A special
SCIENCE
Compendium

Advanced Technology

Edited by
PHILIP H. ABELSON
MARY DORFMAN

American
Association
for the
Advancement
of
Science

Inquiries should be addressed to:

AMERICAN ASSOCIATION for the ADVANCEMENT OF SCIENCE
1515 Massachusetts Avenue, N.W.
Washington, D.C. 20005

Table of Contents

Preface

Young scientists who wish to direct their efforts to a career of maximum significance should consider the important field of materials research. At the moment, energy and molecular biology are in the spotlight, but the long-term values, intellectual challenges, and practical roles of materials combine to make the field very attractive.

Three general groups of materials are of particular interest: (i) polymers, (ii) metals, alloys, oxides, and silicates, and (iii) electronic materials, primarily semiconductors. New kinds of polymers continue to be discovered having special properties such as great strength, high thermal and chemical stability, or electrical conductivity. Fundamental understanding of the behavior of polymers is being obtained through nuclear magnetic resonance spectroscopy and other experimental tools, which give quantitative guidance in efforts to formulate superior products. Major activity, though, is devoted to combining already available monomers and polymers to form objects with desired properties superior to those of a pure polymer. For example, combinations of layers of polymers can lead to containers that are tough, strong, and resistant to passage of oxygen. Mechanical properties can be greatly altered by incorporation of reinforcing fibers, inert materials, or gases. The new products are finding many uses in energy-saving applications.

In studies of materials, chemists and physicists have roamed throughout the periodic table and have made countless combinations of elements and tested them in various proportions. Their work has led to new superconductors and to improved permanent magnets that require less imported cobalt than earlier types. Major advances are being made in improving the strength of materials. One method takes advantage of the fact that some crystals have great unidirectional strength. Another development is the creation of low-alloy, high-strength steels. Even more spectacular has been the development of glassy metals. When liquid mixtures are cooled very rapidly, the resultant solids may have strengths 15 times that of products cooled more slowly. At the same time, other properties such as magnetic permeability and freedom from corrosion may also be greatly improved. Of great importance is the research effort to develop superior specific catalysts. This involves detailed understanding of the interactions among atoms at surfaces. Improvements of as much as a factor of 10^{12} have been obtained in speeds of reaction. When combined with high specificity, such performance leads to major energy savings. The use of catalysts enables reactions to be conducted at lower temperatures. This lessens need for process heat. Catalysts direct the reactions along desired paths so that optimum yields of specific products are obtained. As a result, less energy is expended in such separative processes as distillation. The research effort on catalysts has also led to the development of zeolite cage structures capable of catalyzing the conversion of methanol to gasoline.

During the past decade the most dynamic area of technology has been in exploiting the potential of semiconductors such as silicon. The electronics revolution continues with considerable emphasis on obtaining more transistors per chip and better, lower-cost computer memories. But other frontiers are under scrutiny. Semiconductors such as GaAs (III-V compounds) may be the key to even faster, better computers. Such compounds have already proved useful for lasers and light-emitting diodes. A different approach to increasing the speed of computation is through the development of Josephson-type devices that function at cryogenic temperatures. Another activity is the development of superior photovoltaic materials to capture solar energy. One of the fastest-growing applications of new materials is in prosthetic devices. This year, between 2 million and 3 million such devices will be implanted in humans, creating an interesting set of interactions between living and nonliving substances.

Circumstances have combined to give materials research less visibility than it deserves. This is especially true when the scene is viewed from campus. Federal support for materials research is small in comparison to that given to biomedical activities. The popular media devote far more attention to energy, the environment, and medicine than to materials. Thus young scientists may well receive a distorted picture of the relative significance of and opportunities in various fields and may prepare themselves for careers that offer limited intellectual challenge or limited job opportunities. For those trained in the physical sciences, the major source of employment is industry.

Major industries with total annual sales over $500 billion are intensively engaged in the development of new and better materials. Their efforts are crucial to innovations that will render this nation more energy-efficient and more capable of meeting international competition in the future. In companies that recognize that their future depends on research, scientists enjoy excellent support. Some of their equipment defines the state of the art. In many areas of science pertinent to materials, industrial scientists are the pioneers.

To some degree, companies are a factor in the imperfect picture the academic world has of industrial research. Companies must seek proprietary advantages, and this tends to make them sensitive about releasing information on their technical activities. Nevertheless, a stream of good research papers has come from industry, and several industrial scientists have won Nobel prizes for fundamental discoveries.

During the past decade, the relative circumstances under which scientific research is conducted in academia and industry have shifted. Federal support for university research has declined and has become much more bureaucratic. Funds for instrumentation have not kept up with needs. Openings for young investigators have become scarce.

To have a basis for estimating the quality of opportunities for young scientists and engineers in leading industrial laboratories, I have visited 16 of them, spending 2 days in each and interviewing about 250 staff members. In all of them excellent instrumentation could be seen, and in almost all of them many projects involving materials research were favored with special, very expensive equipment. In addition, a common practice is to maintain a central analytical facility containing as many as 40 instruments, each having a dedicated micro- or minicomputer.

A dominant impression arising from these visits was that materials research enjoys a high priority in industrial laboratories. In addition, it was clear that progress was being made in understanding basic phenomena and in applications of materials. The vitality of the efforts seemed such as to merit special treatment in *Science*. A large number of materials scientists were consulted, and about 75 potential topics were identified. The 20 articles in this volume represent a selection made with the help of experts from the larger group.

The scientists who were asked to contribute were especially cooperative in providing timely articles of high quality. We are indebted to them and to our advisers.

—PHILIP H. ABELSON

Introduction

The modern essays in the 23 May 1980 issue of *Science*, which are reprinted in this compendium, deal opportunely with the advanced technology of materials. For now is the right time to assess national materials capabilities. It is two decades since discussions in the President's Science Advisory Committee led to a national materials science and engineering policy. This enabled designation of the Advanced Research Projects Agency of the Department of Defense, working through the Federal Council for Science and Technology, as sponsor of the Interdisciplinary Laboratories in several versatile American research universities. Accordingly, new discoveries and talented people have been worked into the system of materials makers and users (*1*).

This action was concurrent with the evolution of materials science and engineering as vital industrial elements of the oncoming era of communications, computers, and information technology, now assuming its greatest growth. Similarly, the basis for electronics and photographics in solid-state systems forms part of the essence of the space and rocket epochs. And these, in addition, have demanded extraordinary thermal and physical endurance in yet other structures.

Thus, when the National Academy of Sciences' Committee on the Survey of Materials Science and Engineering (COSMAT) submitted its report a half-decade ago, in September 1975, it was found that the science and technology of materials had developed into a major industrial and, to a lesser but growing degree, academic mission. And in 1978 a successor group to the original committee in the Federal Council for Science and Technology, called the Committee on Materials (COMAT) of the Federal Interagency Coordinating Committee, in cooperation with Battelle Institute and the Industrial Research Institute, reported that U.S. industry devoted about $4.3 billion a year to materials life-cycle research and development. Further, the National Commission on Materials Policy, in its final report published in 1974 (*2*), even then indicated that the application of nonenergy materials and their processing represented 6 to 7 percent of the gross national product of the United States. The large matter of materials interaction with energy usage and the processing requirements for industrial fabrication have, of course, currently enhanced the impact of these figures. They accent the significance of advanced materials science and engineering for overall national progress (*3*).

Now the experts in this volume show how modern physical science has been integrated with age-old techniques applied to metals, ceramics, glasses, cellulose and rubber derivatives, fibers, and composites. But the articles also reflect new paths toward achieving properties of matter unknown or even unimagined before. These mechanical, chemical, and electromagnetic properties may determine the future of great industries such as the automobile, housing, and aerospace industries. Certainly they are crucial to national security capabilities.

Even more stirring is the evidence in this volume that solid-state and materials science is becoming organized and structured so that it joins more traditional science as an intellectual entity. As such, it challenges curiosity and understanding; it provides orderly pedagogy; it thus offers ways to involve new generations of talented people in an especially close and kinetic interaction between basic knowledge and its highly practical applications.

As might be expected from the rapid changes and growth in both materials science and engineering and the world economy, the *record* of such new work has been abundantly complicated by identity with other dominant features of solid-state science and derived technologies. Thus, electronics, communications, and computers have heavily exploited new materials. This is also true of space and missile, as well as aircraft, engineering, and increasing elements of biomedicine (*4*). Similarly, more established fields, such as metallurgy, plastics, fibers, food, lumber, containers, petroleum, stone, clay and glass, and motor vehicles, have participated strongly in other aspects of the new knowledge of matter. So a great deal of current work is associated with these combinations of *makers* of materials and the often quite different *users* of materials, with the common elements of materials progress often being somewhat obscured.

There is another inevitable quality of so rapidly an enlarging domain of science and engineering, which is the sheer burst in volume of published material of all kinds. Thus, since 1950 we find that there have appeared no fewer than 1380 different books on polymers, the basic substances of plastics, rubbers, fibers, coatings, and adhesives. These books are likely to be categorized somewhat in terms of properties, but without general reference to the common features of atoms and molecules, of structure and order, which are expected to govern all materials.

And this is just the issue on which the present collection of papers bears. Here we have an authoritative set of reviews about attractive and compelling roles of materials in most of our major industrial and public technologies. The question then is, How well have we applied the new basic understanding of matter across this growing frontier? Correspondingly, how well have we cross-connected the findings in both technology and science that the hosts of workers in each of the special fields of use of materials are steadily generating? This is the unusual and welcome feature of the present papers—that they stimulate this sort of thinking. And in doing so, it seems very likely that they will lead to new plans and strategies for the next years of materials research and development.

Now it might be asked, why are these questions so compelling, since science will tend to generalize the fundamental properties of matter on the atomic and molecular scale anyway. Further, it might be said that the uses of materials by industries and government are so specialized nowadays that one has to have a self-contained technology for each of them, and cross-linkages do not help much. I believe that these views are invalid. This conclusion is based on several decades of intimate participation in the progress of materials science and engineering, especially electronic materials, polymers, and materials for space and rocket systems.

Instead, the needs for new understanding and ideas about behavior under demanding conditions are such that transfer of knowledge across what have conventionally been impermeable barriers is the best way to advance. This involves especially combining techniques of materials engineering and of design. It offers vast new opportunities for economic gains and help in productivity in industries such as automobiles, other machinery, and energy conversion. But it demands detailed understanding of such solid-state effects as relaxation under stress, creep, fatigue, and many other mechanical and chemical behaviors, which are much affected by design and shapes, thermal and other fabricating dependencies on design, and the like (3, 5).

Strikingly, the atomic and molecular exchangeability among large areas of materials systems (and functions) once typified by basic compositional differences is spreading rapidly. Thus, polyesters govern a variety of both plastics and fibers; increasing numbers of polyolefins, like polyethylene and polypropylene, similarly constitute textiles, sheets, and moldings. And styrene's interaction in synthetic rubbers as well as rigid plastics has become a classic example. While the polyamides and glass remain rather differentiated as principal fiber and sheet materials, respectively, their roles are also being inverted, as glass fibers become reinforcing agents for polymers and nylon becomes a major molded structural element. This interchangeability occurs in many other cases—for example, Pfann and Vogel's discovery of atom dislocations in single-crystal germanium and the transfer of that finding to modifying the strength and processing qualities of a whole range of metals, from aluminum to zirconium.

The sheer quantities involved are indeed symptomatic of the need for unifying principles of both science and engineering in creating and assuring the eventual properties of these classes of matter. For instance, in the 1970's, use of polyamides as rigid thermoplastics grew in the United States from 92,000,000 to 327,000,000 pounds (1969 through 1979), while their use in fibers, including the aromatic polyamides or aramides, went from 1,411,000,000 to 2,720,000,000 pounds. Likewise, the use of olefins in fibers grew from 269,000,000 to 759,000,000 pounds, while their use as plastics grew from 6,580,000,000 to 16,643,000,000 pounds in the same decade. A similar path could be traced for polyesters, which, although they vary in structure, are still basically dominated by the interaction of ester linkages, as shown long ago following Carother's epochal discovery (6).

Overall, in the field of synthetic organic solids and composites, the sharing of knowledge about basic qualities, such as primary valence bonding, interchain dispersion, and dipolar forces between chains, is implicit in the discussion by Anderson, Bartron, and Collette. It is carried further into intra- and interphase, as well as intercomposition or composite, systems, covered by Alfrey and Schrenk. This scope of precepts about matter calls for models of polymer solids comprising larger volumes than the intergroup or "average chain" configurations so far pursued. This is further supported by the growing interest in macromolecular systems qualified by additions of small molecules, which has been done in plasticizing for decades and is used in the induction of electrical and photosensitive behavior, as discussed here by Mort. The latter insights are also related to the borderline between the classic use of carbon in conducting rubber, dominated by interparticle contacts in a dielectric continuum, and intrinsic chain conductivities induced by polar and ionic properties, as in polyamides (7).

Indeed, a prominent quality of these articles, throughout their span from metals to nonmetals and from adhesives to glasses, is the consistency of concepts and experimental observations, even when rigorous analytical treatment is unavailable. The dual behavior of many solids, including the synthetic organics, as insulators and semiconductors or conductors, depending on detailed compositional variations, agrees with the early experiments of Pohl and co-workers (8), whereas the concepts of polyene conductivity advanced by Garrett (9) are affirmed by a series of modern findings.

These widely concordant experimental observations of the electromagnetic and mechanical behavior of materials, under highly divergent conditions of use, indeed encourage us to venture into new dimensions of the science and technology of aggregates. One reason for this is that the supramolecular organizing principles of nature in living matter so vastly exceed the best of our efforts. Thus, although Alfrey and Schrenk bring out elegantly what can be attained through ingenious composites, mixtures, and even molecular combinations, nothing achieved so far comes near the biomorphology that is routine in growing plants and animals. Thousands of researchers over the centuries have marveled at the way bones balance mineral and organic phases. Skins assemble collagen plastics with hydrated rubbers, and feathers incorporate just the right amounts of silica and polypeptides in exquisite combinations. Only recently have we begun to see any path toward comparable processes in making and applying artificial materials. Microfibrils and microtubules are being recognized as widespread units in biomorphology, and modern polymer techniques produce something like the right dimensions of fibrils, but so far we have little idea of how the motility of microtubules arises.

There is good reason to believe that we must now begin to understand ways of generating specified motion—motion such as we have depended on purely thermal agitation to generate on a micro scale (and crude rheology, or at best solution and dissolution, on a larger scale). We really have not joined the phenomenon of Brownian motion in charged and relatively visible colloids with our new understanding of macromolecular qualities, including internal charges, and internal discharging by conduction.

Polypeptides and the cellulose-lignin matrices that are so fabulously effective in plant growth, probably exploit these factors widely. The path to get us started on this new mission of controlled micromorphology brings us back to the comments about simulation of polymer structure under dynamic circumstances, over volumes larger than segments. A vision of how this could be for the system polyethylene is being worked out by Weber and Helfand (10). Also, we should pursue simple boundary controls in the synthesis of macromolecules, as another approach to understanding the structure that nature produces with such facility. Thus, in latex polymerization molecular weight was long ago controlled by the size of the latex particle in the production of microgels (*11*). This method is now widely applied in modifying the structure and properties of natural rubber latices.

We are increasingly recognizing that information transfer and recognition of states is a promising approach to the biomorphology puzzle. Various conferences are emphasizing this with respect to cell composition and function (*12*). It is encouraging that there has been some primitive microtubule assembly in vitro (*13*). Ongoing studies in the recognition and coding potentials for natural and model systems will eventually tie in with these other efforts. Also, in connection with metallo-organic polymers and particular electronic components of large molecules, researchers such as R. P. Blakemore and A. J. Kalmijn several years ago found magnetotactic bacteria whose direction of motion in water is apparently governed by the interaction of the earth's magnetic field with about 20 tiny magnetite crystals forming single magnetic domains, held in a chain 1 or 2 micrometers long within their body. These are again indicative of capabilities for modifying and influencing the assembly of matter as represented in life systems.

Probably, these new kinds of capabilities would first be useful technologically in composites, and other high surface functions such as adhesives. Once more, the new science of surfaces and films is strongly represented in materials science and engineering. Remarkable affirmation of such basic scientific concepts as quantum mechanics appears in this work. Thus, M. B. Panish, his associates A. Y. Cho, J. R. Arthur, and A. C. Gossard, and others have developed molecular beam epitaxy, a method of synthesizing crystals by depositing one layer of atoms after another under precisely controlled conditions. These domains are so carefully arranged, with precise composition, and therefore energy shifts, controlled within 5 angstroms of an interface, that quantum well structures have modeled exactly the properties expected from a two-dimensional electron gas (*14*). The quantum mechanical particle-in-a-box, which most of us learned as an act of faith essential for wave machnics, is now available.

This feature of fundamental science for materials gives hope for much more, but as noted in our statements about polymers, plastics, and adhesives, the questions are still vastly more numerous than the answers. Nowhere is this more apparent than in the use of solids as heterogeneous catalysts. Yet, as Sleight shows in his article, there is extensive use of basic knowledge of structure and bonding which suggests that the new structural probes, such as extended x-ray absorption fine structure (EXAFS), electron impact spectroscopy/microscopy as pursued by David Joy and his associates, and various electron emission and x-ray absorption schemes, will increasingly illuminate this mysterious realm of materials.

Such exact structural analysis has been essential for the admirable progress in superhard materials discussed by Wentorf, DeVries, and Bundy, who have contributed so much to these historic achievements. In the case of diamond, nitrides, borides, and others, there is also an important continuity in relating the bonding of perfect crystals to that of amorphous solids. In turn, the basic theories of P. W. Anderson and N. Mott have opened a new vision of bonding in glasses and other disordered systems. Thus, the finding of Klement *et al.* (*15*) that metals can be vitrified is introduced as a scientific as well as technical resource in Gilman's article.

Indeed, once more the notion of supramolecular assemblies arises in connection with new qualities provided by polyphase ceramics and metals. The articles by Katz and by Kear and Thompson stimulate thought about improved methods of microassembly, perhaps at 2000°C instead of 20°C, but nonetheless with predesigned morphology. The effect is also directly shown in the article on high-strength, low-alloy steels by Rashid, who shows how the age-old art of distribution of particles is being linked with the science of imperfections, found in the totally different arena of semiconductor single crystals. The still more distant realm where nature in living forms distributes the tiny domains and phases we have noted before, and which are the essence of living materials' performance, beckons for exploration.

Many years ago J. D. Fowlkes calculated the positions of veins in plant leaves and showed that nature had distributed the reinforcing members to ensure the greatest photoexposure for a given size and shape. The mystical exercise of living growth and form remains for other science to explain. But the many findings in this volume about composites, combinations of both organic and inorganic substance, impel us to keep seeking new ways of causing dispersion and aggregation. These should be more like nature's and less like those of the conventional milling machine, extruder, or blender.

It is quite likely that some distribution of charges and their changes underlie these dynamics in growth and form. Accordingly, it is interesting that the last ten of the present 20 articles treat solid-state electronics or electromagnetism, as do significant parts of the first ten. This is an accurate reflection of the major technical as well as scientific advances based on a study of solids, as well as a reflection of the vast industrial and technological role of communications and computers, which depend on these materials. Except for the article by Sleight on catalysis, the second half of the volume deals with microscopic qualities of electromagnetics. These range over the hard superconductors treated by Hulm and Matthias, who, with J. E. Kunzler, had practically reorganized parts of the periodic table in terms of these important phenomena. They extend to incisive assessments of our knowledge of magnets of the rare earth–transition metal compounds (Chin). These systems, which amazed me when they were discovered, have intrinsic coercivities of 3 million amperes per meter and can eventually advance the culture of electric motors and magnetic control systems. Similarly, in the other vital technologies of photovoltaic cells (Perez-Albuerne and Tyan), display devices (Kazan), and magnetic memories in bubble systems (Giess), and in the conventional junction device semiconductors (Woodall and Penn), we see again how large and essential technologies have been based on knowledge and control of the pure crystalline state. But it is purity and perfection outside our general experience with matter in the past millennium. The behavior achieved may require particular impurities at less then 10^{12} atoms per cubic centimeter—levels hardly known in any other scientific studies. However, thanks to improved methods of preparation and analysis, in the future we would expect such purities to be widely achieved and employed (*16*). That organic systems as well as the classic inorganic ones are sensitive to impurity levels of this sort indicated in the study of the prostaglandins, whose action on cells, perhaps through membrane effects, is a dramatic reminder of the sensitivity possible in condensed systems.

The message of these articles on materials science and solid-state electronics is not so much the influence of the transistor and its integrated circuit and thin-film derivatives on our lives and economics, for that chronicle is well known, but rather the wider potentials of precise materials processing. The authors have provided a vivid perspective on what modern science and engineering can do in the formation of exact materials structures. These already approximate the dimensions and controls about which I speculated earlier, in discussing the challenge of making material aggregates with efficiencies related to that of living tissues. Thus, experimental thin-film circuits are being achieved with a line width of 2 μm or less in the elements of the circuit. And 1.25-μm dimensions are a reasonable objective for the next few years. Although this is still a long way from the atomic or molecular etchings so deftly employed by nature, which would represent a limit of film control of 3 to 5 nanometers, the interesting point is that the micrometer dimensions already typify oncoming routine manufacturing processes.

Further, an interesting principle of morphology dominates this field. That is, the aim is not so much a very small size for circuit elements such as diodes or transistors; instead, it is downsizing their configuration and interconnection dimensions. The conductors joining them cost about one-hundredth of what they would cost on a printed circuit board of conventional small dimensions. And a modern silicon chip with 150,000 components must contain about 500,000 conductors. Here we are reminded that aggregation of the elements is a basis of efficiency, just as aggregation of various mechanical or chemical elements may turn out to be in materials composites for the future, about which we have speculated. The role of basic science, and particularly of materials research and development, in this extraordinary chapter of electronics progress has been expertly reviewed by Linvill and Hogan (17). They show how the capabilities and styles of civilization have been influenced by the evolution of electronics derived from these materials properties. Earlier, Sir John Cockcroft noted in his address to the British Association for the Advancement of Science that without them, the exploration of outer space and the deployment of missile defenses would have been inconceivable. Thus, as in the chronicle of materials ages in the past—the Bronze, the Iron, and the Steel ages—we can see ahead other opportunities for the dramatic advances through substances and structures.

Indeed, I believe that another one of these is already shaping up. Functionally, it can be termed an epoch of photonics. It is primarily the descendant of the laser, which has stimulated a host of new optics and light energetics. And as lasers were made according to principles set forth by Charles Townes and Arthur Schawlow, materials sciences supported the various solid-state versions. Two spectacular materials developments are extending this field now. One is the glass fibers, which conduct photons thousands of times more efficiently than the clearest glass developed up to now. Thus, losses of 1 decibel or less per kilometer are ensured, commercially, at 0.8- and 1.3-μm wavelengths of the transmitted light. But the remarkable modified chemical vapor deposition synthesis of these lightguides has also resulted in superior mechanical qualities of the fibers. Single fibers with strengths of 800,000 pounds per square inch and higher are readily produced, with exceedingly uniform values for a large number of specimens. These materials properties may eventually enhance the practical use of fiberglass reinforcements in the large realm of composites. Already there is a pronounced effect on improved optics of fiber scopes for physiological, medical, and other instrumental inspection, and of course for communications and instrumentation in aircraft, automobiles, and other vehicles, where light weight and freedom from electromagnetic interference are of special value.

Overall, the study and application of materials are in lively interaction with the other exciting frontiers of science and technology, ranging from biomedicine to nuclear structure. The challenging ways in which the many-body systems in solids can be related to the fundamentals of physics and chemistry—the fundamental principles of nature—have been elegantly expressed by Anderson (18). These concepts of what Anderson calls the "hierarchical structure of science" have guided much of the effort in our laboratories and others for more than four decades.

Thus, we are convinced that such highly practical endeavors as materials technology will continue to be (in fact, will increasingly be) supported by strong science. Striking current advances in the basic concepts of disordered or amorphous matter are good examples. For centuries silica-derived glass was regarded as an indispensable but puzzling kind of witches' brew. Now, theoretical and experimental studies are being stimulated by the materials functions of glass fibers, although sheet glass has been the principal light transmitter for more than 3000 years. In other areas of amorphous structures, carbon has been curiously involved. Diamond is the quintessence of crystalline perfection. Graphite, noted for its crystalline anisotropy, is exploited in lubrication, as an electrode and resistor, as a neutron moderator, and in a host of other materials usages. But what lies in between, in terms of somewhat crosslinked graphite or vastly disordered tetrahedral carbon, has again opened a variety of new systems. The use of graphite fiber reinforcements, discussed by Beardmore et al. and referred to in the biomaterials section by Hench, reflects remarkable hardness and strength. These apparently arise from nongraphitic properties characteristic of what might better be called polymer carbon, since the cross-linkage can readily be established by this macromolecular route (19).

These highly disordered states of matter round out our present perspective on advanced technology materials. Their hardness (surpassed only by that of the most refractory crystals), their corresponding high modulus of elasticity, their versatile electrical properties and response to electrical modification by doping, are all symbolic of the opportunities which lead us into yet new ventures of materials synthesis and processing. High among these is our growing knowledge of surfaces, intrinsic in the epitaxial process reports of Panish, but also applicable to a very wide range of new materials systems. Here the laser is repaying its debt to materials science and engineering by enabling a remarkable energy input. This is virtually exclusively to the electrons of an exposed surface, which then heat up the crystal or glass by electron-phonon collisions. [Yaffa (20) estimates that 1 electron volt of energy can go from electrons to atoms through a series of 20 electron-phonon collisions.]

Indeed, we see now, through pulsed laser energy inputs and the succeeding electronic excitation transformed to heat, ways of processing locally. This occurs especially in surfaces but also in tiny domains for a range of materials that is already producing new alloys, metastable mixtures, and extraordinary crystallite distributions. This may, indeed, turn out to be a step toward the specific internal motion control proposed earlier. In the surface of a silicon crystal, heat will appear in a region about 1 μm thick during a 10^{-8}-second laser pulse of a typically 532-nm, frequency-doubled, Nd:YAG beam. It can be limited with pulses of 10 to 100 nanoseconds to depths of melting of 1 μm, and can then be quenched by the bulk surroundings at about 10^9 K per second.

Work in this very active field is still at a rather early stage, with many conferences being organized internationally. But the point is that a new treatment of surfaces is adding still another process to the ever-widening frontiers of materials science and technology. The articles in this volume indeed assure us that, in materials and the solid-state, science is being related progressively to engineering, now showing a most encouraging consistency of fundamental precepts and practical performance. Thus, with a per capita usage of about 10 tons of matter annually in the United States and with synthetic polymer production, in volume of matter, already greater than our total steel output, we can look forward to creating (with renewable materials, including biomass), extraordinary services of matter, from satellites to paperweights. The patterns of linking novel thought and useful practice as closely as they have been in the modern materials industry may itself be a major stimulus for progress in human affairs (*21*).

W. O. BAKER

Bell Laboratories
Murray Hill, New Jersey 07974

References and Notes

1. W. O. Baker, in *Materials Science and Engineering in the United States*, R. Roy, Ed. (Special Technical Publication 283, American Society for Testing and Materials, Philadelphia, Pa., 1970), pp. 44–65.
2. *Final Report of the National Commission on Materials Policy* (Rep. 06224, Government Printing Office, Washington, D.C., 1974).
3. W. O. Baker, *Metall. Trans.* **8A** (No. 8), 1205 (1977).
4. W. O. Baker, *et al.*, *Engineering and the Life Sciences* (National Research Council, Washington, D.C., 1962).
5. W. O. Baker, *J. Mater.* **3**, 915 (1967).
6. C. S. Fuller and W. O. Baker, *J. Chem. Educ.* **20**, 3 (1943); W. O. Baker, in *Advancing Fronts in Chemistry*, vol. 1, *High Polymers*, S. B. Twiss, Ed. (Reinhold, New York, 1945), p. 105.
7. W. O. Baker and W. A. Yager, *J. Am. Chem. Soc.* **64**, 2164 (1942).
8. H. A. Pohl, in *Progress in Solid State Chemistry*, H. Reiss, Ed. (Macmillan, New York, 1964), p. 316.
9. C. G. B. Garrett, *Semiconductors* (Monograph Series 140, American Chemical Society, Washington, D.C., 1959), p. 634.
10. T. A. Weber, *J. Chem. Phys.* **69**, 2347 (1978); *ibid.* **70**, 4277 (1979); E. Helfand, Z. R. Wasserman, T. A. Weber, *ibid.*, p. 2016.
11. W. O. Baker, *Ind. Eng. Chem.* **41**, 511 (March 1949).
12. *Nature (London)* **285**, 287 (1980).
13. K. Roberts and J. Hyams, Eds., *Microtubules* (Academic Press, New York, 1979).
14. D. C. Tsui, H. Stormer, A. C. Gossard, W. Wiegmann, *Phys. Rev. B* **21**, 1589 (1980).
15. W. Klement, R. H. Willens, P. Duwez *Nature (London)* **187**, 869 (1960).
16. W. O. Baker, *Chem. Eng. News* **57**, 30 (26 November 1979).
17. J. G. Linvill and C. L. Hogan, *Science* **195**, 1107 (1977).
18. P. W. Anderson, *ibid.* **177**, 393 (1972).
19. W. O. Baker and R. O. Grisdale, U.S. Patent 2,697,028 (14 December 1954); F. H. Winslow, W. O. Baker, W. A. Yager, *J. Am. Chem. Soc.* **77**, 4751 (1955); F. H. Winslow, W. O. Baker, N. R. Pape, W. Matreyek, *J. Polym. Sci.* **16**, 101 (1955).
20. E. J. Yaffa, *Laser and Electron Beam Processing of Materials* (Materials Research Society, 1979).
21. W. O. Baker, *Daedalus* **109**, 83 (1980).

Ribbon of metallic glass that was cast directly from liquid alloy at a speed of approximately 900 meters per minute, and ⟶ an average quenching rate close to $10^{6o}C$ per second. The ribbon's nominal composition is $Ni_{70}Cr_6De_2Si_8B_{14}$ (atomic percent). It is 2.5 centimeters wide by 25 micrometers thick. Unlike the crystalline form of the same composition, it is ductile. Such ribbons are useful for brazing pieces of stainless steel together. [Courtesy of Allied Chemical Corporation, Morristown, New Jersey]

Trends in Polymer Development

B. C. Anderson, L. R. Bartron, J. W. Collette

The modern era of polymer research began in the 1930's with pioneering work in the groups of H. Staudinger and W. Carothers. Since that time, many thousands of new polymeric organic compositions have been characterized by researchers in laboratories all over the world. The novelty varies from small differences in structure in well-known families of polymers to compositions representing new families of polymers. The research has been done for many pur-

ered intractable can be formed for study and use. Research on polymer flow, or rheology, has aided understanding of processing requirements.

3) Studies relating structure and processing to final use properties—for example, tensile strength, stiffness, and toughness—have allowed optimization of both polymers and the processes used to prepare them.

4) New synthetic methods provide improved structural control. Many mono-

Summary. Polymer science and technology has flourished as polymers with many new compositions have been synthesized. The range of properties attainable has been continually extended, providing materials with higher strength, better reinforcing capabilities, and greater resistance to extreme thermal and corrosive environments. Examples of evolutionary developments in the polyamides, the fluorocarbon resins, and the aromatic engineering plastics are used to illustrate the trends. It is expected that this process will continue in order to meet changing needs and that emphasis will be put on selective polymer design for specific applications.

poses, ranging from academic investigations of structural possibilities, through subtle modifications to vary processing or physical properties in a useful way, to the preparation of polymers with specific compositions that are expected to have premium properties or combinations of properties.

Interest in polymer research is sustained by steady advances in all phases of polymer science:

1) New and improved analytical techniques are providing more detailed information about molecular structure, molecular weights, end groups, and structural faults in polymers, as well as increasing knowledge of the crystalline and amorphous regions in the supermolecular structure or morphology.

2) Fabrication technology has advanced so that polymers once consid-

mers can be copolymerized either randomly or in blocks as desired. New monomers provide many new structural possibilities. Polymers and copolymers have been developed that contain functional groups needed for cross-linking or other reactions carried out subsequent to synthesis. The chemist can almost routinely prepare structures that seemed impossible only a few years ago.

The field is much narrower when commercially useful, or potentially useful, polymers are considered. Nevertheless, very large technologies involving many structure variations have developed around the polymers made in billion-pound quantities. Many of the commercial polymers made in lower volume are also modified to form numerous new products. These new polymers can range from a polyethylene type useful in cer-

tain packaging applications, to quite new rubbery structures able to withstand extremes of temperatures, to exceptionally strong, stiff organic fibers with many potential uses.

Within the limitations of space and of our own knowledge of the voluminous literature, we have chosen examples that illustrate the extensions of an existing family of polymers, the polyamides; the manipulation of structures in the fluoropolymers; and the development of new aromatic engineering plastics. Our examples are commercial or nearly commercial polymers, particularly those with premium properties or resistance to difficult environments. The reader can consult recent publications for more examples and greater detail (*1, 2*).

Polyamides

Polyamides are a microcosm of the polymer field, illustrating the range of properties attainable by selective changes in structure. After the introduction of nylon 66 in the late 1930's, chemists continued to explore structure-property relationships in other polyamides (*3*). This has recently culminated in the development of the high-strength, high-modulus fibers, probably the most important advance in organic polymers in the last decade.

Nylon 66 is prepared from hexamethylenediamine and adipic acid and is used both as a fiber and as an engineering

$$H_2N(CH_2)_6NH_2 + HOOC(CH_2)_4COOH \rightarrow$$

$$\left[-HN-(CH_2)_6NH\overset{O}{\overset{\|}{C}}-(CH_2)_4\overset{O}{\overset{\|}{C}}- \right]_n$$

Nylon 66

plastic. (The number 66 is a code designating the polymer structure; it refers to the number of carbon atoms in the diamine and the diacid, respectively.)

Nylon 66 is a semicrystalline polymer. Parts of the polymer chain exist in ordered crystalline regions that have definite melting points and parts are in dis-

B. C. Anderson, L. R. Bartron, and J. W. Collette all work in the Research Division of the Central Research and Development Department, E. I. du Pont de Nemours and Company, Experimental Station, Wilmington, Delaware 19898.

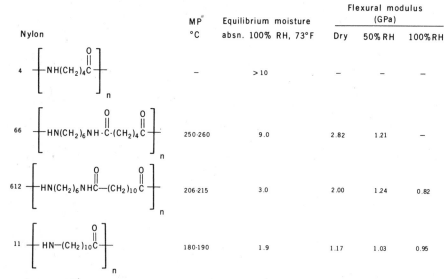

Nylon	MP °C	Equilibrium moisture absn. 100% RH, 73°F	Flexural modulus (GPa) Dry	50% RH	100% RH
4	–	> 10	–	–	–
66	250-260	9.0	2.82	1.21	–
612	206-215	3.0	2.00	1.24	0.82
11	180-190	1.9	1.17	1.03	0.95

Fig. 1. Characteristics of different polyamides. Abbreviations: *MP*, melting point; *absn.*, absorption; and *RH*, relative humidity.

ordered or amorphous regions. The chains in these amorphous regions are rigid if the temperature is below the glass transition temperature (T_g) of the segments, but behave as liquids above T_g. The degree and kind of crystallinity of the ordered phase and the T_g of the amorphous phase profoundly affect the physical properties.

Nylon 66 is moisture-sensitive, primarily because of absorption of water by the amide groups in the amorphous regions; the water acts as a plasticizer, lowering T_g and increasing the mobility of the polymer chains. As a result, physical properties such as modulus, tensile strength, and toughness and structural dimensions in molded parts are sensitive to the relative humidity of the environment. For specialty uses, polyamides that offer a different balance of water absorption, melting points, and crystallinity have been developed; examples are shown in Fig. 1.

Nylon 4, derived by ring-opening polymerization of pyrrolidone, **1**, has a

higher ratio of amide to hydrocarbon groups and absorbs more water. This results in textile characteristics that are more cottonlike than those of nylon 66.

Polyamides with lower moisture sensitivity and greater dimensional stability are also needed. This can be achieved by increasing the ratio of hydrocarbon to amide groups; thus nylon 612 and nylon 11 absorb less water and show less change in modulus than nylon 66 in hu-

mid environments (Fig. 1). The polyamide **2** from bis(4-aminocyclohexyl)-methane and dodecanedioic acid has two structural features that contribute to

dimensional stability: (i) the frequency of the amide group is lower than in nylon 66 and (ii) the large cyclohexyl rings substantially reduce the mobility of the chain, resulting in a much higher T_g, and the polymer is only moderately plasticized by absorption of water (4). Fibers from **2** show little change in properties on exposure to moisture and have unusually good wash and wear properties.

Completely amorphous or glassy polyamides are another structural variation. They can be prepared by using monomers with an irregular chain structure so that the long-range order necessary for crystallinity cannot develop. The use temperature of such amorphous polymers is determined mainly by the T_g. An example is the polyamide from terephthalic acid and mixed trimethylhexamethylenediamines, **3**, which is market-

ed as Trogamid T (5). In common with other glassy polymers, this polyamide is transparent, an important advantage in

many applications, and it has much better retention of mechanical properties at elevated temperatures than many crystalline polyamides of lower T_g.

Aromatic Polyamides and New High-Strength, High-Modulus Fibers

Aromatic polyamides or aramids were introduced in the early 1960's to meet the need for fibers with more heat and flammability resistance. These polymers melt at too high a temperature to be made by the melt polymerization processes used for aliphatic polyamides. New low-temperature polymerization techniques involving solution or interfacial reaction of a diacid chloride and a diamine in the presence of certain salts were developed so that polymers with the high molecular weights required for useful physical properties could be reproducibly prepared (6). The first example is poly(*m*-phenyleneisophthalamide), **4**. The flam-

mability resistance and thermal stability of this polymer have led to its extensive use in protective clothing, electrical applications, and as a replacement for asbestos.

The next step in the evolution of aramids has been the development of high-strength, high-modulus fibers. The fibers are derived from very highly crystalline rigid-chain polymers that form extended chain crystals. These rigid-chain polymers are based on *para*-substituted aromatic polyamides, including **5**, **6**, and **7**.

The commercial material is DuPont's Kevlar aramid fibers. The development of these fibers is a truly revolutionary advance in polymer technology. They greatly extend the range of physical properties attainable in synthetic polymers, provide new insights into polymer crystallinity and properties, and have stimulated research on polymeric liquid crystals (7-10).

The fibers are stiffer—that is, have a higher modulus—than glass or steel and are stronger on an equal-weight basis. Figure 2 compares the strength and modulus of Kevlar to properties of other reinforcing fibers (glass, graphite, and so on) on an equal-weight basis. Modulus values as high as 128 gigapascals (18.5×10^6 pounds per square inch) have been obtained with these *para*-oriented aromatic polyamides. This is more than 90 percent of the calculated theoretical value. Tensile strengths approach 20 percent of the theoretical value—substantially higher than achieved with nylon 66 or polyester fibers (8). The combination of high strength, high modulus, and low density makes these fibers useful for reinforcement of composite structures including tires. Cables as strong as steel with one-fifth the weight of steel are being used to anchor oceanic drilling platforms. Lightweight bulletproof vests have opened a new dimension in protective clothing.

Research to fully understand the structure of these fibers and their structure-property relationships is being actively pursued. Current results suggest that the polymers differ both in their high level of crystallinity and in the arrangements of the crystals. The fiber from poly(*p*-benzamide), 5, is almost completely crystalline and the chains are believed to be extended. A small fraction of randomly oriented but crystalline material is present, which forms defect zones between regular crystalline lamellae. There is a high fraction of extended chains which pass through these defect regions, maintaining a long-range order that is absent in most fibers (11).

Another important characteristic of polymers such as 5 is a propensity to form liquid crystals in concentrated solutions in certain solvents. The polymer chain is sufficiently rigid and the polymer conformation sufficiently rodlike in solution that the chains associate in ordered arrays (liquid crystals) above a critical concentration. The solutions are visually anisotropic and the solution viscosity is substantially lower than expected for random-coil polymer chains. The orientation in these ordered arrays can be maintained during spinning, so that a fiber is obtained with oriented extended-chain crystals without subsequent orientation. This behavior is very important for the economic manufacture of these fibers.

Research in this field is extensive, par-

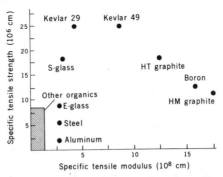

Fig. 2. Specific tensile strength and specific tensile modulus of reinforcing fibers (tensile strength or modulus divided by density).

ticularly on synthetic polymeric liquid crystals. Liquid crystal behavior has been reported in melts of some aromatic polyesters (12), indicating that more examples remain to be discovered.

Fluoropolymers

In the 1940's the preparation of polymers from fluorinated monomers, beginning with polytetrafluoroethylene (PTFE), opened a new area of polymer technology by providing combinations of properties that were unavailable in other materials. Polytetrafluoroethylene resin—for example, Teflon fluorocarbon resin—is a homopolymer of tetrafluoroethylene (TFE), 8; its chemical resist-

$$CF_2{=}CF_2 \qquad\qquad -(CF_2CF_2)_n$$

$$8 \qquad\qquad\qquad PTFE$$

ance is unique and its thermal stability and low-friction characteristics are very useful. The original homopolymer, however, had characteristics that limited its application. PTFE has an extremely high molecular weight and is very difficult to fabricate in complex shapes. It cannot be used under very high loads because of the ease with which the crystals deform and slip.

The use of copolymerization technology to obtain desirable property and processing characteristics in fluoroplastics illustrates the evolutionary modification of this family of plastics very well.

Many monomers can be copolymerized with TFE, but most of the copolymers have reduced thermal stability. Hexafluoropropylene (HFP), 9, was the first fully fluorinated comonomer, which was used to modify the properties of

$$CF{=}CF_2 \qquad\qquad (CF_2CF_2)_x(CF{-}CF_2)_y$$
$$\;|\qquad\qquad\qquad\qquad\qquad\qquad |$$
$$CF_3 \qquad\qquad\qquad\qquad\qquad\quad CF_3$$

$$9 \qquad\qquad\qquad Teflon\ FEP$$

PTFE. Hexafluoropropylene does not copolymerize readily with TFE, so only a limited range of copolymer compositions is readily accessible. A copolymer containing ~10 mole percent HFP was introduced in 1960 as Teflon FEP. Teflon FEP can be melt-fabricated and extruded directly on wires; however, the use temperature is reduced from 250° to 205°C and electrical losses are slightly higher than in PTFE.

Extensive exploratory research at Du-Pont led to the discovery that fluorovinyl ethers, 11, copolymerize with tetrafluoroethylene and related fluoroolefins over a wide range of compositions. This discovery and the existence of a practical synthetic route for the vinyl ethers from hexafluoropropylene oxide, 10, have led to the development of a variety of new perfluoroplastics and elastomers.

$$CF_3CF{-}CF_2 \;+\; R_F\overset{\displaystyle O}{\overset{\|}{C}}{-}F \longrightarrow R_FCF_2OCF{=}CF_2$$
$$\;\;\;\backslash\!/$$
$$\;\;\;O$$

$$\underset{\sim}{10} \qquad\qquad\qquad\qquad\qquad\qquad \underset{\sim}{11}$$

Teflon PFA (perfluoroalkoxy) fluorocarbon resins are copolymers of TFE with small amounts of vinyl ethers such as 11. Because long fluoroalkyl side chains can be introduced, only a small amount of the comonomer is needed to reduce crystallinity and modify the morphology. This gives a melt-processable fluoroplastic that can be injection-molded or extruded and has high-temperature properties very close to or better than those of the original PTFE (13).

The PFA resins are expensive and their use is not commercially justified in many applications. Other approaches to more processable fluoropolymers have involved the preparation of copolymers with simple olefins. Two such materials are copolymers of ethylene with TFE, 8, or chlorotrifluoroethylene, 12. Although

$$CF_2 = CFCl$$

12

these polymers do not have the heat stability or the high use temperature of PTFE, their chemical resistance and electrical properties are remarkably similar. This is undoubtedly because the hydrocarbon units alternate with the tetrafluoroethylene units and these isolated units are less susceptible to oxidative attack. Both materials are less expensive and easier to fabricate than PFA resins. The ethylene-TFE (ETFE) copolymers have found particular use in wire coatings in electrical equipment and computers. In addition to the necessary insulation properties, they have low flammability, and in the event of a fire their combustion products are not as corrosive as those of the materials they replace.

The ETFE copolymers have been known experimentally for many years. Their recent commercialization is mainly due to improvements in several key physical properties. Unmodified copolymers become brittle above 150°C. The problem was managed by careful control of the polymerization conditions and by the introduction of a small amount of a third monomer to modify the crystalline structure. With the increased amorphous content of the polymer, embrittlement was reduced to an acceptable level.

A related melt-processable plastic is derived by copolymerization of vinylidene fluoride, 13, and hexafluoroisobutylene, 14. This is an alternating co-

$$CF_2{=}CH_2 + \underset{\underset{CF_3}{|}}{\overset{\overset{CF_3}{|}}{C}}{=}CH_2 \longrightarrow \left(CF_2{-}CH_2{-}\underset{\underset{CF_3}{|}}{\overset{\overset{CF_3}{|}}{C}}{-}CH_2 \right)_n$$

13   **14**

polymer in which the CH_2 groups are shielded from oxidative attack by the large electronegative CF_3 groups. The copolymer is highly crystalline, with a melting point of 327°C and a use temperature up to 280°C. Like other fluoropolymers it has low surface energy, and compared to PTFE it has higher surface hardness, better abrasion resistance, and better scratch resistance. Because of this combination of properties, uses of this material for hard, low-friction, chemically resistant coatings are being investigated (14).

A recent advance has been the development of a fluoroplastic reinforced with graphite fibers. The composite structure has outstanding rigidity and resistance to creep and can be used in gaskets for extremely corrosive environments.

Fluorocarbon Ionomers

Other new materials that have evolved from the chemistry of perfluorovinyl ethers are copolymers of tetrafluoroethylene with ether monomers containing perfluorinated sulfonyl fluoride functional groups, 15. These polymers can be

$$-(CF_2CF_2)_x{-}(CF{-}CF_2)_y \xrightarrow{NaOH} -(CF_2CF_2)_x{-}(CF{-}CF_2)_y$$

(with pendant chains: $O{-}(C_3F_6){-}O{-}CF_2{-}CF_2{-}SO_2F$ for **15** and $O{-}(C_3F_6){-}O{-}CF_2{-}CF_2{-}SO_3Na$ for **16**)

15   **16**

converted to the sulfonic acid form and then to an acid salt or ionomer, 16.

The ionic groups aggregate in the low-dielectric medium of the fluoropolymer to give highly polar domains that swell in water. These perfluorinated ionomers combine the excellent resistance to chemical attack characteristic of fluoropolymers with selective permeability to certain ions. The sulfonic acid polymers as well as similar ones with carboxyl functional groups are being investigated as membranes in cells for making chlorine and caustic by electrolysis of salt solutions (1, 15).

Fluoroelastomers

The need for thermally and chemically resistant elastomers grew rapidly after World War II, especially for aerospace and military applications. The first fluorinated elastomer was a copolymer, 17, of vinylidene fluoride and chlorotrifluoroethylene, but it was rapidly replaced by superior copolymers of vinylidene fluoride with hexafluoropropylene, 18. In

$$(CH_2CF_2)_x(CF_2\underset{\underset{Cl}{|}}{CF})_y \qquad (CH_2CF_2)_x(CF_2{-}\underset{\underset{CF_3}{|}}{CF})_y$$

17   **18**

each case a large group (Cl or CF_3) was introduced to destroy the crystallinity of the polyvinylidene fluoride. These hydrofluoroelastomers provided a new level of thermal resistance for cured elastomers, having an almost indefinite life at 200°C. Elaboration of the basic structure and extensive curing chemistry has resulted in a family of very useful heat- and solvent-resistant commercial rubbers (16).

Elastomers useful at even higher temperatures have recently become available from copolymers of TFE with the perfluoroalkylvinyl ethers previously discussed. These materials are the most completely fluorinated elastomeric polymers and can be considered elastomeric analogs of PTFE.

$$(CF_2CF_2)_x{-}(CF{-}CF_2)_y\underset{\underset{OCF_3}{|}}{}$$

19

Copolymers, 19, of TFE with 20 to 50 percent (by weight) perfluoromethylvinyl ether are amorphous, again illustrating the general principle of eliminating the crystallinity of PTFE by introduction of chain irregularity. To fully realize the chemical and thermal resistance of this system, special comonomers had to be developed so that

the copolymers could be cured. Examples are shown below.

Terpolymers containing the nitrile 20 can be cross-linked, probably with the

$$CF_2{=}CF{-}O(CF_2)_4CN \qquad CF_2{=}CF{-}O(C_3F_6)_n{-}O{-}C_6F_5$$

20   **21**

ultimate formation of triazines. Terpolymers containing 21 can be cross-linked by reaction with bisphenols. Cured elastomeric parts have a continuous-use temperature of 260°C and are exceptionally resistant to most organic solvents. The parts can be used, for example, in sour gas and oil wells, where temperatures are above 200°C and extremely corrosive environments are encountered.

Chemically resistant elastomers are also needed which are both more flexible at low temperatures than the examples above and stable at high temperatures. The current materials of choice are the fluorosilicones, 22; the silicone chain re-

$$\left(\underset{\underset{(CH_2CH_2CF_3)_2}{|}}{Si}{-}O \right)_x \left(\underset{\underset{(CH_3)_2}{|}}{Si}{-}O \right)_y$$

22

tains flexibility to -100°C and the CF_3 groups confer useful solvent resistance.

A new family of low-temperature rubbers with excellent oil resistance is the polyorganophosphazines. The base polymer is made by polymerization of the cyclic chlorophosphazine 23 to the linear analog 24. The chlorines can be displaced by alkoxy or fluoroalkoxy groups. By use of a mixture of alkoxy substituents with longer alkyl chains, crystallization can be avoided, providing an amorphous rubber, 25, with a low T_g.

$$\text{(cyclic } Cl_2P{=}N\text{ ring, } \mathbf{23}) \longrightarrow \left({=}N{=}\underset{\underset{Cl}{|}}{\overset{\overset{Cl}{|}}{P}} \right)_n \mathbf{24} \longrightarrow \left({=}N{=}\underset{\underset{OR^2}{|}}{\overset{\overset{OR^1}{|}}{P}} \right)_n \mathbf{25}$$

The polymers can be subsequently cross-linked by use of peroxides or radiation (1).

Perhaps the ultimate in a fluorinated rubber with both high- and low-temperature properties will be obtained by polymerization of hexafluoropropylene oxide (HFPO), 10, to poly-HFPO, 26. The oxy-

$$-(CF{-}CF_2O)_n\underset{\underset{CF_3}{|}}{}$$

Poly HFPO

26

gen atom greatly increases the flexibility of the main chain so that the polymer remains flexible at -50°C, and excellent thermal stability has been reported. HFPO polymers with low molecular weights (1 to 2000) are commercially available as thermally stable fluorinated oils and greases. To date, only modest success has been achieved in converting HFPO to a high molecular weight that can be cross-linked (17). Successful completion of this research would extend even further the impressive development of fluorinated elastomers.

Engineering Plastics Based on Aromatics

Engineering plastics are used for structural and mechanical functions, frequently replacing parts once made of metal. They replace metals because they can be fabricated in complex shapes more readily and inexpensively, have better resistance to corrosion, are lighter in weight, or offer a wider range of physical properties. The first and still one of the most important engineering plastics is nylon 66.

The early engineering plastics, however, lacked the strength and stability at higher temperatures to replace metal in many applications. Recent research on the synthesis of aromatic polymers has provided many candidates to meet this need. The aromatic rings confer the structural rigidity, thermal and oxidative stability, and high softening temperatures required for demanding applications.

Examples are the polycarbonates, **27**, introduced in 1959 by General Electric and Bayer; the polyether sulfones, **28**, introduced in 1965 by Union Carbide; and the polyphenylene ethers, **29**, in-

[structure 27]

[structure 28]

[structure 29]

troduced in 1964 by General Electric. All these polymers are glassy or amorphous, with glass transition points between 150° and > 200°C. They are generally more difficult to process than the crystalline plastics because they have higher melt viscosities. Special fabrication equipment and procedures had to be developed for these materials. The poly-

phenylene oxide resins are sufficiently difficult to process that their largest uses are in blends with polystyrene that have lower melt viscosities but also lower use temperatures.

New aromatic polymers continue to extend the range of properties attainable. The aromatic polyesters, **30** (1974), resemble the polycarbonates in chemical

[structure 30]

[structure 31]

stability but add 30° to 40°C to the maximum use temperature. The polyether ketones, **31**, are crystalline and have a wider range of chemical resistance than amorphous polymers.

Polyphenylene sulfides, **32**, combine

[structure 32]

crystallinity and higher use temperatures. Early efforts to prepare useful polyphenylene sulfide by a variety of techniques were unsuccessful; the molecular weights were not reproducible and gelation was a frequent problem. Workers at Phillips Petroleum found that well-defined linear polymers could be prepared by reaction of p-dichlorobenzene with Na_2S in a solvent. The polymer is highly crystalline with a melting point above 288°C and is insoluble in most solvents below 200°C. It is exceptionally thermally stable, showing no weight loss up to 500°C and only a modest change in physical properties after 30 weeks at 204°C. While difficult to melt-fabricate, it is finding uses in corrosive environments where high-temperature solvent resistance is needed.

A partially anisotropic aromatic-aliphatic polyester, **33**, has recently been

[structure 33]

introduced as an experimental high-performance plastic (12). The anisotropy may contribute to easier processing. Broadening of this technology could again increase the maximum use temperature of polyesters.

Probably the ultimate structure in an aromatic polymer is poly(p-phenylene), **34**, which has a theoretical melting point

[structure 34]

above 1000°C. This material can be fabricated by sintering the fine powders slightly below the decomposition temperature of 625°C (18).

An alternative method for achieving high-temperature performance is to create cross-linked or thermosetting polymers. Epoxy plastics are well-known thermosets that are adequate for long-term use at temperatures of 100° to 125°C. For higher use temperatures, polyimides such as **35**, which are ex-

[structure 35]

ceptionally thermally stable, are preferred. Because these polyimides are intractable, two techniques have been used to convert them to usable shapes. In one, a soluble prepolymer is formed into a film or coating and then chemically and thermally converted to the polyimide. In the other, methods developed in powder metallurgy are used. A powder with a high surface area is formed into a part at high pressures and then sintered at high temperatures, during which powder coalescence, imide formation, and cross-linking all occur.

Several modified polyimides are being tested that are designed to combine the thermal stability of the polyimide with greater flexibility in processing. Polyimides **36** and **37** (where Ar rep-

[structure 36]

[structure 37]

[structure 38]

resents an aromatic group) have low enough molecular weights to be melt-processed; they are then cross-linked by addition polymerization of the acetylene or norbornene groups. Polyimide amides, **38**, represent another approach. Although a structure such as **38** can be melt-processed, it also has a reduced high-temperature capability; for example, **38** has a heat distortion temperature (at 1.86 MPa) of 260°C compared to 360°C for **35**.

Conclusion

The field of polymer chemistry has already reached a high degree of excellence and a certain maturity. The available fibers, plastics, elastomers, and coatings provide a wide range of properties at costs competitive with those of alternative materials. Thus it is increasingly difficult to extend the range of properties or find a niche not filled by existing polymers, especially with the current high costs of market introduction.

Nevertheless, some current developments and world trends provide opportunities for further useful invention. These include:

1) The ability to synthesize and engineer new structures with property combinations ideally suited to specific needs.

2) The need to conserve energy by weight reduction, particularly in automobiles.

3) Increasing environmental concern and regulation, which will favor products that can be recycled and are biodegradable.

4) Changing expectations of society, which will favor more durable materials.

5) The development of new raw material sources, which will present opportunities to replace more expensive polymers with less expensive ones.

We foresee polymer science meeting the challenge of the changing future needs. It seems certain that new compositions will ultimately extend the range of properties available in the high-strength, high-modulus polymers, the fluoropolymers, and the high temperature–resistant engineering plastics. In addition, many modifications of existing polymer types can be expected to improve specific characteristics.

References and Notes

1. H.-G. Elias, *New Commercial Polymers, 1969–1975* (Gordon & Breach, New York, 1977).
2. R. D. Deanin, *New Industrial Polymers* (American Chemical Society, Washington, D.C., 1979).
3. M. I. Kohan, *Nylon Plastics* (Wiley, New York, 1973).
4. R. W. Campbell and H. W. Hill, *Macromolecules* **8**, 238 (1975).
5. Produced by Dynamit Nobel A.G. (*1*).
6. P. W. Morgan, *Condensation Polymers by Interfacial and Solution Methods* (Interscience, New York, 1965).
7. _____, *Plast. Rubber Mater. Appl.* (February 1979), pp. 1–7.
8. E. E. Magat and R. E. Morrison, *Chem. Technol.* **6**, 702 (1976).
9. J. Schaefgen *et al.*, in *Ultrahigh Modulus Polymers*, A. Cifferri and I. M. Ward, Eds. (Applied Science, Barking, Essex, U.K., 1979), chap. 6.
10. S. L. Kwolek, P. W. Morgan, J. R. Schaefgen, L. N. Gulrich, *Macromolecules* **10**, 1390 (1977).
11. P. Avakian, R. C. Blume, T. D. Gierke, H. H. Yang, M. Panar, *Polymer Prepr. Am. Chem. Soc. Div. Polym. Chem.*, in press.
12. W. C. Wooten *et al.*, in (*9*), chap. 8.
13. R. L. Johnson, in *Encyclopedia of Polymer Science and Technology* (Wiley, New York, 1976), Suppl. 1, pp. 266–267.
14. F. Petruccelli, *Am. Chem. Soc. Div. Org. Coat. Plast. Chem. Prepr.* **35** (No. 2), 107 (1975).
15. H. Ukihashi, *Polym. Prepr. Am. Chem. Soc. Div. Polym. Chem.* **20** (No. 1), 195 (1979).
16. R. G. Arnold, H. L. Bounez, D. C. Thompson, *Rubber Chem. Technol.* **46**, 619 (1973); H. E. Schroeder, *Rubber Plast. News* (September 1978), pp. 21–22.
17. J. T. Hill, *J. Macromol. Sci. Chem.* **8** (No. 3), 499 (1974).
18. D. M. Gale, *J. Appl. Polym. Sci.* **22**, 1955 (1978).
19. We acknowledge the assistance of R. E. Putscher in assembling the information for this article.

Multipolymer Systems

Turner Alfrey, Jr., and Walter J. Schrenk

The study of structure and the development of structure-property relationships are central issues in materials technology. In the case of organic polymers, structure embraces two distinctly different aspects—molecular structure and supramolecular structure. Molecular structure refers to the macromolecules—their chemical compositions, molecular weights (average molecular weights and distributions of molecular weight), extent of branching or cross-linking, and so on. Supramolecular structure refers to the geometric arrangement of these macromolecules in a specimen. In many cases, the molecular structure of a polymer is established during its synthesis,

A polymer which is employed as a semicrystalline solid can exhibit a variety of different crystalline morphologies, depending on the thermal and mechanical history encountered in its previous mechanical processing (2, 3). For example, linear polyethylene, the simplest prototype of a linear polymer, can exist in any of the following morphologies:
1) Spherulitic.
2) Drawn fibrillar.
3) "Shish kebab."
4) Extended-chain crystals.
5) Oriented extended-chain crystals.
Spherulitic morphology develops when an unstressed melt is simply cooled. Nucleation of crystallization re-

gree of molecular orientation. Polystyrene (PS), a typical glassy noncrystalline polymer, is brittle when unoriented. Uniaxially oriented PS is strong and tough in the direction of orientation, but weak and fragile in the transverse direction. Biaxially oriented PS can be strong and tough in all directions in the plane of orientation (4).

The range of properties obtainable with a given polymer through control of crystalline morphology and molecular orientation can be greatly broadened by the use of different materials in combination: polymers with polymers, polymers with glass or metals, polymers with gases (polymeric foams). The geometric arrangements of the various phases in such composite systems constitute another level of supramolecular structure. Several comprehensive reviews (5, 6) describe the vast scope of polymer-polymer composite materials. Two solid materials can be combined in an infinite number of geometric arrangements: dispersion of particles in a continuous medium, fibers in a continuum, parallel lamellar phases, two interpenetrating continuous phases, and so on. The properties of a particular combination depend on:
1) Properties of the individual materials.
2) The geometric arrangement.
3) The character of the interfaces.
One very important class of polymeric composite materials is that of fiber-reinforced plastics; these are discussed elsewhere in this issue. We will briefly discuss multilayer plastic sheets and films, and polymer blends, and list miscellaneous other examples. First, however, we will examine some basic principles common to all these multipolymer systems.

Summary. Different polymers can be combined to yield a wide variety of composite materials: layered sheets and films, homogeneous and heterogeneous blends, interpenetrating polymer networks, bicomponent fibers, and others. Some properties of a multipolymer material are roughly additive, but synergistic interactions can yield properties and performances superior to those of the individual constituents. Consequently, the use of polymers in combination is a rapidly growing component of polymer materials technology.

whereas the supramolecular structure is established during subsequent mechanical processing into a shaped article. The properties of a polymer depend on both its molecular structure and its supramolecular structure (1).

At high temperatures, a typical linear polymer molecule is in a state of rapid segmental motion, wriggling about from one three-dimensional conformation to another (by means of rotation about covalent bonds within the chain). Under these conditions, the material is a viscoelastic fluid, which can be pumped through channels and formed into desired shapes. Upon cooling, the polymer can solidify by either of two mechanisms: crystallization (ordered packing of chains in a crystal lattice) or vitrification (formation of a glassy amorphous solid). The crystalline melting point, T_M, and the glass transition temperature, T_G, of a given polymeric species are important determinants of its potential applications as a material.

sults in a spherical growing cluster of radially oriented, folded-chain ribbons; mutual impingement of these growing spherulites results in a set of polyhedral grains in the final specimen. Drawing of such a spherulitic structure, below the crystalline melting point, results in a structural reorganization in which chunks of the folded-chain ribbons are oriented into a drawn fibrillar morphology. Crystallization from an oriented melt can result in the shish kebab morphology, consisting of long central fibrils with lateral folded-chain growths. Crystallization under high hydrostatic pressure can lead to extended-chain crystals. The properties of linear polyethylene depend strongly on which of these various morphologies has been developed, as shown in Table 1.

Some polymers are used as glassy amorphous (noncrystalline) solids. Their properties also depend markedly on the supramolecular structure developed during processing—in particular, on the de-

Miscibility, Immiscibility, and Interfaces

Different low-molecular-weight organic liquids are often miscible in all proportions, but if immiscible, exhibit sharp boundaries between the separate phases. Organic high polymers of differing structures are usually thermodynamically immiscible; even a very small unfavorable heat of mixing per segment adds up to a large value for an entire polymer molecule. Krause (7) lists 17 polymer pairs which appear to be miscible in all proportions at room temperature and 25 pairs which appear to be conditionally miscible, but such miscible systems constitute a small minority among the my-

The authors are research scientists at the Dow Chemical Company, Midland, Michigan 48640.

Table 1. Young's modulus of linear polyethylene.

Linear PE	Young's modulus (psi)
Transverse to drawn direction	50,000
Spherulitic (isotropic)	100,000
Drawn fibrillar	600,000
"Superdrawn"	9,000,000
Theoretical upper limit	23,000,000

riad possible combinations. Some polymer pairs are miscible at elevated temperatures, but immiscible at room temperature. Other polymer pairs are miscible at room temperature, but immiscible at elevated temperatures.

The phase boundary between two immiscible polymers is more diffuse than boundaries between low-molecular-weight phases; considerable local interpenetration occurs. Although entire polymer chains do not invade the interior of the foreign phase, sections of polymer molecules (loops and tails) can penetrate across the boundary from both sides. This results in a diffuse boundary zone of varying composition which may be 20 or 50 angstroms thick. Two molten polymers which are nearly compatible will have a more diffuse interfacial boundary than a pair of strongly incompatible polymers.

When a two-phase polymer melt is cooled and solidified, the interpenetrating chains may hold the two phases firmly together in the solid state; however, in many cases the phase boundary becomes a site of mechanical weakness—the two solid phases exhibit poor interfacial ad-

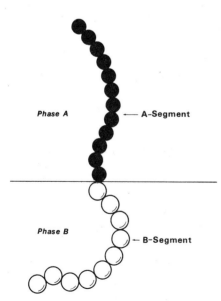

Fig. 1. An A-B block copolymer molecule located at the interface between two immiscible polymers, poly-A and poly-B. High-molecular-weight copolymers would have many more units than are shown here.

hesion. (For example, the act of crystallization may forcibly withdraw some polymer tails which in the molten state penetrated across the boundary, and thus alter the character of the diffuse zone.) A pair of polymers thus may be (i) miscible (a minority of cases), (ii) immiscible, but with strong interfacial adhesion in the solid state, or (iii) immiscible, with poor adhesion.

Two polymers which normally do not adhere well can sometimes be caused to do so by the addition of a third polymer which adheres well to both. Such an agent may be introduced as a discrete "glue layer" between the two nonadhering phases, or may be localized at the interface, with sections of individual molecules penetrating into both phases (8). Figure 1 is a schematic diagram of a block copolymer of A-units and B-units at the interface between the two immiscible phases poly-A and poly-B. Such an additive can tie together two immiscible polymers which would otherwise exhibit poor interfacial adhesion.

These features of compatibility or incompatibility, interfacial adhesion, and "compatibilizing" additives are critical in all types of multipolymer systems—layered films, polymer blends, and fiber-reinforced polymers.

Multilayer Plastic Sheets and Films

The use of polymers in layered combination is quite old. The utility which could be gained by combining layers with different desirable properties often justified the expense of multiple operations—laminating films together, applying coatings, and so on. In recent years, an inexpensive method of producing such laminar structures has led to a rapid growth in the number and quantity of such products. Two or more molten polymers are combined as fluid layers within a die, and the layered product is produced directly in one step. Today, many layered plastic sheets and films are manufactured by melt coextrusion (9). Figure 2 is a schematic diagram of one method of producing coextruded laminar products. Individual streams of molten polymers are introduced through feed ports and joined together in a parallel array. This multilayer fluid passes through a transition zone in which the layers are extended laterally and thinned down, while remaining parallel. Finally, the multilayer fluid exits through a wide, shallow die opening and is cooled to yield the solidified layered product. Films containing hundreds of layers can be produced in this manner; most com-

mercial coextruded products contain from two to seven layers. Figure 3 illustrates a few typical examples. Often a desired laminar structure contains adjacent layers which exhibit poor interlayer adhesion; this compels the introduction of a thin glue layer of a polymer which will adhere well to both (9).

The properties of a multilayer polymer sheet depend on:

1) The properties of the individual polymers.

2) The layer geometry.

3) The character of the interfaces.

Some sheet properties turn out to be derived from those of the individual layers by simple addition or averaging. In other cases, synergistic interactions result in sheet behavior which is significantly different from the sum of its parts. Finally, some multilayer films exhibit characteristics which are unique.

Barrier to diffusion—an additive property. An important group of applications for layered polymer products is in the packaging of food, medical products, chemicals, and so on. Many of these applications demand low permeability to oxygen, water vapor, or other migrating species. Normally, the diffusion barrier provided by a multilayer sheet is the simple sum of the barrier values of the individual layers ("resistances in series") (10). Polymers differ greatly in their permeabilities to small molecules; for example, the permeability of low-density polyethylene to O_2 is more than 1000 times that of Saran. Thus, a relatively thin barrier layer of one polymer can be combined with other layers chosen for mechanical toughness, heat sealability, or other required properties, to yield a superior packaging film or a sheet which can be formed into containers.

Mechanical properties. Some mechanical properties of multilayer sheets are similarly additive. However, it is often observed that a thin layer in a laminate behaves much differently from a free film of the same material. Particularly striking effects can be observed in laminar composites containing a hard component and a soft component, or a brittle component and a tough component. Sword makers of early times hammered down alternate layers of hard and soft steel, obtaining blades that would take a fine cutting edge and yet were strong and tough (11). Similar effects can be observed with coextruded multilayer polymer systems (10).

Consider a laminar composite containing a layer of high-modulus, low-elongation material flanked by layers of more extensible materials. When the composite is tested to failure in tension, the adhering high-elongation layers may act to prevent transverse crack propagation in the hard layer. With crack propagation so blocked, the hard layer may be able to reach its ductile yield stress, and the entire composite can stretch in a ductile manner to high elongation. This toughening effect has been termed mutual interlayer reinforcement (10). Interlayer interaction can sometimes have the opposite effect; a normally tough polymer may act in a brittle manner in a laminar composite. In this case the high-elongation material is not able to block crack propagation in the brittle layer; a crack forms in the brittle layer, and continues into and through the tough layer, causing a localized failure at low overall elongation. For example, a skin of general-purpose PS on a core of high-impact PS can force the entire composite to undergo brittle failure on bending. This behavior has been called mutual interlayer destruction (12). Such interlayer interactions, whether of a beneficial or a deleterious nature, require some degree of adhesion at the interfaces. However, the thinner the layers, the lower will be the degree of adhesion required for one layer to alter the response of its neighbor.

The mechanical properties of each layer in a laminar composite depend on the molecular orientation present. As with homogeneous sheets, uniaxial orientation can increase tensile strength in one direction while reducing it in the other direction. Biaxial orientation can provide desirable mechanical properties in all directions in the plane. In a multilayer polymer film, an appropriate amount of biaxial orientation can optimize the mechanical properties of the individual layers, and thereby the likelihood and extent of interlayer reinforcement. For example, biaxially oriented 125-layer films of polypropylene and PS have been reported to exhibit high tensile strengths and high ultimate elongations over a wide composition range (10).

Optical properties. Finally, multilayer films made of alternating layers of two transparent polymers with differing refractive indices can exhibit vivid optical effects. In nature, the iridescent colors of many butterflies, beetles, birds, and fish result from selective reflection from layered microstructures (13). Similar effects are observed with coextruded multilayer polymer films (14, 15). Figure 4 is a schematic drawing of a 125-layer film in which the odd layers have a refractive index of 1.6 and the even layers a refractive index of 1.5. When a beam of light of wavelength λ shines on the film, partial reflection occurs at each of the many in-

terfaces; the reflections at the interfaces of increasing refractive index suffer a phase reversal. Since the layers differ by only 0.1 in refractive index, the individual reflections are weak. However, if reflections from the different interfaces all leave the film in phase with each other, the constructive interference yields a high-intensity reflection. The wavelength of the first-order reflection (for normal incidence) is given by

$$\lambda_I = 2(n_{odd}d_{odd} \times n_{even}d_{even})$$

where n and d are refractive index and thickness. For example, if the thicknesses of the odd and even layers are set at 700 and 746.5 Å, respectively

$$n_{odd}d_{odd} = 1.6 \times 700 = 1120$$
$$n_{even}d_{even} = 1.5 \times 746.5 = 1120$$
$$\lambda_I = 2(1120 + 1120) = 4480 \text{ Å}$$

In addition to the first-order reflection, λ_I, higher-order reflections occur at wavelengths λ_{II}, λ_{III}, . . . , which are 1/2,

1/3, 1/4 . . . of the λ_I value. The relative intensities of these higher order reflections depend on the ratio of optical thicknesses of the two kinds of layer. For example, if the odd and even layers have equal optical thickness, the second-order and fourth-order reflections are suppressed. On the other hand, if $n_{odd}d_{odd} = 2n_{even}d_{even}$, the third-order reflection is suppressed.

In the illustration above, where the layer thicknesses are uniform through the film, the reflection at 4480 Å would involve only a narrow band of wavelength. The reflected color would be "pure," but "pale." If the layer thicknesses are appropriately varied through the film, a broader band of wavelength will be reflected. Theoretically, the most vivid color impression, called a C-color, corresponds to strong reflection extending over a spectral region bounded by two complementary wavelengths; this provides the optimum compromise between intensity and color purity (16).

Fig. 2. Schematic diagram of the feedblock method of coextruding multilayer polymer sheets and films.

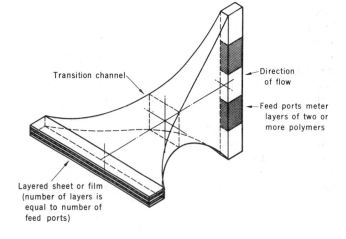

Transition channel

Direction of flow

Feed ports meter layers of two or more polymers

Layered sheet or film (number of layers is equal to number of feed ports)

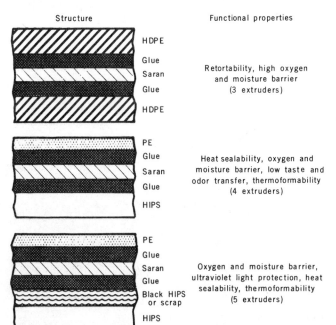

Structure

Functional properties

HDPE
Glue
Saran
Glue
HDPE

Retortability, high oxygen and moisture barrier (3 extruders)

Fig. 3. Schematic diagram of some multilayer plastic sheets with functional properties developed by the combinations of layers.

PE
Glue
Saran
Glue
HIPS

Heat sealability, oxygen and moisture barrier, low taste and odor transfer, thermoformability (4 extruders)

PE
Glue
Saran
Glue
Black HIPS or scrap
HIPS

Oxygen and moisture barrier, ultraviolet light protection, heat sealability, thermoformability (5 extruders)

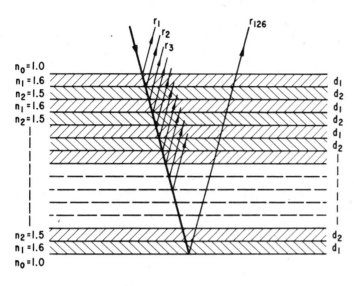

Fig. 4. Reflection of light by a multilayer film made up of two transparent polymers with different refractive indices (n = refractive index; d = layer thickness).

The Mearl Corporation has commercialized an iridescent microlayer film which consists of more than 100 layers, each less than 1000 Å thick; the film is coextruded from two polymers with different refractive indices.

Lamellar combinations of polymers with other materials. Although our primary concern is with multipolymer systems, the widespread use of polymers in laminar combination with other materials deserves mention. Safety glass used in automobile windshields employs a layer of energy-absorbing plastic between glass layers. Metal-plastic laminates with metal skins and plastic cores provide low-weight rigidity; laminates with plastic skins and metal-foil cores are used in packaging. Metallized plastics employ a thin metal layer on the surface of a plastic part; polymeric surface coatings on metal surfaces are the reverse of this. The mechanical properties of these laminar composites depend on the individual materials, the layer geometry, and the interfaces. Interlayer reinforcement and destruction are sometimes encountered. For example, Bhateja and Alfrey (*17*) studied the tensile behavior of sheets with aluminum cores and Mylar skins (bonded with an adhesive). With good adhesion, the Mylar skins prevent tensile fracture of the aluminum core, and the entire composite can exhibit ductile stretching to more than 100 percent elongation.

Polymer Blends

If two thermoplastic polymers are mixed together mechanically in the molten state, and then cooled and solidified, a new "hybrid" material is produced. Ordinarily this will consist of two separate phases, since most pairs of polymers are thermodynamically immiscible. Often the blend will prove to be worthless; many pairs of polymers not only are immiscible but also exhibit poor interfacial adhesion. But some polymer blends exhibit excellent physical properties and may offer advantages over either of the individual polymers. The properties of a polymer blend depend on the properties of the individual constituents and their respective volume fractions, the phase geometry ("morphology"), and the character of the interfaces. When a mixture of two immiscible fluid polymers is vigorously stirred, many different morphologies can be developed. One of the polymers may constitute a continuous phase, with the second polymer dispersed as a discontinuous phase. The dispersed particles may be spheres, ellipsoids, lenticular ribbons, or fibrils. As mixing proceeds, the morphology changes; droplets of one phase may be drawn out into thin lamellae or fibrils, or may be broken up into smaller droplets. Alternatively, both phases may be continuous, forming an interpenetrating structure which may be coarse or fine in texture.

Some blend properties (such as density) are insensitive to these morphological features, being dependent primarily on the volume fractions of the two phases. Other properties are highly sensitive to morphology. Consider, for example, a 50/50 blend of a hard, glassy polymer, A, and a soft rubbery polymer, B. A dispersion of spherical B-particles in a continuous A-phase is a hard blend; a dispersion of A-particles in a continuous B-phase is soft. The elastic modulus (stiffness) of the former could be 1000 times that of the latter. The strength and toughness of a polymer blend are strongly influenced both by the morphology and by the degree of adhesion between the polymers (*6*).

In 1977, *Modern Plastics* printed a news item with the provocative heading:

"Yes, new plastics are coming; not from chemistry but from *alloying*" (*18*). In 1979, a subsequent paper (*19*) described the growing use of polymer blends as engineering thermoplastics (materials with outstanding toughness and temperature resistance) and listed 21 commercially available engineering blends. At the other end of the property spectrum, rubber blends are commonly employed in the manufacture of articles such as automobile tires. Polymer blends are also employed in textile fibers (see below). Polymer "alloys" can also be formed chemically, by polymerization of one polymer in the presence of another—or by the combination of chemical and mechanical processing. The range of possibilities—modes of formation, identities of the constituents, morphologies, and resulting properties—is too great to cover adequately here. The following specific examples have been selected to provide an indication of this diversity.

Noryl, a miscible blend. Polystyrene is a hard, glassy, brittle polymer with good melt-processing behavior. Polyphenylene oxide (PPO) is a tough, temperature-resistant engineering polymer, but is difficult to process. Polystyrene and PPO are miscible in all proportions, and blends of the two exhibit useful combinations of properties—easier processability than PPO alone and higher toughness and heat stability than PS alone. Some PS-PPO blends have a higher tensile strength than either component by itself. Introduced in 1966 by General Electric Company under the trade name Noryl, PS-PPO blends are the most important commercial example of miscible blends (*20*).

Blends of immiscible polymers. Polystyrene and polyethylene (PE) not only are immiscible, but exhibit very poor adhesion in the solid state. When these two polymers are melt-blended, the resulting product is weak and "cheesy." Paul (*21*) describes the improvement which can be obtained by the addition of a "compatibilizer" to such a blend. In the case of PS and PE, a "graft copolymer" containing both PS and PE sections acts as a compatibilizer. The graft copolymer plays an interfacial role, acting as a surfactant during the melt-blending operation and thereby affecting the morphology which is developed, and acting as an interfacial adhesive during subsequent mechanical testing of the blend. Paul also describes immiscible blends of PE with polyvinyl chloride (PVC) and the use of chlorinated polyethylene (CPE) as a compatibilizer for this polymer pair. The CPE molecule is believed to contain chlorinated sections which are miscible with PVC and unchlorinated sections miscible with PE. Finally, some immiscible polymer pairs exhibit strong interfacial adhesion without the need for any compatibilizer.

Interpenetrating polymer networks (IPN's). Manson and Sperling (*5*, chap. 8) have described a broad class of polymer structures involving two intimately interlocked polymer networks. One way to produce such a structure is to start with a lightly cross-linked polymer, A, swell it with a second monomer, B (plus cross-linker), and polymerize B. A second polymer network (B) forms within the space enveloped by the first. The two polymer networks are distinct, but physically interpenetrating. Another way to prepare an IPN structure is to simultaneously polymerize two mixed monomers with their cross-linkers, by separate non-interfering polymerization mechanisms (for example, styrene plus divinylbenzene polymerized by a free radical mechanism, in the presence of a polyurethane network being formed by polycondensation). Whether formed simultaneously or sequentially, IPN's commonly undergo restricted phase separation into microdomains rich in the respective polymer species. An IPN of polyethylacrylate (PEA) and polymethylmethacrylate (PMMA), which are nearly miscible polymers, has phase domains smaller than 100 Å. In the case of PEA-PS, a more immiscible pair, a cellular structure about 1000 Å in size develops. The properties of an IPN depend on the properties of the individual polymers, the phase morphology, and interactions between the phases (*5*, chap. 8). Some IPN properties are simple averages of the individual polymer properties; other properties may be synergistic. Practical exploitation of this broad class of materials can be anticipated.

High-impact polystyrene (HIPS). Polystyrene is a useful and important thermoplastic material which is rather brittle. This glassy polymer can be toughened by appropriate molecular orientation, or by the introduction of dispersed rubber particles. HIPS is produced by dissolving a rubber in styrene monomer, which is then polymerized. Since PS and rubber are immiscible, phase separation accompanies polymerization. Initially, the PS phase is dispersed in a continuous rubber phase (both highly swollen with styrene monomer). Mechanical agitation causes a phase inversion to occur, yielding a continuous PS phase with dispersed rubber particles. Further formation of PS occurs in both phases, the final product having a continuous PS phase and dispersed rubber particles, within which PS inclusions

are embedded—an A-in-B-in-A dispersion. The properties of this dispersion, produced by combined chemical and mechanical means, are different from those of a simple melt blend of the same amount of rubber with PS. HIPS is much tougher than PS, but less transparent.

Multicomponent Textile Fibers

Textile products containing more than one polymer are now common. Different polymers can be used in combination at various levels. A garment might be made from sections of nylon cloth and polyester cloth. Nylon yarn could be interwoven with polyester yarn to yield a composite fabric. Nylon fibers could be spun together with polyester fibers to yield a composite yarn. Finally, molten nylon and polyester polymers can be coextruded to yield composite fibers. At some point in this series, it becomes appropriate to regard the composite as a new material, with its own distinctive material properties.

Although the textile industry combines polymers at the level of garment, fabric, yarn, and fiber, this discussion will be limited to the coextrusion of two

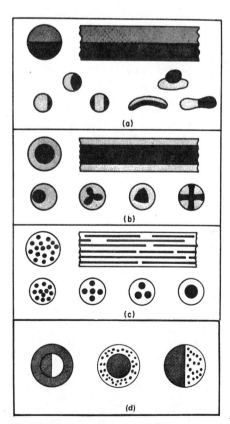

Fig. 5. Possible bicomponent and biconstituent fiber structures: (a) side-by-side, (b) sheath-core, (c) dispersed fibrillar, and (d) combinations. [After Paul (*6*, vol. 2, pp. 171 and 173]

different polymers to form bicomponent or biconstituent fibers. The term bicomponent is applied to skin-core fibers and side-by-side two-component fibers, prepared by supplying two separate feed streams to each die opening. When two molten polymers are blended to disperse the minor component as droplets, and then melt-spun and drawn to produce a dispersion of parallel, oriented fibrils within each fiber, the term biconstituent fiber is used. Bicomponent fiber manufacture has many parallels to the coextrusion of multilayer sheets and films (see above), and biconstituent fibers represent a special case of polymer blend technology.

Figure 5 illustrates the wide range of phase arrangements within multicomponent textile fibers which can be developed by such coextrusion processes. Some of these represent commercial products; others are taken from patents. Still another possible phase morphology is a bicontinuous structure similar to the interpenetrating polymer networks discussed earlier; Paul (*22*) indicates that the mechanical interlocking of the two phases in such a fiber reduces the need for adhesion between the phases. Finally, the fiber composition or morphology can be deliberately varied along the length in order to develop desired properties.

Some properties of bicomponent and biconstituent fibers are essentially additive with respect to the individual polymer constituents, but some properties are synergistic or even unique. One important practical phenomenon often observed with bicomponent fibers is self-crimping. If one side of a bilateral fiber tends to shrink more on heating than the other side, the fiber can be made to develop a spontaneous helical crimp. Fiber crimp is important in the spinning of staple yards and in the mechanical behavior of the eventual fabrics. (High bulk and high stretchability are examples of fabric properties which can be developed by use of self-crimping bicomponent fibers.)

Other applications of multipolymer fibers include antistatic fibers, nylon tirecord with reduced "flat spotting," flame retardance, improved dyeability, and bondable fibers for nonwoven fabrics.

Bicomponent fibers with low interfacial adhesion can be fibrillated (split apart) to yield blends of ultrafine fibrils. In this case, poor adhesion is an exploitable advantage rather than a deficiency. Rasmussen (*23*) developed a process for producing bicomponent fibers from multilayer film. Three polymers are coextruded into the layer sequence AB-

13

CABC . . . , where B adheres to A and C but the A-C interfaces can delaminate. The multilayer melt stream is extruded through orifices, drawn into thin ribbons, and mechanically delaminated to form very thin bicomponent fibers (components A and C held together by the adhesive B).

Expanded Polymers (Foams)

Still another class of two-phase polymer structures, which further broadens the range of useful material properties, is the versatile family of foamed polymers. Nature produces and man utilizes many lightweight polymeric materials which contain large amounts of distributed void-space (cork, balsa, sponges). These natural materials have been joined by a wide variety of synthetic polymers in expanded or foam form. A few examples are foamed rubbers, expanded PS, and rigid and flexible polyurethane foams. These expanded polymers find wide and growing applications in thermal insulation, cushioning (furniture, automobiles), packaging, and construction.

Expanded polymers can be rigid or flexible, open-cell or closed-cell, crystalline or amorphous, thermoplastic or thermosetting, high-density or low-density. The range of possible structures (and properties) is very great. The compressive softness of an open-cell elastomeric foam is an obvious characteristic. Less obvious is the fact that rigid foams can answer structural needs in applications requiring high rigidity. The bending rigidity of a plate or shell increases with the cube of its thickness; consequently, a plate or shell composed of a rigid plastic foam has a much greater bending stiffness than a (thinner) plate made from the same weight of unfoamed polymer (even though the foam modulus is low).

This capacity for lightweight rigidity can be augmented by preparing sandwich structures with high-modulus skins and rigid foam cores (a structural principle long utilized by nature in the load-bearing bones of birds). Plastic foam cores can be combined with sheet-metal skins, or plywood skins, to yield such structural panels. Alternatively, a foaming polymer can be molded in such fashion as to directly produce a finished shaped article with a foamed core and a dense, unfoamed skin; such systems are referred to as structural foams. The properties of an expanded polymer depend on the properties of the polymer (or polymers) and the foam morphology. The versatility of polymer foams can be further expanded by the incorporation of additional phases, such as reinforcing fillers.

Industrial Growth of Multipolymer Materials

During the past few decades, the production of synthetic polymers has grown at a greater rate than that of most other materials. The production of combined polymeric materials is growing faster yet—coextruded multilayer sheets and films, polymer blends, bicomponent fibers, foams (including structural foams), and fiber-reinforced plastics. This rapid industrial growth is largely due to the enhanced utility which can be achieved with polymers by varying and controlling the supramolecular structures as well as the molecular structures of these materials.

References and Notes

1. T. Alfrey, in *Applied Polymer Science*, J. K. Craver and R. W. Tees, Eds. (American Chemical Society, Washington, D.C. 1975), chap. 5.
2. D. R. Uhlmann and A. G. Kolbeck, *Sci. Am.* **235**, 96 (December 1975).
3. E. S. Clark and C. A. Garber, *Int. J. Polym. Mater.* **1**, 31 (1971).
4. L. S. Thomas and K. J. Cleereman, *SPE (Soc. Plast. Eng.)* **28** (No. 4), 2 (1972).
5. J. A. Manson and L. H. Sperling, *Polymer Blends and Composites* (Plenum, New York, 1976).
6. D. R. Paul and S. Newman, Eds., *Polymer Blends* (Academic Press, New York, 1978).
7. S. Krause, in (6), vol. 1, p. 36.
8. D. R. Paul, in (6), vol. 2, p. 135.
9. W. J. Schrenk, *Plast. Eng.* **30** (No. 3), 66 (1974).
10. ___ and T. Alfrey, Jr., *Polym. Eng. Sci.* **9**, 393 (1969).
11. C. S. Smith, *Endeavour* **16** (No. 64), 199 (1957).
12. W. J. Schrenk and T. Alfrey, Jr., in (6), vol. 2, p. 129.
13. H. Simon, *The Splendor of Iridescence* (Dodd, Mead, New York, 1971).
14. T. Alfrey, Jr., E. F. Gurnee, W. J. Schrenk, *Polym. Eng Sci.* **9**, 400 (1969).
15. J. A. Radford, T. Alfrey, Jr., W. J. Schrenk, *ibid.* **13**, 216 (1973).
16. P. J. Bouma, *Physical Aspects of Colour* (Elsevier, New York, 1947), p. 140.
17. S. K. Bhateja and T. Alfrey, Jr., *J. Compos. Mater.* **14**, 42 (1980).
18. *Mod. Plast.* **54** (No. 11), 42 (1977).
19. A. S. Wood, *ibid.* **56** (No. 12), 44 (1979).
20. A. S. Hay, *Polym. Eng. Sci.* **16** (No. 1), 1 (1976).
21. D. R. Paul, in (6), vol. 2, p. 48.
22. ___, in (6), vol 2, p. 167.
23. O. B. Rasmussen, *Am. Chem. Soc. Div. Org. Coatings Plast. Chem. Pap.* **32** (No. 1), 264 (1972).

Conductive Polymers

J. Mort

When polymers were first developed, their most obvious characteristics were their mechanical and chemical properties. The onset of World War II and the need for synthetic rubber and natural product substitutes determined the course of development of the polymer industry. Almost in parallel with this development, the exploitation of solids in electronics centered on elemental inorganic materials such as silicon and germanium and subsequently on inorganic compound semiconductors. As a consequence, only scattered studies of the electronic properties of organic solids in general, and polymers in particular, were made up to 1970. Thus, although the origins of a definable plastics technology can be traced back about 50 years, it was not until very recently that polymers have been viewed as a potential source of electronic materials.

In the 1970's there has been increased attention to disordered materials and organic solids. One reason for this is interest in the effects of disorder on solid-state properties of solids and in the crystalline and amorphous forms of molecularly bonded or organic solids (*1*, *2*). A second motivation has been a growing need for large-area, low-cost elements in electronic and photoelectronic devices. The need of the electrophotography industry for such elements provided opportunities for the use of amorphous inorganic and polymeric materials. This is now a multibillion-dollar industry, at the technological heart of which are amorphous materials.

Work in the past few years has included substantial studies of the electronic properties of polymers, since they combine a number of advantageous features such as good mechanical properties, ease and diversity of synthesis, and the potential for molecular engineering. These studies have utilized and coordinated the skills of chemists, materials scientists, and physicists. Sophisticated experimental and theoretical approaches are being applied to establish a firm understanding of the basic phenomena and to elucidate structure-property relationships (*3*, *4*).

Although significant progress has been made, there is much that is still not understood. My purpose in this article is to review our present understanding of the electronic properties of organic polymers and their actual or possible technological applications. The discussion will be limited to key phenomena and materials involving bulk transport of charge and thus excludes interesting phenomena such as triboelectricity and materials such as electrets whose properties are determined by bound charge.

Summary. Recent research has shown that polymers, normally thought of as being insulators, exhibit a wide range of electrical conductive properties.

Conduction Processes

It is convenient here to take a broader than usual interpretation of the concept of conductivity. The strong interaction between the atomic building blocks and their ordered arrangement on a three-dimensional lattice in a crystalline, covalently bonded solid lead to the existence of bands of allowed energy. The bandwidths are related to the strength of the interaction between the building blocks, which in organic and molecular solids are molecules. In this class of solids intermolecular interactions are typically weak, so bandwidths even in crystals are generally very much smaller than for covalently bonded solids. Within the bands, electrons are not identified with a particular site, but have an equal probability of being associated with any site. This phenomenon of delocalization, in which the electronic wave function is extended in space, has led to the states in these energy bands being called extended or conduction states. The bands of interest correspond to the highest fully occupied states (valence band) and lowest empty states (conduction band). These energy bands are separated by a range of forbidden energies which defines the band gap, E_g, characteristic of a particular solid.

In such an idealized crystal, the material at the absolute zero of temperature is an insulator since all states in the valence band are occupied and all states in the conduction band are empty. As the temperature is raised, there is an increasing probability (defined by Fermi-Dirac statistics and the position of the Fermi level) that some electrons in the valence band can acquire sufficient thermal energy to make the transition to states in the conduction band. Partial occupation of the conduction band can occur, and in an applied electric field the electrons can accelerate and move into higher energy states within the conduction band. Since the valence band is also no longer full, electrons in this band can experience a similar effect, although this is visualized as being associated with the motion of the vacated states, called holes.

The conductivity σ is proportional to $ne\mu_e + pe\mu_h$, where n and p are the equilibrium concentrations of free electrons and holes and μ_e and μ_h are the electron and hole mobilities or velocities per unit field. The mobility therefore determines the facility with which the carriers can move through the crystal lattice and for a material such as crystalline silicon is about 1000 square centimeters per volt per second. Such motion is hindered by collisions or interactions with the vibrating atoms of the crystal at finite temperatures. Thermal vibrations of the lattice are termed phonons. The electrical conductivity is thermally activated and, because of the increasing free carrier population, increases with temperature. This type of solid, where the electrons and holes are produced in equal numbers, is called an intrinsic semiconductor. In the case of metals the conduction band is already partially occupied because of the number of valence electrons per atom and the filling of available states. Since the number of free carriers is therefore essentially temperature-independent, the conductivity of metals decreases as the temperature rises because the carrier mobility is reduced by enhanced scattering of electrons by lattice imperfections or phonons.

A major advance in the field of semiconductors was the discovery that the

The author is manager of the Physical Sciences Laboratory, Webster Research Center, Xerox Corporation, Webster, New York 14580.

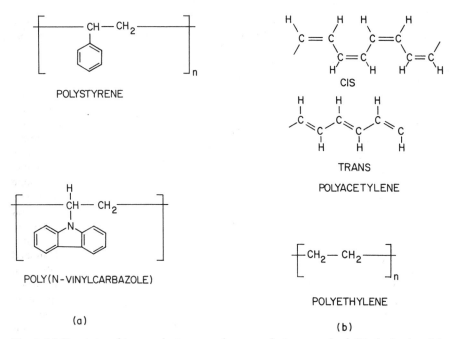

POLYSTYRENE

POLY(N-VINYLCARBAZOLE)

CIS

TRANS

POLYACETYLENE

POLYETHYLENE

(a)

(b)

Fig. 1. (a) Structures of two pendant-group polymers, polystyrene and poly(*N*-vinylcarbazole). (b) Structures of a saturated-backbone polymer, polyethylene, and a non–pendant-group polymer with an unsaturated chain, polyacetylene.

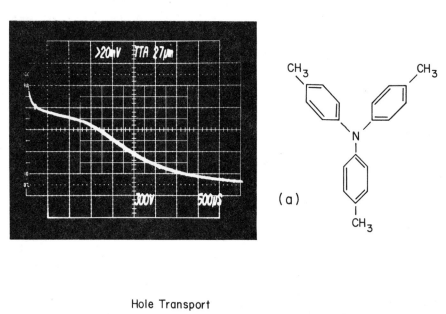

(a)

Hole Transport

m   ⊕ ← e
   m m   m   m

Electron Transport

      e
m   ⊖ → m   m
  m

(b)

Fig. 2. (a) Current transit pulse in a molecularly doped polymer, polycarbonate Lexan doped with molecules of tri-*p*-tolylamine (TTA), shown at the right. The transient photoconductivity is induced by a 10-nanosecond 3371-Å flash strongly absorbed by the TTA molecules. The motion of the induced charge sheet across the sample exhibits considerable dispersion, as discussed in the text. Undoped Lexan exhibits no electronic activity under these conditions. (b) Schematic representation of the basic transport process; *m* stands for molecule, *e* for electron.

addition of a small concentration of impurities could drastically modify the electrical conductivity at room temperature. Such doping can lead to *n*-type extrinsic or impurity-controlled conduction since the excess electrons can be thermally excited out of a localized or donor level, resulting from the introduction of the impurity, into the conduction band. Similarly, certain impurities can lead to localized states in the forbidden gap just above the valence band. These unoccupied states can accept electrons and lead to an increase of holes in the valence band. This type of doping gives *p*-type material (the Fermi level lies closer to the valence band). By controlled doping of materials in this manner, it is possible to form contiguous regions of *p*- and *n*-type materials, which results in a *p-n* junction.

There are solids with sufficiently large band gaps that the thermal generation rate and equilibrium carrier concentration are so low that the material is an insulator even at room temperature. This can be the case even though any thermally excited carriers would move with high mobility in the available conduction states. This quiescent, low, dark conductivity can be changed by photoexcitation of carriers into the high-mobility states (photoconductivity) or by injection of carriers from contiguous solids such as metal electrodes (5). In either case, the introduction of nonequilibrium carriers leads to their transport in already existing transport states, and the solids can be highly insulating although electronically active. Some materials remain insulating and electronically inactive because their insulating nature is associated with a lack of adequate accessible transport states. In this case doping can transform an electrically inactive material into an electronically active but insulating material by the provision of otherwise absent transport states. Additional chemical modification can make these materials semiconducting. All of these approaches lead to conductivity in polymers, and the remainder of this article will be devoted to representative examples.

Electronic States of Polymers

Figure 1 shows examples of two classes of polymers that have quite different optical properties—a difference that is reflected in their electrical properties. These are (i) polymers with saturated backbones and appended aromatic chromophores (Fig. la), of which poly-

16

styrene and polyvinylcarbazole are examples, and (ii) non–pendant-group polymers (Fig. 1b) with saturated or unsaturated backbones, such as polyethylene and polyacetylene. A third class of polymers with unsaturated backbones and appended chromophores has not received much attention and will not be discussed further.

It is characteristic of the pendant-group polymers (the first class) that they are remarkably similar in their electronic states to the isolated pendant aromatic molecule. Their electronic interactions are essentially those of a random and therefore disordered ensemble of the constituent pendant molecules. This led to the concept of molecular engineering of electronically active polymers. In non–pendant-group polymers, the electronic properties are very much controlled by the states associated with the strong intrachain covalent bonding. As with inorganic covalent solids, one expects wide conduction bands because one-dimensional order is substantially maintained along the chain even in otherwise largely amorphous polymers. Indeed, this is the most unusual aspect of such polymers. By analogy with the inorganic solids, appropriate doping with donors and acceptors should lead to extrinsic semiconductors. Although this has been observed, recent studies suggest that the analogy cannot be drawn too closely and the situation is more complex than was first thought.

The electronic absorption spectra of pendant-group polymers are similar to those of the polymers in solution or of effectively isolated chromophores in the gas phase (6). It has long been recognized intuitively that this is evidence that the weak interaction between the pendant molecules leads to small exchange energies and bandwidths much smaller than the energy fluctuations associated with the disorder. In general, such electronic states will be localized. The pendant molecules have a much lower ionization potential than the saturated polymer backbone and are expected and observed to dominate the low-energy transitions and to play an integral role in their electrical properties. The energy required to separate a negative and a positive charge for an isolated molecule (as in gas-phase ionization) is $I_g - A_g$, where I_g is the gas-phase ionization energy and A_g is the energy gained by binding the ionized electron to a neighboring homolog molecule. In a solid the energy ΔE to produce these ion states is diminished by the existence of the molecules within a polarizable medium; thus

$\Delta E = I_g - A_g - 2P$, where P is the polarization energy of the "free electron" and, to a good approximation, is the same for electrons and holes. In this sense, one can speak of conduction states for electrons and holes separated by a band gap ΔE.

Polyethylene is a non–pendant-group polymer with a saturated backbone which has been extensively studied over the past decade (see Fig. 1b). The early studies were of its ultraviolet absorption, and the dominant feature is an absorption threshold extending from 7.2 to 11 electron volts. Energy band calculations for a single polyethylene chain predict a large energy gap between filled and vacant states. The strong covalent intrachain bonding leads to energy bands of considerable width with mobility estimates of ~25 cm²/V-sec. In this picture, the photoconductivity edge should coincide with the absorption edge; that is, the solid-state properties would be expected to be similar to those of a covalent inorganic semiconductor. In fact, the photoconductivity measurements show a threshold at 8.8 eV, which is significantly higher than the optical absorption threshold at 7.2 eV. Inclusion of the effects of highly localized excitons (electron-hole pairs bound by their mutual Coulomb attraction), which dominate the absorption threshold, helps to reconcile the photoconductivity results with the broad energy bands dictated by the structure of the polyethylene chain.

The most studied polymer with an unsaturated backbone is polyacetylene, $(CH)_x$, whose two possible structural forms are shown in Fig. 1b. It is the simplest unsaturated organic polymer, having a backbone composed of a linear chain of carbon atoms alternately singly and doubly bonded (7). The π electron is in a p orbital perpendicular to the plane of the chain. Within each C_2H_2 unit, the p orbitals combine to form a filled π and empty π^* orbital. In an extended polyene chain such as $(CH)_x$, these molecular orbitals broaden into filled and unfilled bands. If the π electron delocalization was complete and the carbon-carbon bond length was constant, the separation between the π and π^* bands would be zero and intrinsic metallic behavior would be observed. In practice, however, bond alternation persists and results in a finite gap between the π and π^* bands. Band structure calculations indicate that conduction and valence bandwidths are essentially the same for the cis and trans forms. The general feature of the alternating conjugation is the appearance of an absorption threshold in the visible region of the spectrum, and $(CH)_x$ at room temperature behaves as a semiconductor. It is this absorption that

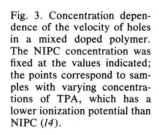

Fig. 3. Concentration dependence of the velocity of holes in a mixed doped polymer. The NIPC concentration was fixed at the values indicated; the points correspond to samples with varying concentrations of TPA, which has a lower ionization potential than NIPC (14).

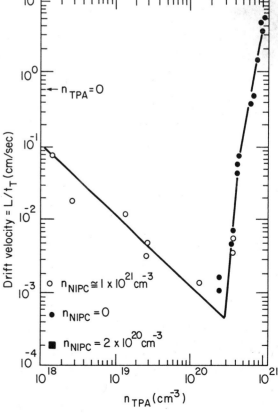

is responsible for the dichroism of the K polarizer invented by Land (8); this is formed by the dehydration of oriented polyvinyl alcohol, which results in oriented polyene chains.

Conduction in Pendant-Group Polymers and Molecularly Doped Polymers

A few highly insulating pendant-group polymers do exhibit significant conductivity when excess carriers are produced—for example, by light. Poly(N-vinylcarbazole), PVK, is a well-known example (see Fig. 1a). The facility with which charges move through such an insulator has been measured by a technique, widely applied to insulating solids, called time of flight (5). A sheet of charge, defined both spatially and temporally, is produced by a short pulse of light strongly absorbed in the film. The drift of the resultant photocarriers across the sample can be timed electronically to determine their velocity or mobility. The transport in PVK, in common with inorganic amorphous solids (9), is dispersive in the sense that although all the carriers begin to move simultaneously, they have a wide spread of arrival times at the other side of the sample. In addi-

tion, the mobilities, which are only observable for holes, are typically many orders of magnitude smaller than those in inorganic crystals.

Although initially this dispersive transport was thought to be due to ionic drift or impurity effects, results in the past few years have proved that it is associated with basic intermolecular electronic transfer interactions (10, 11). The hole transport in PVK is now known to involve electron hopping from one carbazole chromophore to another without involvement of the vinyl backbone. Since films of PVK are amorphous, the pendant carbazole units are in effect randomly oriented in space. The random distribution of distance and relative orientation between the planar carbazole chromophores results in a wide distribution of hopping probabilities between molecular sites and thus a wide distribution of effective velocities.

These ideas are also involved in the molecular doping of polymers—an approach that can be accurately described as molecular engineering of electronically active, insulating polymers. Otherwise electrically insulating and inactive polymers such as polycarbonates or polyesters can be doped with organic molecules such as N-isopropylcarbazole (NIPC),

tri-p-tolylamine (TTA), or trinitrofluorenone (TNF) (11–13). It is possible to form solid solutions with dopant polymer concentrations of about 10^{21} per cubic centimeter. In such a molecular dispersion, the molecules are separated on the average by ~10 angstroms from their nearest neighbors. Figure 2a illustrates the transport of holes produced by photoexcitation of the dopant molecules by a 10-nanosecond ultraviolet light pulse in a polycarbonate film doped with TTA. The dispersive nature of the transport is evident, with the transit time indicated by the shoulder corresponding to a hole drift mobility of ~2×10^{-5} cm²/V-sec.

Figure 2b indicates schematically the basic transport process. A neutral dopant molecule is photooxidized to produce a molecular cation, followed by electron transfer from a nearby neutral molecule to this cation. The "free hole" propagates through the film by successive reversible oxidation-reduction (electron loss–electron gain) reactions. Transport is therefore typically unipolar: hole transport occurs if the neutral dopant molecule is donor-like (for example, NIPC or TTA), electron transport if it is acceptor-like (TNF) (11, 13). (Thus, in an interesting conjunction of physics and chemistry, an effect that a solid-state physicist perceives to be macroscopic electrical conductivity could be described by a chemist as reversible solid-state chemistry.)

Detailed aspects of the transport process have been unraveled by changing the concentration of the dopant molecules in predetermined ways. This results in controlled, variable intermolecular separations, and shows clearly that the transport process is governed by electronic wave function overlap between neighboring molecules with any excess charge being highly localized on the host molecule. The high degree of localization is the determinant of the electronic properties of these systems and of the analogous pendant-type polymers. As a consequence, electronic transport in these systems is decidely non-band-like and proceeds by an intermolecular hopping process. Concomitantly low mobilities with values $\approx 10^{-3}$ cm²/V-sec are expected. This imposes considerable and definable constraints on the range of potential technological applications of these systems, although the applications that remain, as will be discussed later, can be of major consequence.

In principle, the broadest objective of science is that of harnessing nature's properties through understanding. In this

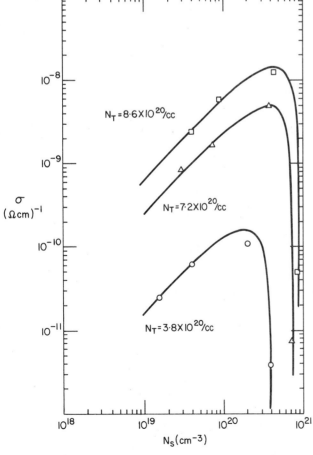

Fig. 4. Plot of the logarithm of the d-c conductivity of Lexan doped with three different concentrations (N_T) of TTA molecules against spin density, N_s. The spin density measures the concentration of the paramagnetic molecular cation TTA⁺ (15).

$N_T = 8 \cdot 6 \times 10^{20}$/cc

$N_T = 7 \cdot 2 \times 10^{20}$/cc

$N_T = 3 \cdot 8 \times 10^{20}$/cc

σ (Ωcm)⁻¹

N_s(cm⁻³)

respect, doped polymers provide a model system in which to test our understanding of the basic electronic properties of systems involving electron transfer between molecules. Such processes are basic to many phenomena both in nature and in potential technical applications, ranging from photosynthesis to photochemical photovoltaics. Figure 3 shows the results of experiments that confirm intuitive expectations of how admixtures of different molecules will act. In these experiments (14) controlled small quantities of a molecule, triphenylamine (TPA), with a lower ionization potential than a second molecule (NIPC), present in very high concentrations, are introduced. In the absence of TPA, the holes normally propagate by hopping between the closely spaced NIPC molecules. The introduction of TPA, with its lower ionization potential, results in the trapping of a hole and its temporary removal from the transport channel. As the concentration of TPA rises, the average carrier velocity falls. This is an unambiguous example of a trap-controlled hopping process. A concentration of TPA is ultimately reached at which (because of sufficient overlap between TPA molecules) it becomes an alternative and preferred transport channel, so the carrier velocity rises again.

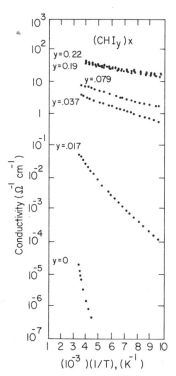

Fig. 5. Plot of the logarithm of conductivity against the reciprocal of temperature for polyacetylene for various concentrations (y) of iodine. The sharp rise in conductivity and related decrease to a weak activation energy is clearly seen (16).

In an extension of these ideas, chemical control of dark conductivity by the introduction of finite equilibrium concentrations of free carriers has been demonstrated (12). (The carriers are free in the sense that in an applied field they determine the conductivity; they are, of course, highly localized on the constituent molecules and are not free in the sense of carriers in a metal or a crystalline semiconductor.) The concept employed here can be understood in the following way. As discussed before, transport in electronically active but otherwise insulating systems is observed and measured by the transient photooxidation of some of the dopant molecules by suitable exposure to light. This photoinduced conductivity is nonequilibrium in that it decays with the removal of the illumination. By analogy, it is possible to effectively chemically oxidize a controlled, equilibrium fraction of the neutral molecules.

Figure 4 shows results of a study where, in addition to doping the polymer matrix with the molecule TTA, variable amounts of the molecular cation TTA$^+$ are incorporated so that the total number of molecules, neutral and oxidized, is constant. The TTA$^+$ is in the form TTA$^+$SbCl$_5^-$ (SbCl$_5^-$ is the average composition), prepared by reacting neutral TTA with the strong oxidizing agent SbCl$_5$. In this way different fractions of the TTA molecule can be oxidized to the molecular cation TTA$^+$ and constitute free carriers. The conductivity thus initially increases as the fraction of oxidized species is increased. Since the transport process requires electron transfer from a neighboring neutral molecule, if all the molecules are oxidized no transport can occur and the conductivity must go through a maximum, as observed. The maximum value of the dark conductivity achieved is $\sim 10^{-7}$ (ohm-cm)$^{-1}$, reflecting the intrinsically low mobility associated with hopping between localized states; the maximum also shows that in this system free carriers are ultimately formed at the expense of transport states. Interestingly, these molecular ion states that play an important role in the dark conductivity are, in the case of TTA, also paramagnetic, so opportunities exist to study correlations between electrical conductivity and magnetic phenomena. Recent coordinated studies (15) have revealed that in such systems exchange between two paramagnets (TTA$^+$ species) can be mediated by the presence of a diamagnetic, neutral TTA molecule; this effect is called superexchange.

Conduction in Non–Pendant-Group Polymers

This class embraces polymers with saturated backbones, such as polyethylene, which exhibit no significant electronic activity either intrinsically or after photoexcitation or chemical doping, and those with unsaturated backbones, such as polyacetylene. Recent work (16) on doping such an unsaturated polymer indicates that materials can be produced with conductivities close to those of conventional metals and with intermediate conductivities by controlling the doping concentrations. In fact, a remarkable range spanning 12 orders of magnitude in conductivity is possible. The materials have been known for several years, but the systematic study of their electrical properties is a relatively recent development, which has been made possible by advances in the production of high-quality materials and films. Thus the production of high-quality polyacetylene films of useful thickness and good mechanical flexibility by catalytic polymerization of acetylene gas (17) has led to extensive studies of their electronic properties. Much of this work is current, and so many of the mechanisms are not fully understood. Some general features do seem clear, however. In contrast with the pendant or doped polymers, the electronic states in these systems are associated with the in-chain bonded atoms. In this sense, within the single dimension of the chain the strong covalent bonding between the constituent atoms results in extended or bandlike states. Thus, con-

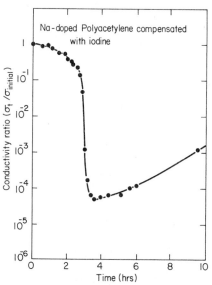

Fig. 6. Compensation curve for Na-doped polyacetylene, showing the conductivity ratio versus time. The sample was initially doped n-type and subsequently exposed to iodine vapor (18).

19

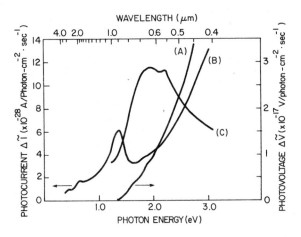

Fig. 7. (Curve A) Open-circuit photovoltaic response of a Schottky barrier junction formed between p-type polyacetylene and a metal with a low work function, normalized to the incident photon flux density and the transmittance of the indium contact. (Curve B) Photoconductivity similarly normalized. (Curve C) Absorbance spectrum of undoped $(CH)_x$ measured at room temperature (19).

ceptually one-dimensional order exists in an otherwise disordered or amorphous polymer.

As polymerized, undoped polyacetylene is a semiconductor with a conductivity of $\sim 10^{-5}$ to 10^{-6} (ohm-cm)$^{-1}$. Thermopower measurements indicate that the conductivity is p-type. The activation energy of the conductivity is much lower than the band gap deduced from optical absorption—0.3 versus 1.4 eV. By doping with the electron acceptor AsF$_5$ or with halogens, p-type conductivities as high as 1000 (ohm-cm)$^{-1}$, comparable to that of mercury, have been achieved in unoriented films. Doping with electron donors such as sodium naphthalide results in n-type material with comparable conductivity. The concentration dependence of the doping indicates that an abrupt semiconductor-metal transition occurs at ~ 1 to 3 atomic percent (Fig. 5). Below the transition the conductivity is strongly thermally activated, as expected for a semiconductor; above the transition, the conductivity approaches a temperature dependence like that for a metal. In the regime of high dopant concentrations, additional strong electronic absorption is observed in the infrared that is interpreted as being due to free carrier absorption. The ability to control conductivity by compensation, analogous to the technique familiar in inorganic semiconductors, has also been demonstrated. This is illustrated in Fig. 6, where the n-type conductivity produced by sodium doping is reduced by incorporation of acceptor-type states due to iodine (18). In addition, recent measurements by workers at IBM (19) indicate the onset of a photovoltaic response at ~ 1.48 eV in $(CH)_x$ (Fig. 7). The photoconductive quantum efficiencies reported to date are extremely low, of the order of 10^{-8} carriers per absorbed photon.

One of the more remarkable aspects of the work on polyacetylene is the fact that

such high conductivities are observed experimentally. The films consist of highly disoriented fibers (average diameter, 200 Å) and are reported to have a density only one-third the theoretical density determined from x-ray diffraction studies. Indeed, orientation of the films by stretching does result in further significant increases in the conductivity.

Considerable debate is taking place about the detailed interpretation of the observations in polyacetylene, particularly with respect to the semiconductor-metal transition. There are various possible explanations for this effect, but only two will be mentioned as illustrative of the kinds of ideas being considered. According to the first mechanism, the transition could be the result of a percolation phenomenon in which the segmented $(CH)_x$ chains constitute essentially unconnected metal wires as a result of doping. The problem is akin to dispersing small metal needles in an insulating medium and asking at what concentration sufficient interconnection will occur for the system to appear like a metal. Such a percolation phenomenon does exhibit a sharp conductivity threshold transition (20). A more fundamental explanation involves excitations known as solitons. In the case of polyacetylene, solitons are kinks in which domains to the left and right of the kink exhibit bond alternations that are out of phase with each other. Calculations (21, 22) suggest that at low doping levels, it is energetically more favorable for the charge from the dopant molecule to bind itself to the kink rather than enter the conduction or valence bands. In this model the precipitous drop in conductivity below the transition could be associated with a change in the transport mechanism (22). Carrier mobilities would therefore be highly concentration-dependent and would be relatively low in the semiconducting regime. The two effects discussed could both be

involved, of course, since they are not mutually exclusive.

Work is also under way on other materials such as polyparaphenylene, polypyrrole, and pyrolyzed polymers. Polyparaphenylene is a polymeric chain of phenyl rings, which exists in the form of a thermally moldable powder but has not, to date, been produced in thin flexible films. The band gap in undoped polyparaphenylene lies in the near ultraviolet. Doping with alkali metals or AsF$_5$ results in broad absorption from the far infrared to the visible and n- and p-type conductivities with relatively high values, but significantly less than those achieved in polyacetylene (23).

Highly stable flexible films of polypyrrole with p-type conductivities of ~ 100 (ohm-cm)$^{-1}$ have been prepared (24) by electrolytic oxidation of appropriate pyrrole monomers. The preparation of films by electrolysis of mixtures of pyrrole and N-methylpyrrole dissolved in acetonitrile, using tetraethylammonium tetrafluoroborate electrolyte, leads to conductivities that can be systematically varied over five orders of magnitude to the maximum of 100 (ohm-cm)$^{-1}$. This process avoids the diffusive and often nonuniform doping of films by exposure to oxidizing or reducing chemicals. Very little is known about the fundamental processes controlling the electronic properties of this material. The polymer is believed to be formed by the linkage of pyrrole units through α-carbon atoms and is therefore not a pendant-group polymer; Raman studies confirm the maintenance of the pyrrole rings in the polymer.

Technical Applications

The most extensive technological application of conductive polymers is in the electrophotographic industry. Polymers that are electrically insulating in the dark but can transport nonequilibrium carriers produced by light are well suited to function as photoreceptor elements (25). Polymers such as PVK or molecularly doped polymers have strong intrinsic optical absorption only in the ultraviolet, and since visible light is employed in practice, the polymer photosensitivity must be extended into the visible. This can be done by (i) formation of a charge-transfer complex with absorption in the visible, (ii) dye sensitization with an appropriately absorbing dye, or (iii) use of a thin contiguous sensitizing layer such as amorphous selenium. The ability to make large-area flexible polymer films at relatively low cost by solution coating accounts for the application of these materials in electrophotography. Although pendant-group polymers or polymers doped with aromatic molecules have very low mobilities $\gtrsim 10^{-3}$ cm²/V-sec, this is not a limitation for electrophotographic usage. In this process the most important parameter, within limits, is how far the carriers move before they are immobilized rather than how fast they travel (26). For most process speeds such mobilities are adequate provided the photogenerated carriers can transverse the total device thickness; that is, $\mu E \tau >$ sample thickness, where μ is the mobility, E the electric field, and τ the lifetime of carriers with respect to deep traps. Remarkably, such polymers do exhibit large carrier ranges because of their very large deep-trapping lifetimes.

For systems where the photogeneration occurs by photoexcitation within a polymer, such as the charge-transfer complexes or dye-sensitized systems, the photogeneration efficiency can be controlled by a geminate recombination mechanism. In such a process, the initial photoexcitation leads to electron-hole pairs, which diffuse within their mutual Coulomb well. There is a finite probability, which decreases with applied field, that the pair will recombine on the initial excitation site rather than thermally dissociate into a free electron and a hole (27). Carrier recombination in molecular systems can lead to quantum efficiencies for photogeneration that are substantially less than unity and are strongly field-dependent. This may result in a photosensitivity limitation, depending on the particular system, and can be overcome to some degree by increasing the light exposure in a machine.

For other electronic applications the magnitudes of the carrier mobilities are of paramount importance, since they determine the frequency response of devices or, through the related diffusion lengths, determine ultimate collection efficiencies in devices such as photovoltaic cells. Pendant-group polymers and disordered molecularly doped systems are not likely to find applications in these areas because of their low mobilities and the probable importance of geminate recombination processes. No such assessment can yet be made for materials such as polyacetylene or polyparaphenylene, since questions remain to be answered regarding many fundamental parameters. The ability to produce p- and n-type materials in these solids by appropriate doping holds some promise. Questions remain about the long-term chemical stability of individual and contiguous layers. Preservation of high mobility associated with the one-dimensional order along the chains in the semiconducting regime would be an enormously important feature, but thermopower and conductivity measurements suggest that this may not be the case. Very little information is available regarding localized gap states and minority carrier parameters. Initial photoconductivity studies (19) indicate extremely low quantum efficiencies for photogeneration in currently available materials, but it is not clear whether this is intrinsic to the materials.

Conclusion

The current status of investigations of polymers as electronic materials has been briefly reviewed. Despite many gaps in our understanding, some general codification of properties according to the electronic states of polymers is possible. The remarkable spectrum of conductive behavior observed should provide further stimulation for fundamental studies and the exploration of possible technological applications. This mutual interplay of scientific and technological motivation can be expected to lead to improved understanding, new materials, and new concepts.

The materials are quite complex, and the many questions concerning chemical reactivity and materials stability are further complicated by the morphological features of polymers. Eventual practical application will hinge on the tractability of materials from the perspective of materials processing. Offsetting such difficulties is the diversity of molecular and materials architectures inherent in organic chemistry which can be brought to bear on these problems. Thus the field, still very much in its infancy, poses experimental and theoretical challenges to experimentalists, theorists, synthetic chemists, and materials scientists alike.

Attempts to predict specific applications for the semiconductive and metal-like polymers are not warranted at this stage, since the definition of the full scope and limitations of the phenomena and materials has just begun. The situation is quite different for electronically active insulating polymers, which have already made the difficult transition from laboratory to marketplace. Electronically active polymers function as electronic materials with major industrial applications in the electrophotographic industry. As such, they constitute a tangible demonstration of the promise of appropriate polymers as electronic materials for specialized applications where low cost and large area are important.

References and Notes

1. N. F. Mott and E. A. Davis, *Electronic Processes in Non-Crystalline Materials* (Oxford Univ. Press, New York, 1979).
2. N. Karl, *Festkörperprobleme XIV*, H. J. Queisser, Ed. (Pergamon, New York, 1974).
3. F. Guttman and L. E. Lyons, *Organic Semiconductors* (Wiley, New York, 1967).
4. J. Mort, *Adv. Phys.*, in press; M. Stolka and D. M. Pai, *Adv. Polym. Sci.* **29**, 1 (1978).
5. J. Mort and D. Pai, Eds., *Photoconductivity and Related Phenomena* (Elsevier, Amsterdam, 1975).
6. J. Ritsko and R. Bigelow, *J. Chem. Phys.* **69**, 4162 (1978).
7. P. M. Grant and I. P. Batra, *Synth. Met.* **1**, 193 (1980).
8. E. M. Land, *J. Opt. Soc. Am.* **41**, 957 (1951).
9. J. Mort and G. Pfister, *Polym. Plast. Technol. Eng.* **12**, 89 (1979); H. Scher and E. W. Montroll, *Phys. Rev. B* **12**, 2455 (1975).
10. W. D. Gill, in *Proceedings of the 5th International Conference on Amorphous and Liquid Semiconductors, Garmisch-Partenkirchen 1973*, J. Stuke and W. Brenig, Eds. (Taylor & Francis, London, 1974), p. 901.
11. J. Mort, G. Pfister, S. Grammatica, *Solid State Commun.* **18**, 693 (1976).
12. J. Mort, A. Troup, S. Grammatica, D. J. Sandman, *J. Electron. Mater.* **9**, 411 (1980).
13. W. D. Gill, *J. Appl. Phys.* **43**, 5033 (1973).
14. G. Pfister, J. Mort, S. Grammatica, *Phys. Rev. Lett.* **37**, 1360 (1976).
15. A. Troup, J. Mort, S. Grammatica, D. J. Sandman, *Solid State Commun.* **33**, 91 (1980).
16. D. J. Berets and D. S. Smith, *Trans. Faraday Soc.* **64**, 823 (1968); C. K. Chiang, Y. W. Park, A. J. Heeger, H. Shirakawa, E. J. Louis, A. G. MacDiarmid, *J. Chem. Phys.* **69**, 5098 (1978).
17. T. Ito, H. Shirakawa, S. Ikeda, *J. Polym. Sci. Polym. Chem. Ed.* **13**, 1943 (1975).
18. C. K. Chiang, S. C. Gau, C. R. Fincher, Jr., Y. W. Park, A. G. MacDiarmid, A. J. Heeger, *Appl. Phys. Lett.* **33**, 18 (1978).
19. T. Tani, P. M. Grant, W. D. Gill, G. B. Street, T. C. Clarke, *Solid State Commun.* **33**, 499 (1980).
20. R. Zallen, *Statphys. 13, Proceedings of the 13th IUPAP Conference on Statistical Physics*, D. Cabib, C. G. Kuper, I. Riess, Eds. (Hilger, Bristol, 1978), pp. 309–321.
21. M. J. Rice, *Phys. Lett. A* **71**, 152 (1979).
22. W. P. Su, J. R. Schrieffer, A. J. Heeger, *Phys. Rev. Lett.* **42**, 1698 (1979).
23. D. M. Ivory, G. G. Miller, J. M. Sowa, L. W. Shacklette, R. R. Chance, R. H. Baughman, *J. Chem. Phys.* **71**, 1506 (1979).
24. K. K. Kanazawa, A. F. Diaz, R. H. Geiss, W. D. Gill, J. F. Kwak, J. A. Logan, J. F. Rabolt, G. B. Street, *J. Chem. Soc. Chem. Commun.* (1979), p. 854.
25. J. Weigl, *Angew. Chem. Int. Ed. Engl.* **16**, 374 (1977).
26. J. Mort and I. Chen, *Appl. Solid State Sci.* **5**, 69 (1975).
27. P. M. Borsenberger, L. E. Contois, D. C. Hoesterey, *Chem. Phys. Lett.* **56**, 574 (1978).
28. I have benefited from numerous discussions with colleagues, both at Xerox Corporation and in the wider scientific community, whose collective work and insight have made the scientific and technological progress described in this article possible.

Biomaterials

L. L. Hench

As living beings get older, they begin to wear out. Although many factors responsible for aging are not understood, the consequences are quite clear. Our teeth become painful and must be removed, joints become arthritic, bones become fragile and break, the powers of vision and hearing diminish and may be lost, the circulatory system shows signs of blockage, and the heart loses control of its vital pumping rhythm or its valves become leaky. Tumors appear almost randomly in bones, breast, skin, and vital organs. And, as if these natural processes did not occur fast enough, we have achieved an enormous capacity for maiming, crushing, breaking, and disfiguring the human body with motor vehicles, weapons, and power tools or as a result of our participation in sports.

A consequence of these natural and unnatural causes of deterioration of the human body is that some 2 million to 3 million artificial or prosthetic parts are implanted into individuals in the United States each year. A list of some of the devices and their function is given in Table 1. More than 50 implanted devices made from more than 40 different materials are included alone and in various combinations. Although many materials appear several times in this table there is no apparent commonality of microstructure, atomic structure, composition, or surface features.

The challenge of the field of biomaterials is that all implant devices replace living tissues whose physical properties are a result of millions of years of evolutionary optimization, and which have the capability of growth, regeneration, and repair. Thus, all man-made biomaterials used for repair or restoration of the body represent a compromise. The relative success or failure of a biomaterial reflects the scientific and engineering judgment used in achieving this compromise. The interaction of many complex physical, biological, clinical, and technological factors must be considered.

The author is professor and head, Ceramics Division, Department of Materials Science and Engineering, University of Florida, Gainesville 32611.

For example, consider the following characteristics of a natural tooth that must be satisfied in some measure to achieve a successful tooth implant: a tensile strength of 15,000 to 20,000 pounds per square inch in flexure; a biologically bonded interface with epithelial skin cells, gingival tissues, and bone, which results in a difference of more than 10^3 in elastic moduli across the various interfaces in contact with a tooth; and an attachment structure (the periodontal ligament) that converts compressive stresses applied to the tooth to tensile stress within the jawbone. Although many materials have the requisite flexural strength, no material known today can reliably achieve the stable interfacial attachments required to mimic a natural tooth. Is it any wonder that few extensive clinical studies show more than 50 percent success rates for long-term (> 5 years) dental implants?

Control over the biomaterials-tissue interface is the paramount problem in this field of materials science (1). The physical properties of most tissues can be matched within engineering limits by careful selection of metals, ceramics, or polymer materials singly or in specially designed combinations (Table 1) (1, 2). Even the requirements that the biomaterial be nontoxic to the host tissues can be achieved relatively easily by screening of the materials with tissue culture tests or short-term implants. But, achieving the necessary match or gradient in physical properties across the interface between living and nonliving matter is a formidable scientific challenge. Part of the difficulty is that the science of adhesion of biological interfaces is still being developed. Until cell biologists and biochemists discover which molecular species control the bonding of cells to each other, the understanding of adherence or lack of adherence of tissues to implant devices will remain incomplete.

The field of biomaterials developed historically so as to achieve a suitable combination of physical properties to match those of the replaced tissue with a minimal toxic response to the host. This

approach has led to a reasonably large catalog of "nearly inert" biomaterials (most of those in Table 1) that comprise the bulk of the 2 million to 3 million devices implanted yearly. A common feature of these materials is that they initiate the growth of a thin, fibrous capsule which separates the normal tissue from the implant.

Figure 1A shows an example of the thin fibrous capsule formed between a nearly inert biomaterial, a copolymer of methylmethacrylate and hydroxyethyl methacrylate, and the subcutaneous tissue of a rat 8 weeks after the material was implanted. The muscle and subcutaneous connective tissue are normal; the thin fibrous capsule is the only evidence that the implant has been present in the host tissue.

In contrast, a reactive material such as an acrylic acid and methylmethacrylate copolymer (Fig. 1B) produces a very thick fibrous capsule which extends throughout the subcutaneous region 8 weeks after implantation in a rat. Even the muscle tissue shows extensive inflammation resulting from the reactivity of the implant.

There is little, if any, adhesion between the implants and the fibrous capsules. Consequently, movement of the implant within the capsule can occur when stress is applied with the following possible results. (i) The capsule may increase in thickness. A thick capsule may interfere with the local blood supply to tissues (3) or provide a site for accumulation of biochemical by-products perhaps associated with formation of tumors (3, 4). (ii) The capsule may calcify and harden. Progressively stiffening capsules around devices such as silicone breast implants produce pain and deterioration of underlying tissues because of the mismatch of mechanical properties (5). (iii) Localized concentrations of stress may result. Mechanical damage to the host tissue, such as the microfracture of boney spicules adjacent to the stem of a hip or knee implant, can cause pain around and progressive loosening of the implant (6). More motion results, and even larger stress concentrations occur until either the bone or the implant fractures. (iv) The effects of infection may be magnified (7). An infection may occur or persist at an implant site because there is not a normal blood supply to the capsule. The lack of blood prevents the invasion of white cells necessary to attack the infection and retards the transport of cell debris away from the site of infection. (v) The capsule may separate from the device. Spalling of fragments of a poorly adherent layer from the surface of cardiovascular implants such as heart valves, arterial or venous grafts, or from the walls of an artificial heart can result in fatal emboli (8). (vi) Corrosion products may accumulate. Again because of the lack of circulation, products from the corrosion of metals or deterioration of polymers can accumulate within the capsule or at the capsule-implant interface (9).

Methods of Controlling
Biomaterials Interfaces

Because of the interfacial problems associated with nearly inert biomaterials, much research during the last decade has been directed toward stabilizing the tissue-biomaterial interface by controlling either the chemical reactions or the microstructure of biomaterials. That the microstructure can be controlled is based on the hypothesis that tissue can grow into pores or surface depressions if the pores are big enough and if the tissue can maintain a vascular supply and tissue vitality (10, 11). A fibrous membrane will still interpose itself between the surface and the pores and the infiltrating tissue. However, the mechanical keying caused by the interdigitation of the living and inanimate material serves to inhibit growth of a fibrotic capsule by retarding motion and by distributing stress over a large interfacial area. Because of the large interfacial area exposed to tissue and tissue fluids it is important that porous biomaterials be especially resist-

ant to corrosion and deterioration in the body. Bioceramics, especially aluminum oxide, have potential for microstructural control of the interface without formation of potentially toxic corrosion products. Only a few metals, that is, titanium and cobalt-chromium alloys, exhibit sufficient corrosion resistance to be considered for use in porous implants (12).

Quantitative analyses of the growth of tissues into pores of different sizes show that soft connective tissue will grow into pores of greater than 50 micrometers in diameter and remain healthy over periods of at least several years; bone will grow into pores bigger than 100 μm (13, 14). There is debate about whether bone will develop as fully mineralized tissue within a porous structure under the continual stress and micromovement associated with a functional load-bearing implant (15). Most studies of hard tissue ingrowth into porous materials have been done with the use of nonfunctional plugs in the long bones of dogs, and it is not known how much stress is transmitted to the plug. When stress transfer to bone via porous interfaces does occur a high modulus of elasticity by the implant can cause extensive bone resorption (16). Design of porous or porous surface implants must take these factors into consideration.

Another factor still open to debate is the mechanical fatigue resistance of an interface composed of a large number of thin webs of tissue and material. Figure 2 shows the microstructure of bone ingrowth into 200-μm (average) pores of an aluminum oxide ceramic implanted in a rabbit femur for 8 weeks (17). Since there is a distribution of the pore area filled with dense bone there must also be distribution of the localized strength, elastic moduli, and fatigue resistance of such an interface. Current studies involving modeling of microstructurally controlled interfaces with finite element analyses coupled with postmortem stress-strain testing of functional implants put into animals, should help determine the long-term utility of this type of interface control (18).

Microporous structures have been successfully used as vascular prosthetic devices (19). To ensure that blood will continue to flow without clotting through such a replacement artery it is essential that the flow surface, originally of "biomaterial," be lined by a natural lining—the so called "neointima." Once laid down this lining must be maintained and one way to do this is to establish transmural tissue growth from the outer scar tissue that includes blood vessels to nourish the inner lining. Alternative methods

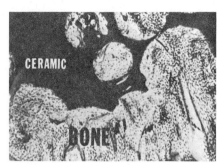

Fig. 2. Thin section of Al_2O_3 with 200-μm pores with rabbit femoral bone ingrowth in 8 weeks. [Photo courtesy of J. J. Klawitter and S. Hulbert]

of maintaining this neointima, by simple diffusion of oxygen from the blood passing through and by growth of thin blood vessels into the neointima from the ends where the biomaterial joins the natural artery, are sufficient only over relatively short distances. Transmural growth seems to be desirable when longer lengths of vessel need replacement.

A second method of manipulating the biomaterials-tissue interface is controlled chemical breakdown, that is, resorption, of the material. Resorption of biomaterials appears a perfect solution to the interfacial problem because the foreign material is ultimately replaced by regenerating tissues (20). Ideally, there is eventually no discernible difference between the implant site and the host tissue. This method comes closest to the grafting of a patient's own tissue when

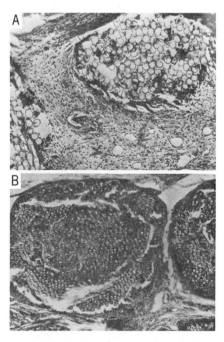

Fig. 3. Effect of time on resorption of a Dexon (PGA) suture implanted subcutaneously in rat ($\times$ 250). (A) After 2 weeks; (B) after 8 weeks. [Photo courtesy of A. Reed, J. Wilson, and K. Gilding]

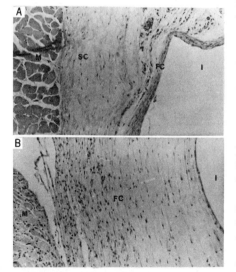

Fig. 1. Fibrous capsule (FC) developed between an implant (I) and the subcutaneous tissue (SC) and muscle (M) of a rat 8 weeks after implantation. (A) Nearly inert biomaterial: hydroxylpolymethylmethacrylate; (B) reactive biomaterial: acrylic acid-methylmethacrylate copolymer ($\times$ 100). [Photo courtesy of J. Wilson]

Table 1. Implant devices in use or test today, their function, and the biomaterials used.

Device	Function	Biomaterial
Sensory and neural systems		
Artificial vitreous humor	Fill the vitreous cavity of the eye	Silicone Teflon sponge; polyglycerylmethacrylate (PGMA)
Corneal prosthesis	Provide an optical pathway to the retina	Polymethylmethacrylate (PMMA); hydrogels
Intraocular lens	Correct problems caused by cataracts	PMMA (lens); nylon, polypropylene, Pt, Ti, Au loops
Artificial tear duct	Correct chronic blockage	PMMA
Artificial eustachian tube	Provide clear ventilation passage	Silicone rubber, Teflon
Nerve tubulation	Align severed nerves	Silicone membrane, porous surgical metals
Middle ear prostheses	Replace diseased bones of the middle ear	PMMA; metallic wire, Proplast (PTFE + carbon fiber); Bioglass
Percutaneous leads	Conduct power to electrical sensory devices	Nylon or Dacron velour, PMMA
Auditory prostheses, visual prostheses	Restoration of hearing and vision	Pt and Pt-Ir wires and electrodes; $Ta-Ta_2O_5$ electrodes, stainless steel, Elgiloy wires; silicone rubber; PMMA
Electrical analgesia	Eliminate chronic pain	Pt and Pt-Ir wires and electrodes, $Ta-Ta_2O_5$ electrodes, stainless steel, Elgiloy wires, silicone rubber, PMMA
Electrical control of epileptic seizure	Conduct electrical signals to brain	Same
Electrophrenic stimulation	Control breathing electrically	Same
Bladder control	Stimulate bladder release	Same
Heart and cardiovascular system		
Myocardial and endocardial stimulation (heart pacer)	Maintain heart rhythm	Stainless steel, Ti cans; silicone rubber, wax epoxy encapsulants; Pt or Pt-Ir alloy-electrode, Elgiloy wire
Chronic shunts and catheters	Assist hemodialysis	Polyethylene, hydrophilic coatings
Cardiac heart valves	Replace diseased valves	Co-Cr alloys; low-temperature isotropic carbon; porcine grafts; Ti alloy with Silastic or pyrolytic carbon disks or balls
Arterial and vascular prostheses; artificial heart components; heart assist devices	Replace diseased arteries and blood vessels; replace the heart; augment diseased heart	Segmented polyurethanes; silicone rubber or pyrolytic carbon mandrels with Dacron mesh sheaths; heparin + GBH or TGBH coupled coatings on Teflon or silicone rubber; poly-HEMA-coated polymers; Dacron velours, felts, and knits; textured polyolefin (TP), TP with cross-linked gelatin surface; Teflon (PTFE) alone
Skeletal system repair and replacement		
Artificial total hip, knee, shoulder, elbow, wrist	Reconstruct arthritic or fractured joints	Stems: 316L stainless steel; Co-Cr alloys; Ti and Ti-Al-V alloy; Co-Cr-Mo-Ni alloy cups: high-density, high-molecular-weight polyethylene; high-density alumina; "cement" PMMA; low-density alumina; polyacetal polymer; metal-pyrolytic carbon coating; metal-Bioglass coating; porous polytetrafluroethylene (PTFE); and PTFE-carbon coatings on metal; PMMA-carbon fibers, PMMA-Ceravital powder composite; porous stainless steel; Co-Cr; Ti and Ti alloys
Bone plates, screws, wires	Repair fractures	316L stainless steel; Co-Cr alloys; Ti and Ti alloys; polysulfone-carbon fiber composite; Bioglass-metal fiber composite; polylactic acid-polyglycolic acid composite
Intramedullary nails	Align fractures	Same
Harrington rods	Correct chronic spinal curvature	Same
Permanently implanted artificial limbs	Replace missing extremities	Same plus nylon or Dacron velours on Silastic for soft tissue ingrowth
Vertebrae spacers and extensors	Correct congenital deformity	Al_2O_3
Spinal fusion	Immobilize vertebrae to protect spinal cord	Bioglass
Functional neuromuscular stimulation	Control muscles electrically	Pt, Pt-Ir electrodes; silicone; Teflon insulation
Dental		
Alveolar bone replacements, mandibular reconstruction	Restore the alveolar ridge to improve denture fit	PTFE carbon composite (Proplast); porous Al_2O_3; Ceravital: hema hydrogel-filled porous apatite; tricalcium phosphate; PLA/PGA copolymer; Bioglass
Endosseous tooth replacement implants (blades, anchors, spirals, cylinders—natural or modified root form)	Replace diseased, damaged, or loosened teeth	Stainless steel, Co-Cr-Mo alloys, Ti and Ti alloys, Al_2O_3, Bioglass, LTI carbon, PMMA, Proplast, porous calcium-aluminate, $MgAl_2O_4$ spinel, vitreous carbon, dense hydroxyapatite
Subperiosteal tooth replacement implants	Support bridge work or teeth directly on alveolar bone	Stainless steel, Co-Cr-Mo alloy, LTI carbon coatings
Orthodontic anchors	Provide posts for stress application required to change deformities	Bioglass-coated Al_2O_3; Bioglass-coated Vitallium

possible. Practically, there are severe limitations on achieving the combination of the requisite physical properties and resorption into chemical constituents that can be processed by the metabolic system. Polymer systems based on polyglycolic (PGA) or polylactic acids (PLA), or both, decompose into CO_2 and H_2O and are used clinically as resorbable sutures and implantable drug delivery systems (21). Copolymer systems composed of PGA and PLA also show some promise for use as resorbable fracture fixation devices. Resorbable calcium phosphate or apatite ceramics break down to soluble calcium and phosphate salts which are able to be metabolized within hard tissues and are being tested for use in dental reconstruction (22).

A problem with the chemical breakdown method is that the strength of the resorbable biomaterial decreases as resorption occurs. Unless there is close matching of the reduction in implant strength with the increase in strength of the healing tissues the implant-tissue system will fail. The physicochemical reactions associated with resorption are complex and this makes it difficult to match time-dependent physical properties. Figures 3 and 4 show that the strands of a PGA suture, Dexon (American Cyanamid), become progressively infiltrated with tissue during an 8-week period in rat (23). Although one can still

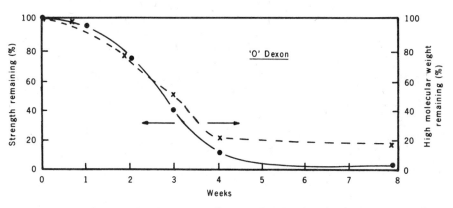

Fig. 4. Effect of implantation time on tensile strength (●) and molecular breakdown (X) of Dexon (PGA) suture in rat subcutaneous tissue in vivo. [Based on Reed (23)]

see the suture after 8 weeks, its tensile strength has diminished to only a few percent of the original value. A progressive breakdown of the higher molecular-weight polymerized structure occurs during this period. In contrast, experimental PLA implants show no detectable degradation within the 8-week period (23). Thus, it might seem that copolymers of PGA and PLA could be used to carefully control time-dependent changes of strength. However, recent studies show that PGA/PLA copolymers produce wide variations in rates of resorption that depend on both the degradation of the end-members and the degree of crystallinity in the copolymer (23). Prediction of the time dependence

of properties is still not possible without knowing more about the mechanisms of chemical breakdown of copolymers in the body and how molecular structure is related to strength.

Resorbable polymers are too weak to be used as replacements for bones and joints. Tricalcium phosphate and calcium-aluminate-phosphate bioceramics show promise for use as hard tissue replacement (22), but more research is required to get the right rate of loss of strength (24), ensure the nontoxic metabolism of the large concentration of mineral salts, and relate the effects of microstructure to the rates of resorption for this class of biomaterials (22).

The third approach to control of the

Table 1 (continued).

Device	Function	Biomaterial
Space-filling soft tissue prostheses		
Facial contouring and filling prostheses (nose, ear, cheek)	Replace diseased, tumorous, or traumatized tissue	Silicone rubber (Silastic), polyethylene, PTFE, silicone fluid, dissolved collagen fluid, polyrane mesh
Mammary prosthesis	Replace or augment breast	Silicone gel and rubber, Dacron fabric; hydron sponge
Cranial boney defects and maxillofacial reconstruction prostheses	Fill defects	Self-curing acrylic resin; stainless steel, Co-Cr alloy, Ta plates; polyethylene and polyether urethane–coated polyethylene terephthalate–coated cloth mesh
Artificial articular cartilage	Replace arthritis deterioration cartilage	Crystallized hydrogel-PVA and polyurethane polymers; PFTE plus graphite fibers (Proplast)
Miscellaneous soft tissues		
Artificial ureter, bladder, intestinal wall	Replace diseased tissue	Teflon, nylon-polyurethane composite; treated bovine pericardium; silicone rubber
Artificial skin	Treat severe burns	Processed collagen; ultrathin silicone membrane polycaprolactone (PCA) foam-PCA film composite
Hydrocephalus shunt	Provide drainage and reduce pressure	Silicone rubber
Tissue patches	Repair hernias	Stainless steel, Marlex, Silastic, Dacron mesh
Internal shunt	Provide routine access to dialysis units	Modified collagen; Silastic
External shunt	Provide routine access to dialysis	Silastic-Teflon or Dacron
Sutures	Maintain tissue contact to aid healing	Stainless steel, silk, nylon PGA, Dacron, catgut, polypropylene
Drug delivery systems	Release drugs progressively; immobilize enzymes	Silicone rubber, hydrogels ethylene-vinyl acetate copolymer, PLA/PGA polysaccharides-vinyl polymers
Artificial trachea	Reconstruct trachea	Porous Dacron-polyether urethane mesh, Ta mesh, Ivalon sponge and polypropylene mesh

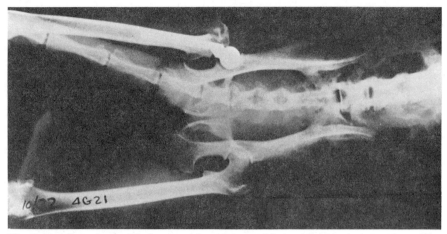

Fig. 5. X-radiograph of a Bioglass-coated stainless steel femoral head replacement in monkey after 40 weeks without use of a polymer "bone cement." [Photo courtesy of W. Petty and G. Piotrowski]

materials interface is to use biomaterials with controlled surface reactivity. In this class of biomaterials the composition is designed such that the surface undergoes a selected chemical reactivity with the physiological system and thereby establishes a chemical bond between tissues and the implant surface (25, 26). The desired bonded interface protects the implant material from further deterioration with time, that is, it self-passivates. Thus the potential of this approach is to combine the high strength or flexibility of nearly inert biomaterials with the surface chemical reactivity needed for tissue adherence and bonding. Ideally, interface stabilization by surface reactivity produces more flexibility in device design and fabrication than does mechanical interlocking or resorption. Practically, it is difficult to get the requisite mechanical and surface chemical properties in the same material. Certain compositional ranges of soda-calcia-phospho-silicate glasses (27), the glass-ceramics (Bioglass, Bioglass-Ceramic, Ceravital) (28) and dense biologically reactive hydroxylapatite (Durapatite) (29) develop a chemical bond with living bone. The mechanical limitations of these compositions have required the development of a number of means of using the controlled surface-active materials as a component in a composite system, such as coatings on dense high-strength alumina (30) or surgical alloys (31), as an active filler in bone cement (32), or with metal fiber reinforcement (33).

Figure 5 shows an x-radiograph of a Bioglass coated 316L stainless steel partial hip replacement in a monkey 40 weeks after implantation. Mechanical testing of the implant in tension showed that it resisted a fracture load nearly equivalent to that of the bone of the op-

posite leg. Thus, a viable chemical bond had been established at the interface between the bone and the implant via the reactive glass coating.

A scanning electron micrograph of the bonded interface (Fig. 6) shows continuity between the metal, the reactive glass coating, and the bone. Energy dispersive x-ray analysis at various points across the interface shows the characteristic secondary x-ray peaks for calcium and phosphorus in the bone and silicon and calcium in the reactive glass, and intermediate concentrations in the 100-μm-thick bonding zone. The mechanism of formation of the bond has been shown to be the development of a biologically reactive hydroxylapatite and silica-rich layers on the glass surface (34). This active surface incorporates metabolic constituents such as collagen within itself as polymerization and crystallization of the inorganic phases proceeds. A scanning electron micrograph of collagen bonded within such a surface after just 2 weeks in vitro (Fig. 7), shows that mechanically and chemically graded interfaces between physiologically derived substances and prosthetic devices can be achieved (35).

Importance of Reliability

Regardless of whether a biomaterial responds to the body as: (i) nearly inert, (ii) porous microsurface, (iii) resorbable, or (iv) surface-reactive material, the central issue today is the reliability of the biomaterial and the devices made from it. The use of even more prostheses is likely in the future, for several reasons. The high (85 to 98 percent) success rates reported for many short-term (< 5 years) implants encourages patients to seek

physicians who will use prostheses and, at the same time, encourages surgeons to use devices in patients with a wide range of symptoms and in younger patients. Since more surgeons are gaining confidence in the use of implants, and since prosthetic devices are being used in new clinical applications, longer and more severe service is imposed on the implants, which they do not always withstand. For this and other reasons (36) the number of reoperative cases involving implants is steadily increasing. Most reparative operations are more difficult technically than the initial operation, less amenable to the use of generalized procedures and devices, and are complicated by the extensive tissue damage resulting from the implant failure. These factors, the increased age of the patients, and the negative psychological consequences of a previous implant failure all reduce the probability of subsequent success. Reoperative cases generally require the services of the more specialized and capable implant surgery teams, and can consume a progressively larger fraction of their time and facilities. Obviously, the reliability of prostheses must be improved.

I recommend that reliability oriented biomaterials research be directed toward three areas.

1) Composite biomaterials systems.

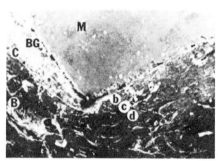

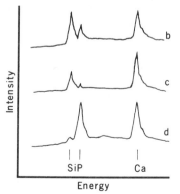

Fig. 6. (Top) Scanning electron micrograph of an interface (C) of monkey femoral bone (B) bonded to Bioglass (BG) coated to a stainless steel metal (M) prostheses. (Bottom) Compositional analysis to show tissue at spot (d), bonding interface (c), and Bioglass coating (b) (×20).

Many of the devices listed in Table 1 are constructed of more than one material or include modifications of the surface of a material. However, the unique combinations of biological and physical properties that can be achieved in this way have only begun to be exploited.

2) Mechanisms of interfacial reactions. The interface between tissue and implant surface and that between phases in a composite, such as a coating and substrate, are potential weak links in long-term reliability. Only by a thorough study of the mechanisms and kinetics of interfacial reactions will it be possible to determine why failure occurs. Knowledge of failure mechanisms is essential in designing better devices. Likewise, predicting the reliability of a biomaterial or device in service requires understanding the modes of failure of the tissue-implant system.

3) Performance prediction for long-term service (36). To develop endurance tables for prostheses will require (i) an understanding of the mechanisms and kinetics of interfacial reactions; (ii) use of mathematical techniques, such as finite element stress analysis, and biomechanics to describe the anticipated stress-time cycles to be applied to prostheses and to define reasonable safety margins of stress; (iii) expansion and application of appropriate theories of the fracture mechanics of brittle materials to predict the expected lifetimes of implants loaded at given stress levels (37); (iv) collaboration with the fracture mechanics R & D community to extend life prediction theories to include viscoelastic materials such as bone and polymeric materials, plastically deforming metallic components and combinations thereof, with variable degrees of interfacial attachment between the materials and between materials and tissue; (v) developing accelerated fatigue tests of simple sample configurations and devices that are representative of in vivo conditions; and (vi) establishing methods to correlate predictive relationships with data on implants that have been removed from patients.

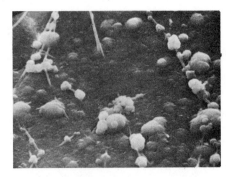

Fig. 7. Scanning electron micrograph of collagen fibers attached to a Bioglass surface after exposure in vitro at 37°C for 10 days (×5000). [Photo courtesy of C. G. Pantano]

An effort has been made to predict the long-term reliability of a Bioglass coated-alumina system by using some of these approaches (38). A potential problem was identified in the use of this composite for load-bearing orthopedic implants, and research has been redirected to try to solve it.

Conclusions

The impact of advanced materials technology on the biomaterials field is threefold. New types of composite materials can be created with previously unobtainable combinations of biological and physical properties. Some are already in clinical trials. New techniques for characterizing biomaterials and their interfaces are now available. Methods are becoming available for predicting service lives of materials and prostheses. Use of these new capabilities in concert should produce better prostheses in the decade ahead.

References and Notes

1. L. L. Hench and E. C. Ethridge, *Biomaterials: An Interfacial Approach* (Academic Press, New York, 1980).
2. J. B. Park, *Biomaterials: An Introduction* (Plenum, New York, 1979).
3. I. Elsebai, *Cancer J. Clin.* 27, 100 (1977).
4. C. B. Ripstein, D. M. Spain, I. Bluth, *J. Thorac. Cardiovasc. Surg.* 45, 362 (1968).
5. J. Smahel, *Plast. Reconstr. Surg.* 61 (No. 1), 80 (1978).
6. P. Ducheyne, A. Kagan, J. A. Lacey, *J. Bone J. Surg.* 60-A (No. 3), 384 (1978).
7. J. Calnan, *Br. J. Plast. Surg.* 16, 1 (1963).
8. S. D. Bruck, *Ann. N.Y. Acad. Sci.* 283, 332 (1977).
9. L. L. Hench, H. A. Paschall, W. C. Allen, G. Piotrowski, *Natl. Bur. Stand. U.S. Spec. Publ.* 415, 19 (1975).
10. S. F. Hulbert, S. J. Morrison, J. J. Klawitter, *J. Biomed. Mater. Res.* 6, 347 (1972).
11. P. Predecki, B. A. Auslaender, J. E. Stephan, V. L. Mooney, C. Stanitski, *ibid.*, p. 401.
12. J. Galante, in *Mechanical Failure of Total Joint Replacement* (Document 916-78, Steering Committee, American Academy of Orthopedic Surgeons, Chicago, 1978), p. 107.
13. S. F. Hulbert and J. J. Klawitter, *J. Mater. Res. Bull.* 7, 1239 (1972).
14. J. J. Klawitter and A. M. Weinstein, *Acta Orthop. Belg.* 40, 755 (1974).
15. P. Ducheyne, P. DeMeester, E. Aernoudt, M. Martens, J. C. Mulier, *J. Biomed. Mater. Res.* 11, 811 (1977).
16. R. M. Pilliar, H. U. Cameron, A. G. Binnington, J. Szinek, I. Macnab, *ibid.* 13, 799 (1979).
17. J. J. Klawitter and S. F. Hulbert, *J. Biomed. Mater. Res. Symp.* 2, 231 (1971).
18. R. A. Brand, in *Mechanical Failure of Total Joint Replacement* (Document 916-78, Steering Committee, American Academy of Orthopedic Surgeons, Chicago, 1978), p. 81.
19. C. D. Campbell, D. H. Brooks, M. W. Webster, H. T. Bahmson, *Surgery* 79, 485 (1976).
20. S. N. Bhaskar, J. M. Brady, L. Getter, M. F. Cromer, T. Driskell, M. O'Hara, *Oral Surg.* 32, 336 (1971).
21. T. M. Jocknicz, H. A. Nash, D. L. Wise, J. B. Gregory, *Contraception* 8 (No. 3), 227 (1973).
22. K. de Groot, *Biomaterials* 1, 47 (1980).
23. A. M. Reed, thesis, University of Liverpool (1978).
24. J. T. Frakes, S. D. Brown, G. H. Kenner, *Am. Ceram. Soc. Bull.* 53, 183 (1974).
25. L. L. Hench, R. J. Splinter, W. C. Allen, T. K. Greenlee, Jr., *J. Biomed. Mater. Res. Symp.* 2, 117 (1972).
26. L. L. Hench and H. A. Paschall, *ibid.* 4, 25 (1973).
27. L. L. Hench, R. W. Petty, G. Piotrowski, *An Investigation of Bonding Mechanisms at the Interface of a Prosthetic Material* (Summary Report to U.S. Army Medical R & D Command, Contract DAMD17-76-C-6033, 1979).
28. B.-A. Blencke, H. Bromer, E. Pfeil, H. H. Kas, *Langenbecks Arch. Klin. Chir.* 116, 119 (1973).
29. M. Jarcho, J. F. Kay, K. I. Gumaer, R. H. Doremus, H. P. Drobeck, *J. Bioeng.* 1, 79 (1970).
30. D. C. Greenspan and L. L. Hench, *J. Biomed. Mater. Res.* 10, 503 (1976).
31. D. E. Clark, M. C. Madden, L. L. Hench, in *An Investigation of Bonding Mechanisms at the Interface of a Prosthetic Material* (Annual Report No. 8 to U.S. Army Medical R & D Command, Contract DAMD17-76-C-6033, 1977). pp. 67-77.
32. H. Bromer, K. Deutscher, B. Blencke, H. Pfeil, *Sci. Ceram.* 9, 94 (1978).
33. P. Ducheyne and L. L. Hench, in preparation.
34. L. L. Hench, in *Proceedings of Surfaces and Interfaces of Glass and Ceramics*, V. D. Frechette, W. C. LaCourse, V. Burdick, Eds. (Plenum, New York, 1974), pp. 265-283.
35. C. G. Pantano and L. L. Hench, in preparation.
36. L. L. Hench, *Biomater. Med. Devices Artif. Organs* 7, 339 (1979).
37. J. E. Ritter, Jr., in *Fracture Mechanics of Ceramics*, R. C. Bradt, D. P. H. Hasselman, F. F. Lange, Eds. (Plenum, New York, 1978), vol. 4.
38. J. E. Ritter, Jr., D. C. Greenspan, R. A. Palmer, L. L. Hench, *J. Biomed. Mater. Res.* 13, 251 (1979).

One of the Intelsat V series of advanced communication satellites. The first satellite in this series is scheduled to be launched by NASA for the International Telecommunications Satellite Organization later in 1980. The satellite will carry telephone calls, television, telex, and telegrams; it can simultaneously handle 12,000 telephone calls and two TV channels. Graphite fiber-reinforced composites are used for the truss structure and large dish antennas of the satellite. [C. K. H. Dharan, Ford Aerospace and Communications Corp., Western Development Laboratories, Palo Alto, California]

Fiber-Reinforced Composites: Engineered Structural Materials

P. Beardmore, J. J. Harwood, K. R. Kinsman, R. E. Robertson

We are on the verge of witnessing the widespread introduction of a new major class of engineering materials in the industrial marketplace. It has long been the aspiration of materials scientists and engineers to reproduce in tailor-made, engineered materials the intricate structural networks so often found in nature's products and the human body which give rise to such remarkable functional characteristics. This long-sought goal seems to be approaching realization through the incorporation of a variety of fine fibers or filaments into appropriate matrices for the achievement of engineering properties not easily attained, if at all, in conventional bulk materials. In particular, continuous fiber-reinforced plastic (FRP) composites are close at hand as competitive classes of materials for structural and engineering applications.

Over the past decade there has been a remarkable growth in the use of plastics in the United States and in the rate of their acceptance for a broad range of industrial and consumer products. The polymer science and technologies which underlay these developments are described by Anderson et al. (1) and by Alfrey and Schrenk (2). But it has been claimed that a "second revolution" in the plastics industry and technology will come from the large-scale industrial applications of FRP's. The features of the fiber composites that make them so promising as industrial and engineering materials are their low density, high specific strength (strength/density), and high specific stiffness (modulus/density), and the opportunities to tailor the materials properties through the control of fiber and matrix combination and fabrication processing.

A most familiar use of FRP's is in recreation products involving largely glass fiber–resin systems in a variety of forms typically assembled by hand-labor–intensive processes. Fiberglass boats and other recreation vehicles, either all or in part, are commonly constructed of chopped glass on woven glass fiber mats held in resin matrices. But applications involving more demanding performance and more innovative use of continuous fiber reinforcement are evolving. An early example is rod forms, most familiarly fiberglass fishing rods. Today, even lighter and more responsive rods are filament wound with graphite and Kevlar fibers. Golf club shafts, tennis rackets, skis, ship masts are other products in which the advantages of lightness and the ability to tune the response of the component through manipulation of fiber placement are being tested in the marketplace.

But the most sophisticated use of the advantages offered by composites always has been pursued in high-performance aircraft and in military and space vehicle applications, where the premium on weight reduction supported the associated high cost. This application experience fostered sufficient confidence in design, engineering, performance, and reliability of certain composites that selected control surfaces in current-generation commercial aircraft are being retrofitted with these lighter materials in the interest of increased fuel efficiency. Boeing recently publicized plans to use significant quantities of advanced graphite and graphite/Kevlar composites in their next generations of planes, such as the Boeing 757 and Boeing 767 (Fig. 1). Advanced communication satellites, such as Intelsat V (Fig. 2), make extensive use of graphite fiber-reinforced composites for the truss structure and the large dish antennas. Not only are these designed for strength, stiffness, and lightness, but the composites provide the dimensional (thermal) stability critical to the communications task performed.

Of more impact on the consumer marketplace are the composite materials developments taking place in the automotive industry. Some 600 million pounds of fiberglass-reinforced plastics were used by the transportation industry in 1979. It has been projected that this amount may grow to more than 1 billion pounds per year over the next 5 years. Moreover, the intensive commitment on the part of the automotive industry to develop and produce new generations of vehicles with significantly improved fuel economy has brought lightweight, high-strength materials into key roles. Analyses based on expectations of technological and manufacturing developments indicate that fiber-reinforced polymer composites have a real potential as a next-generation class of materials for vehicle applications. Within this class, hybrid graphite fiber composite systems offer dramatic weight-saving opportunities—perhaps up to 70 percent weight reduction on a materials basis and more than 30 percent reduction in vehicle weight. These are the stimuli that have led to aggressive exploration and development of new classes of composite materials for advanced vehicle applications.

In this article we will review briefly the nature of fiber-reinforced polymer composites, the types of fibers currently being utilized for reinforcement, the requirements and functions of the resin matrix, the manufacturing and fabrication processes involved in composites production, and the properties and engineering advantages of composites. Using the automotive industry as an example, we will indicate the striking potential of composites as structural materials and the issues involved in bringing this potential into reality.

Summary. Fiber-reinforced composites are an emerging new class of engineering materials. The ability to tailor-make composite materials and structures offers exciting opportunities for a broad spectrum of industrial applications. This article reviews the nature of fiber-reinforced polymer composites, their characteristics and properties, and the manufacturing and fabrication processes involved in composite production. The automotive industry is used as an example of the striking potential of composites as structural materials and the issues involved in bringing this potential into reality.

J. J. Harwood is director of the Materials Sciences Laboratory; P. Beardmore is principal research scientist, and K. R. Kinsman and R. E. Robertson are staff scientists in the Metallurgy Department, Ford Research Laboratory, Dearborn, Michigan 48121.

Fiber Composites

Is a fiber composite simply a mass of fibers dispersed in some polymer or metal matrix? Polymer engineers have viewed composites as fiber-reinforced plastics—an upgrading of the polymer's stiffness and strength by fibers without a significant loss in resilience or toughness, and often without a large decrease in the ease of processing. Ceramists have viewed composites as a clever way (and it is) to utilize the high stiffness and high intrinsic strength of lightweight, brittle ceramics. The cleverness is in isolating the surface flaws that are catastrophic in a monolith by subdividing the ceramic into a very large number of fibers separated by a crack-stopping matrix. And metal physicists have viewed composites as a way of utilizing metal whiskers so defect-free that the strengths approach theoretical estimates.

For structural designers and materials engineers, fiber composites have become more than any of the above concepts. With the advent of graphite (carbon) and aramid (Kevlar) fibers, fiber composites have become a means for tailor-making "engineered," mechanically anisotropic materials that have superior properties in selected directions. With the reinforcement of plastics with glass fibers, for example, the composite is often anisotropic but at least the constitutents are isotropic. With Kevlar and graphite fibers, the molecular arrangement within the fiber is

Table 1. Fiber properties.

Fiber	Density (g/cm³)	Tensile strength (MPa)	Modulus (GPa)	Cost ($/lb)
Graphite-I	1.75	2760	235	20–32
Graphite-II	1.85	2415	220	75
S-Glass	2.63	3450	90	2.3
E-Glass	2.63	2415	72	0.50
Kevlar	1.45	2760	130	9

highly oriented, with the strongest covalent bonds along the fiber axis and weaker covalent and even van der Waals bonds in the transverse direction. The fibers, then, mirror internally the properties of the composite in which they are used—high stiffness and strength, including fatigue and creep resistance, along the fiber axis, with much lower transverse stiffness and strength. Optimum utilization of such oriented materials requires matching the direction-controlled properties of the composite to the service load conditions.

Fibers

A variety of fibers is available for use in composites. Table 1 summarizes the major types currently used in engineering structural plastics. Not included are boron fibers, which are considered too expensive for anything but aerospace and military applications, and alumina

and silicon carbide fibers, which are newer types being developed for composite reinforcements. Some of the fibers, such as graphite, can be fabricated with a wide range of properties. Table 2 lists mechanical properties of fiber-reinforced composites with epoxy matrices. (Also included in Table 2 are comparative properties for steel and aluminum.) It is the unique combinations of properties available in these fibers that provide the outstanding structural characteristics of fiber-reinforced composites. The key features of low density and high strengths and moduli give rise to high specific strengths and specific stiffness properties. Graphite fiber composites are particularly outstanding in this respect (Fig. 3).

All the fibers listed in Table 1 are used commercially, the selection of any particular fiber being dependent on the required combination of properties in the composite and cost considerations. The use of these fibers stems, of course, from the capability of producing the outstanding properties in the fiber form—it is important to emphasize that equivalent properties cannot be produced in the bulk form. Thus, it is the specific conditions used in generating the fiber that dictate the properties; development of successful fiber technology has been the key to advanced composite technology.

The fiber-producing process is unique for each different fiber. For example, for aramid fibers such as Kevlar the polymer is drawn into fiber form under the appropriate set of conditions (stress, temperature, and so on) to generate and retain an extremely high degree of orientation in the polymeric chains. For graphite fibers, a two-stage heating process is involved, as summarized in Fig. 4. The major manufacturing process for graphite fibers involves the generation of the fibers from a precursor fiber; the predominant precursor fiber is polyacrylonitrile (PAN), although other polymeric fibers such as rayon have been used. The PAN fiber is pulled under tension through the first heating stage, at 250° to 400°C, in which it is oxidized to produce a stable state. The fiber is then pyrolized to drive off all the atoms except a carbon ladder backbone; in this stage the temperature varies between 1500° and 2500°C depending on the degree of graphitization required. The properties of the resulting graphite fibers depend on the amount of graphite crystal structure in the fiber and the degree of orientation of the basal plane of the graphite crystallites along the fiber axis. Both of these are a function of the ten-

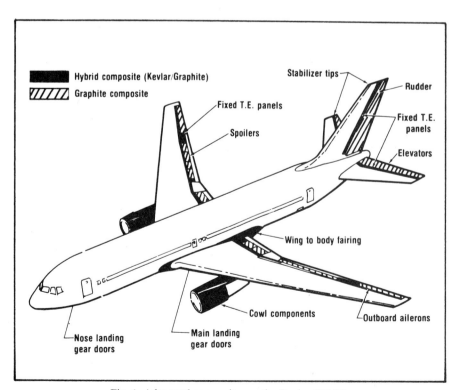

Fig. 1. Advanced composites on the Boeing 767 (5).

sion applied during heating and the temperature of conversion. Control of these two variables allows a family of graphite fibers to be commercially produced with a range of properties.

Also under industrial development is a new technology for making graphite fibers using pitch as the precursor material. The pitch process has the potential for significantly reducing the cost of graphite fibers.

Matrix

The resin matrix performs several functions in FRP composites. (We use the term matrix here to include the material between the fiber surfaces, both the matrix resin and any coating previously applied to the fibers.) First, the matrix must keep the fibers separated so they cannot abrade each other during any periodic straining or deflection of the composite. Second, the matrix must be mechanically coupled to the fibers so that it can bear loads. Ideally, the fibers should carry the total load. The fibers are generally much stiffer, stronger, and more fatigue and creep resistant. But even in ideal cases when the fiber orientations are well matched to the applied stress, the matrix is needed to transfer externally applied loads into and out of the fiber composite as well as to transfer internal loads around fiber breaks in continuous fiber composites and between fibers in chopped fiber and whisker composites. In less ideal cases where loads are complex, the composite may even have to bear loads transverse to the fiber axis. Not only must the matrix resin and its coupling to the fibers be able to bear these loads, the loads often must be sustained under conditions of changing temperature and moisture.

Although it is clear that the matrix must be coupled to the fibers, there has been considerable discussion over the years about just how much strength and stiffness is desirable for the matrix and coupling. If they approach those of the fiber too closely, the composite will behave like the brittle monolithic material. Although there is still no completely satisfactory answer, an empirical rule that is followed for polymeric matrices is a strength and coupling to the fibers that allows stubs of fibers roughly five to ten times their diameter to be pulled from the matrix when the composite is stretched to failure along the fiber axis.

There are many combinations of fibers and matrices that yield desirable composite properties. The key to the appro-priate combination is the ability to manufacture it. Composites, perhaps more than other structural materials, are manufacturing process–dominated; that is, the properties are a direct consequence of the method of preparation. In general, the matrix should be derivable from a liquid precursor. The transverse strength requirement means that the liquid precursor must wet the surface of the fibers. Wetting is needed to bring the matrix and fiber into proximity for covalent, dipolar, or van der Waals bonding to occur, as desired. Although various flow patterns and pressure applications can be used to speed the mixing and wetting of the fibers by the matrix resin, the local rate of wetting is governed by the parameter $\gamma \cos \theta / \eta$, where γ is the surface tension of the matrix precursor liquid, θ is the advancing contact angle of the liquid on the fiber surface, and η is the viscosity of the liquid (3). Since $\gamma \cos \theta$ is usually small, the viscosity must also be small for reasonable processing speeds. General methods for increasing the matrix liquid viscosity after wetting are cooling the liquid, thereby inducing crystallization; inducing a chemical reaction like polymerization; or evaporating a low-viscosity solvent. In certain processes for manufacturing fiber composites, such as pultrusion (like extrusion in metals), some filament winding, and matched die molding from preforms, the low-viscosity resin is brought to the rigid state in a single step by polymerization. In other processes, an intermediate matrix state is utilized, obtained by one or more of the viscosity-increasing mechanisms above. The intermediate viscous or leathery state is desired because it makes the handling of the composite before final curing much simpler and cleaner.

Coatings are often applied to the fibers prior to mixing with the matrix resin to

Table 2. Typical properties of fiber-reinforced composites.

Material	Weight of fiber (%)	Tensile strength (MPa)	Modulus (GPa)
Graphite-I/epoxy	65 continuous	1380	124
Graphite-II/epoxy	65 continuous	1100	180
S-glass/epoxy	65 continuous	1660	48
E-Glass/epoxy	65 continuous	1100	35
Kevlar/epoxy	65 continuous	1240	62
SMC-30	30 chopped	140	10
SMC-65	65 chopped	280	13
Steel (5160)		1380	200
Aluminum		560	70

Fig. 2. Intelsat V. [C. K. H. Dharan, Ford Aerospace and Communications Corp.]

prevent the fibers from abrading one another during subsequent processing and to aid adhesion between fiber and matrix. When $\gamma \cos \theta$ for the matrix liquid against the fibers is particularly small, the rate of wetting, or the quality of the wetting in a fixed process, can be greatly enhanced by spreading a thin layer of the matrix on the fibers. Sometimes a monolayer bridging link is employed between the specific chemical functionalities of the fiber and matrix. Thus, practically all glass fibers for composites are treated with silanes, in which one end of the molecule can form up to three Si-O bonds with the glass surface while the other end can chemically join to the matrix.

A list of polymeric matrix resins of current automotive interest is given in Table 3 along with the maximum continuous-use temperatures. The resins of greatest current interest are the polyesters. Although they are called *poly*esters, the ester linkage is used to produce only short chains of roughly 20 monomers. The monomers consist of dibasic acids and glycols. Typical acids include maleic anhydride, fumaric acid, phthalic anhydride, and isophthalic acid. Typical glycols include ethylene glycol, propylene glycol, and diethylene glycol. These are then combined to give oligomers with molecular weights of 1500 to 3500. To reduce the viscosity and enhance the subsequent reactivity of the unsaturated backbone, these polyesters are dissolved in unsaturated monomers like styrene. Concentrations of 30 to 50 percent styrene are common. It should be clear that the term polyester refers to a class of materials and not to a particular compound. After the polyester-monomer mixture is applied to the fibers and they are shaped in the desired form, the resin can be cured by an addition-type polymeriza-

Table 3. Organic matrices and maximum use temperatures.

Matrix	Maximum use temperature (°C)
Thermosetting	
Polyester	95
Vinyl ester	95
Epoxy	175
Polyimide	315
Thermoplastic	
Nylon 66	140
Poly(butylene terephthalate)	180
Polysulfone	150
Poly(amide-imide)	260

tion in about 2 minutes at 150°C, using a peroxide initiator such as *t*-butyl perbenzoate. The final polymer network thus consists of the original polyester oligomer intersecting or interconnecting polystyrene chains, which typically consist of two styrene monomers between polyester backbone junctions. Polyester matrix formation is seen to involve a two-step polymerization: an esterification or condensation polymerization followed by an addition polymerization.

A three-step polymerization is used for the commercially important SMC process to avoid handling problems. In this process, the mixture of resin, fibers, and a particulate filler, if desired, is held between sheets of polyethylene film until it thickens to a leathery sheet molding compound (SMC). Thickening is obtained by making the polyester with excess acid and including in the fiber mixture a divalent metal oxide like magnesium oxide. The reaction between these produces, after a day or two, the desired change from a soft, sticky mass to a handleable sheet. The original formulation for SMC, and perhaps still the most important commercially, is 30 percent chopped glass fiber, 30 percent ground limestone, and the rest resin, by weight. The same process is useful for many different combinations of chopped and continuous fibers.

The fiber composites exhibiting perhaps the best mechanical performance to date are those made from epoxy resins. Most of the aerospace composites have epoxy matrices, although these matrices are less attractive for most automotive applications because of their long cure times. Although there are a few processes in which the liquid resin applied to the fibers is cured in a single step to a rigid matrix, most processes, including most filament winding, use "prepregs" as an intermediate state. That is, the fibers are

impregnated by the resin and the resin thickened to facilitate subsequent handling. In some cases, a solid novolac epoxy resin is applied to the fibers in a solvent and then dried. In other cases, part of the epoxy curing reaction is allowed to occur. Epoxy resins often used for composites are the diglycidyl ether of bisphenol A, dimers of this, the tetraglycidyl ether of tetraphenolethane, tetraglycidylmethyldianiline, and the epoxies derived from the novolacs. Most epoxy systems for prepregging involve mixtures of the epoxy resins to achieve a balance of properties. There are many agents for curing at elevated temperatures; important agents are diaminodiphenyl sulfone and dicyanodiamide, which with accelerators cure with the diglycidyl ether of bisphenol A after 30 to 60 minutes at 165° to 175°C (4).

The good mechanical performance of epoxy matrices arises from their general toughness and their good adhesion to the fiber surface through hydroxyl groups formed when the oxirane ring is opened on curing. As a way of taking advantage of the mechanical performance of epoxies and the rapid cure of polyesters, unsaturated acids such as methacrylic acid can be reacted with the glycidyl ethers of bisphenol A to form vinyl esters. Like the polyesters, monomers such as styrene are added to the vinyl ester to reduce viscosity. As a result, the curing behavior of the vinyl ester is essentially the same as that of the polyes-

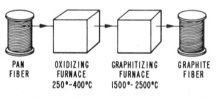

Fig. 4. Schematic of graphite fiber production from polyacrylonitrile.

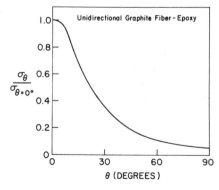

Fig. 5. Decrease in tensile strength away from the fiber direction in a continuous fiber-reinforced composite. θ is the angle with the fiber alignment direction.

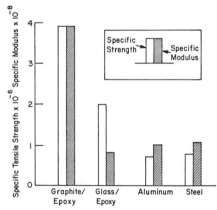

Fig. 3, Comparison of specific tensile strengths and specific moduli for different materials.

ters. In general, the mechanical properties of composites with vinyl ester matrices are intermediate between those with polyester and epoxy matrices.

Thermoplastic matrix resins are of interest because they should be formable by a hot stamping process. This would allow shorter processing times than are possible even with the vinyl esters and polyesters. Except for two stampable materials of moderately low stiffness and strength, PPG's Azdel with polypropylene and Allied Chemical's STX with nylon 6, reinforced thermoplastic composites are still in an early stage of development.

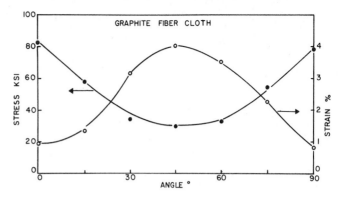

Fig. 6. Variation in properties as a function of testing angle in a graphite fiber cloth in which alternate layers of fibers are aligned in the 0° and 90° directions.

Types of Fiber Composites

The useful properties of the fiber-reinforced composites are the end result of the combined properties of the fiber and the supporting matrix. Key features of composite technology are the availability of matrix resins and fabrication processes that do not significantly degrade the intrinsic properties of the fiber. A realistic target is to have the strength and modulus properties of the composite roughly follow the law of mixtures of the fiber/resin two-component system.

In principle, there is an infinite gradation of fiber composite types ranging from chopped fiber composites at the low properties end to a continuous, unidirectional fiber composite at the high properties end. Essentially, the composites differ in the amount of fiber, fiber length, and fiber type (Table 2). In general, the chopped fiber systems are used in lightly loaded or, at best, semistructural applications, whereas the continuous fiber–reinforced composites are high-performance structural materials. By their very nature, continuous fiber–reinforced composites are highly anisotropic. If all the fibers are aligned in one direction, maximum properties are achieved in that direction. However, the properties decrease rapidly in directions away from the fiber direction, as illustrated in Fig. 5. To generate more orthotropic properties, alternate layers of fibers are frequently alternated between 0° and 90° directions, resulting in less directionality but at the expense of absolute properties in the 0° direction (Fig. 6). Because of such marked directionality of properties, it is extremely important to have a good definition of the total stress environment under which a composite component operates. Accurate knowledge of the variations and types of stresses acting on the component enable the rational selection

of design methodologies and the critical orientation of laminate and fiber arrays to provide for structural design optimization.

The types of composites and composite design technologies employed by different industries can be quite specific to the particular requirements and practice of the individual industry. For aerospace, the criticality of minimizing weight at (virtually) any cost combined with low-volume production allows the most exotic and expensive fibers to be used, combined with long matrix cure times and hand-layup fabrication procedures. In mass-production, consumer-oriented industries such as the automotive industry, the high volume and high production rates demand automated fabrication, short matrix cure times, and minimization of cost. Such demands suggest that for highly stressed structural application, fiber hybrids are the most likely candidates, with a large fraction of the fibers being continuous. As the name suggests, hybrid fiber composites contain mixed fibers, tailored to optimize properties and minimize cost. The hybrids of perhaps greatest promise are E-glass composites containing appropriate amounts of graphite fiber, but other combinations are being considered. Particular hybrid combinations will be determined by the appropriate property-cost trade-offs for any given application. It is significant that cost pressures are also stimulating the development of hybrid composites for aeronautical applications.

Composite Processing

There are basically two types of processing for converting the raw fiber into the finished product:

1) Direct processes. In these processes, the fiber is combined with the matrix polymer and formed into the finished part in one continuous operation.

2) Indirect processes. In these processes, the combining of the fiber and matrix polymer takes place in a separate, preliminary stage, and the conversion of this preform to the finished product takes place in a subsequent, physically separate fabrication procedure.

Direct processes. The attraction of direct-process fabrication lies in the simplicity and cost effectiveness of such operations. Starting with the raw fiber, the fabrication process is continuous through the final product. The disadvantage of such processes lies in the restricted geometric shapes that can be fabricated. Examples of such processes are injection molding, pultrusion, and filament winding. Pultrusion is particularly applicable to parts that are straight or only slightly curved. It consists of pulling a bundle of fibers through a resin bath and subsequently through a heated die in which the matrix is cured. Production rates in the range of 2 to 5 feet per minute are currently achievable.

Filament winding is an important, high-speed process for manufacturing tubes. The fibers are prewet by a technique similar to pultrusion and then wrapped around a mandrel at precisely determined orientations and subsequently cured either in a continuous or batch-heating process. Production rates greater than 6 feet per minute are achievable in this process. Filament winding is also suitable as a high-speed fabrication process for making preforms with controlled orientations and amounts of fiber. Large-diameter filament-wound tubes can be slit before curing into preform sheets for fabrication into shapes by other methods.

Perhaps the most rapid polymer processing is injection molding of thermoplastic parts, both short glass fiber–reinforced and unreinforced. Molding cycle times are measured in fractions of minutes and multiples of the part are made at the same time. The fluid polymer is injected into a closed, cooled mold. After

33

Table 4. Summary of weights of GrFRP components. Vehicle weights: Ford LTD 1979, 3740 pounds; lightweight concept vehicle, 2504 pounds.

Component	Weight (pounds)		
	In steel	In graphite	Savings
Body-in-white*	461.0	208.0	253.0
Frame	282.8	207.2	75.6
Front end	96.0	29.3	66.7
Hood	49.0	16.7	32.3
Deck lid	42.8	13.9	28.9
Bumpers	123.1	44.4	78.7
Wheels	92.0	49.3	42.7
Doors	155.6	61.1	94.5
Miscellaneous†	69.3	35.8	33.5

*Complete body structure without closure panels (doors, deck lid, etc.) and trim (seats, etc.). †Bracketry, seat frame, and other items.

the polymer has cooled enough for the part to become rigid, the mold opens for the part or parts to be ejected and then closes again to begin the next cycle. The process is rapid and inexpensive enough to be used for many small parts on the car as well as for a wide variety of everyday consumer plastic items.

Indirect processes. The indirect processes involve an intermediate step in the transition from the raw fibers to the finished product. All the processes involve the preparation of some kind of preform in which the fiber is intermixed with the matrix resin and treated so that the preform is handleable. For thermoset matrices, such a step involves partial curing to allow handling of the preform; the preform must then be storable for up to 6 months without a further significant amount of curing. For thermoplastic matrices, the preform is rigid and stable at room temperature with infinite storage time.

Examples of indirect processes are compression molding and vacuum bag curing for the thermoset matrix composites. In compression molding, the preform is pressed into final shape in a matched metal die and heated in the die to fully cure the matrix. The cycle time depends on the thickness of the part and the particular resin employed, but typically varies between 2 and 3 minutes for the faster-curing polyester systems. Vacuum bag curing is generally a much slower process in which the preform is "sucked" against a female mold by evacuation of a plastic bag around the mold, and subsequently cured in an autoclave. This process is normally restricted to low-volume applications, such as in aerospace use, and tends to use longer-cure, epoxy-type matrices.

Warm stamping of thermoplastic matrix composites is not as well developed as the thermoset matrix procedures, but is appealing because of the infinite shelf life and handleability. Basically, it consists of preheating the sheet, forming to final shape in a die, and then cooling to retain the shape before removal from the die. The rheological behavior of fiber-containing thermoplastic resin and the mobility of the fibers during the stamping operations are key process features that require control for the successful development of thermoplastic stamping as a composite material fabrication process.

Composites in the Automotive Industry

Except for a few specialized applications, the development of glass fiber-reinforced plastics (GlFRP) for the auto industry has concentrated primarily on manufacturing techniques that can economically turn out large numbers of parts and only secondarily on strength properties. However, attempts are now being made to improve properties, by replacing particulate fillers with glass fibers and by lengthening the fibers, while trying to retain some of the rapid fabricating techniques.

Current and near-term automotive uses for glass fiber-reinforced plastics. The best known example of GlFRP usage in automobiles is the Chevrolet Corvette. Since its introduction in the mid-1950's, the Corvette has had a GlFRP exterior body fabricated by compressive molding of SMC. Practically all grille opening panels on American-made cars are fabricated from glass-reinforced SMC materials; in such applications, the final component is less expensive to manufacture from SMC because numerous steel stampings can be integrated into one molding with a consequent reduction in labor and assembly costs. A continuing problem that has compromised other sheet metal replacement (outer panels of doors, hoods, and so on) has been the inability to produce a high-quality exterior surface in SMC, especially where there is much variation of thickness. This results in an undesirable finished paint appearance. It seems likely that a newly developed in-mold urethane coating process may solve this problem and open the way for large-scale application of SMC in exterior components.

To date, most automotive usage of plastics has been in relatively non-load-bearing applications. The next step is to apply composites in components bearing moderate loads. For example, a radiator support with an integral fan shroud is

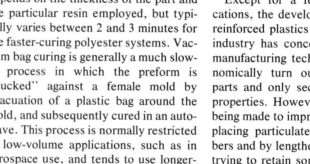

Fig. 7. Typical GrFRP drive shaft for a medium-sized automobile. The tube was manufactured by filament winding, using an epoxy matrix, and was adhesively bonded onto steel yokes. The weight saving compared to the same component made with steel was 5 pounds.

Fig. 8. Automobile leaf spring made of GrFRP. This component weighs 7 pounds, compared to 28 pounds in steel.

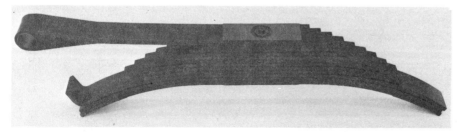

Fig. 9. Experimental GrFRP heavy truck spring. The weight is 30 pounds compared to 125 pounds in steel. With four springs per rear truck suspension, the potential weight saving per truck is nearly 400 pounds.

being evaluated for automotive companies. To meet the load requirements, an SMC formulation with increased glass fiber content (60 percent by weight) must be used and the glass fibers must be a mix of continuous or relatively long (4 to 12 inch) unidirectionally oriented fibers and short (1 inch) randomly oriented fibers. GlFRP composite bumper reinforcements and leaf springs are also beginning to appear in production vehicles. GlFRP composites are competing effectively in the heavy truck market because of their weight-saving and cost features. Heavy truck cabs and related exterior components are becoming frequent choices for composite applications.

Advanced composites—graphite and hybrid graphite composite. Intensified concern about energy and current shifts in the marketplace have accelerated efforts to reduce vehicle size and weight, and applications of lightweight materials have assumed a key role in advanced fuel-efficient vehicle designs. Materials are critical to the design of vehicles with optimum capacity and utility at minimum weight. In looking beyond 1985, it has become clear that the dramatic weight reduction potential of the advanced graphite fiber composites (developed initially for military and aerospace applications) offers opportunities to achieve further improvements in fuel economy. This recognition has been a driving force for the exploration and development of graphite fiber-reinforced plastic (GrFRP) composites for future vehicle application that is now taking place almost worldwide.

Engineering data on the characterization and engineering properties of the GrFRP's are being generated, and numerous prototype components are being extensively evaluated. However, the design methodology needed for continuous fiber composite components is radically different from that for isotropic metals. The fiber composite components consist of discrete layers, or laminae, and the properties of individual layers may be completely different, depending on the material (type of fiber) and fiber orientation. The resulting laminate is usually anisotropic, and the design analysis must examine every layer for potential failure. Finite-element analysis, computer-aided design, and computer graphics and modeling techniques have become essential features of the development and design of these new materials.

The principles of this design methodology have been used in the development of a number of GrFRP experimental components. In particular, drive shafts and leaf springs were used as initial examples of the potential for GrFRP composites in automobiles. Among the more promising of these experimental components are a typical automobile drive shaft (Fig. 7), an automobile leaf spring (Fig. 8), and a truck spring (Fig. 9).

The Ford GrFRP Concept Vehicle

This promising developmental experience with graphite fiber composite components led Ford Motor Company in 1977 to undertake the building of an experimental car with body, chassis, and power train components to be made of graphite fiber composites to the maximum extent possible. The 1979 Ford was selected as the design to demonstrate the potential of graphite fiber composite technology for making a lightweight six-passenger car with good fuel economy while retaining the performance and characteristics of larger vehicles. The project was intended to demonstrate concept feasibility and design and materials feasibility and to identify the critical issues related to production feasibility for future vehicles. Manufacturing and cost feasibility were not program objectives.

A schematic of the GrFRP vehicle is shown in Fig. 10, where the shaded areas represent the parts of the vehicle fabricated in GrFRP. The completed experimental vehicle with a 2.3-liter engine weighed 2504 pounds—1246 pounds less than 1979 production Ford LTD with the standard 5.0-liter engine. The fuel economy was projected to be 33 percent, or 6 miles per gallon, better than the base production car. Virtually all of the body, frame, and chassis parts—about 160 parts—are made of graphite fiber-reinforced composites. Only the power train, trim, and some chassis components were not converted. However, most of these (for instance, engine, brakes, transmission) were downsized or downgaged for secondary weight reductions.

Some 600 pounds of graphite composites was used, containing about 400 pounds of graphite fiber. The primary weight saving—by direct materials substitution—was 706 pounds; the remaining 540 pounds saved was the result of secondary weight reductions. A summary of the weight savings achieved in key components is given in Table 4. For

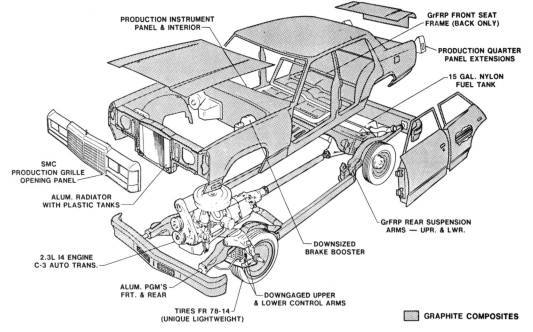

Fig. 10. Schematic of experimental car with body, chassis, and power train components made of GrFRP.

the hood, front end, deck lid, bumpers, and door, part weight savings ranged from 61 to 69 percent—gratifyingly close to our design expectations. For the wheels and miscellaneous brackets, a 45 to 50 percent weight reduction was achieved.

The weight and design targets were successfully achieved and, in its preliminary evaluation, the experimental vehicle exhibited an initial ride quality equivalent to that of production vehicles in the same stage of development.

The experience gained in this car program underscored the two key issues that will determine large-scale future applications: materials cost and manufacturing feasibility. The fiber industry indicates that the price of graphite fibers will come down dramatically as capacity and markets increase. A price of $6 per pound of fiber in the middle to late 1980's appears to be a reasonable projection, compared with the current price of $20 per pound. However, on a materials basis, all-graphite composites will not approach the more favorable economics of automotive materials such as steel or aluminum. The automotive industry, together with the composites and resin industry, must develop hybrid composites containing graphite, glass, and other fibers, to improve the economic effectiveness of such composites.

Equally important is the need for low-cost manufacturing processes for producing both preforms and components compatible with automotive industry practice. Hand-layup techniques obviously are not economically feasible. High-speed manufacturing processes, perhaps involving new resins and processing techniques, are required.

The automotive industry needs to have in hand well-characterized and tested classes of materials, more understanding of performance characteristics in service and failure modes, design methodologies compatible with new materials and manufacturing processes, bonding and joining methods, and methods of nondestructive evaluation and quality control.

One long-range issue critical to the future automotive application of composites is related to crashworthiness and durability. The ability to design and integrate composite materials and composite structures to ensure vehicle safety and integrity must be fully demonstrated before we can move toward large-scale production considerations.

In our view, the graphite fiber composite experimental car marks the end of a phase in graphite fiber composite technology—a phase in which conceptual feasibility was the predominant focus. We now face the hard tasks of bringing to reality the potential demonstrated to date. Hybridization, formability, and low-cost, high-rate manufacturing processes emerge as the targets for the next round of materials manufacturing R & D programs.

Future

The future picture for composites will involve the aggressive pursuit of every application where they are thought to be of advantage. This spans a wide range of materials and economics of application. New product opportunities will continue to be coupled to developments in the technology of composite fabrication and design. Larger-scale structural use of composites will be constrained by the challenges of rapid and automated processing, the reliability of mass-produced structures, material cost, and more efficient design.

Aerospace use of composites will continue to be the proving ground for high-performance and exotic fiber-matrix combinations and for design methodology. Such structures have been labor-intensive handcrafted articles. But the development of next-generation combat aircraft, which could be 50 percent composite materials, will also be more cost-constrained and automation will be required to adequately fabricate the increasingly larger and more complex components while remaining competitive in cost.

The advantages of tailored high-performance products will always support specialized applications, as in sports equipment and in certain advanced technologies. But composite use continues to expand into more highly cost-constrained sectors. Perhaps the automobile industry's need for high-rate automated processing will encourage development of more know-how and the ability to control the compromises between performance and reliability. As in aerospace, progress will likely include a tailoring of material form and shape to an automated industry, but it will take a different form and include more variety. Optimized design will afford cost-effective use, which will probably include hybrid composites and composite structures of various forms. Reliable prediction and control of fiber distribution in complex structures formed by machines will open new opportunities for optimal use of composite materials.

If indeed the R & D activities during the next decade realize the promise of composite materials in consumer market applications, we will have provided a stimulus for a quantum increase in the output of the plastics industry for the 1990's and perhaps a realistic basis for the coming of a second revolution in plastics.

References and Notes

1. B. C. Anderson, L. R. Bartron, J. W. Collette, *Science* **208**, 807 (1980).
2. T. Alfrey and W. Schrenk, *ibid.*, p. 813.
3. H. van Oene *et al.*, *J. Adhes.* **1**, 54 (1969).
4. J. Delmonte, in *Epoxy Resins: Chemistry and Technology*, C. A. May and Y. Tanaka, Eds. (Dekker, New York, 1973), pp. 589–631.
5. R. H. Wehrenberg II, *Mater. Eng.* (January 1980), p. 33.

High-Temperature Structural Ceramics

R. Nathan Katz

Accelerated research activity over the past decade on the ''new'' high-temperature structural ceramics may lead to one of those rare instances in the history of technology when the appropriate materials and engineering technology come together at the right time to meet urgent societal needs. Availability of energy, availability of critical materials, and maintaining a livable environment are, perhaps, the three most urgent problems facing the industrialized world. It is widely recognized that the new high-

tive engines. The universal abundance and inherent low cost of the elements silicon, carbon, and nitrogen contrast sharply with the increasingly short supply and increasing costs of critical metals for high-temperature alloys such as chromium, nickel, cobalt, and tungsten. Table 1 gives some perspective on the fuel savings that the use of these ceramics could produce in several applications (6–10). Recent estimates (10, 11) indicate that fuel savings of 1/2 billion barrels of oil per year, amounting to $17.5 billion at

Summary. The unique properties of ceramics based on silicon carbide and silicon nitride make them prime candidates for use in advanced energy conversion systems. These compounds are the bases for broad families of engineering materials, whose properties are reviewed. The relationships between processing, microstructure, and properties are discussed. A review and assessment of recent progress in the use of these materials in high-temperature engineering systems, and vehicular engines in particular, is presented.

temperature structural ceramics based on silicon carbide, SiC, and silicon nitride, Si_3N_4, offer the possibility of helping to resolve the first two of these problems (1–4), with no negative effects [and maybe some slight positive ones (5)] on the third.

The silicon carbide- and silicon nitride–based ceramics have unique combinations of properties that make them very attractive to designers of vehicular engines, energy conversion systems, and industrial heat exchangers. In particular, they provide high strength at high temperature, good thermal stress resistance, and excellent oxidation, corrosion, and erosion resistance. This combination of properties can be used to increase the operating temperature, or reduce the heat lost to cooling, in gas turbine, diesel, and Stirling engines and in industrial heat exchangers, thereby yielding more power per unit of fuel. The low density of these materials ($\sim$ 40 percent that of high-temperature superalloys) may offer components of lower weight and inertia, which would translate into improved performance for conventional automo-

current prices, can be realized in the areas of highway vehicles and industrial heat exchangers if this technology is fully exploited. Thus there is considerable incentive to apply these ceramics in high-temperature energy conversion technology.

So far, I have made silicon carbide and silicon nitride ceramics sound like panaceas. If this class of materials can do all of the above, why are they not used in today's engines, other energy conversion devices, and industrial heat exchangers? The answer lies in the brittleness of ceramic materials—but this is a bit too simplistic an answer. All engineering design represents a trade-off in which (we hope) the best compromise in the use of materials and design emerges. The practice in engineering design to date has been to take advantage of the fact that metals are forgiving of contact stresses and stress concentrations (that is, they relieve local overstresses by local yielding). This simplifies the design process. The trade-off is that metals soften or melt and must be cooled for high-temperature use. Management of the

cooling fluid adds design complexity and weight and drains power. Up to now, the compromises in design that have been required to use metals have had a net payoff. However, in either economic or actual physical terms, the end of this line of development is in sight. Further development of heat engines, in particular, requires higher temperatures with reduced cooling and at acceptable costs. Thus if a combination of improved ceramic materials and the ability of designers to create compliant designs to accommodate the noncompliant nature (brittleness) of ceramics exists, then the materials-design trade-offs will favor ceramics. Partly because of the brittleness problem, there is very little design experience with such structural applications of ceramic materials. This lack of experience has itself served to restrict the use of ceramics in these applications.

Designers have been intrigued by the possibility of applying ceramics to turbines and diesels for more than 40 years. However, it was not until about 10 years ago that the two prerequisites were present: thermal shock-resistant structural ceramics, such as SiC and Si_3N_4, developed to the point where they could begin to be considered as engineering materials, and, equally important, sufficiently developed computer capability to handle the complex job of detailing stresses with the degree of refinement required in brittle materials design. The fact that the materials and design technology were advancing at the same time that the need for this technology was being widely perceived led to a large number of technology demonstration programs in the mid-1970's (2, 3). I will return to some of these programs later in this article in order to assess where this technology stands.

The SiC and Si_3N_4 ceramics that are being developed for engine and heat exchanger use are actually families of ceramic materials, in much the same way as steels and brasses are alloy families. In materials families, as in all families, certain characteristics are shared. In the case of the carbides and nitrides of silicon the shared characteristics include strongly covalent bonding, dissociation rather than melting at high temperatures ($>$ 1900°C), low coefficients of thermal expansion (which lead to the excellent thermal shock behavior of these materials), and resistance to oxidation provided by a spontaneously forming surface oxide film. These similarities aside, each individual member of these families

The author is chief of the Ceramics Research Division, U.S. Army Materials and Mechanics Research Center, Watertown, Massachusetts 02172.

Table 1. Examples of projected energy savings from use of ceramic technology.

Technology	Systems configuration	Reduction in fuel use (%)	Reference
Gas turbines			
Automotive	~ 150 hp, regenerated, single-shaft engine, 1370°C turbine inlet temperature	27	(6)
Truck	~ 350 hp, regenerated, two-shaft engine, 1240°C turbine inlet temperature	17	(7)
Industrial/ship	~ 1000 hp, simple-cycle, three-stage engine, 1370°C turbine inlet temperature	10	(8)
Diesel engines			
Truck	~ 500 hp, adiabatic-turbocompound, 1210°C maximum component temperature	22	(9)
Industrial heat recovery			
Recuperators	Silicon carbide recuperated slot forging furnace operating at ~1300°C	42	(10)

of ceramics is a unique material made by a unique process. For example, within the Si_3N_4 family we have hot-pressed, slip-cast sintered, injection-molded sintered, and injection-molded reaction-bonded materials. These materials have very different properties as a consequence of their processing. Also, the ease and cost of producing useful engineering shapes differ widely from one processing method to another. In ceramic materials, unlike metals, one cannot readily manipulate the microstructure, and hence the properties, after the primary processing of the material. Therefore, processing science and technology is central to the development of improved ceramics. The progress during the past decade on the carbides and nitrides of silicon has largely been the result of increased understanding of processing and the relationship between processing, microstructure, and properties. Therefore, to have a perspective for evaluating the progress in the development of these materials, it is important to briefly review the processing of ceramics in general.

Ceramic Processing

Virtually all high-temperature engineering ceramics are formed by the consolidation of powders. This is the case for the silicon-based structural ceramics under consideration here. The application of heat with or without pressure will, in general, cause consolidation of a ceramic powder, resulting in a strongly bonded polycrystalline body. This is known as sintering (or, if pressure is involved, hot-pressing).

The driving force for sintering is the reduction of the surface free energy of the powder aggregation. In general, sintering will occur when the surface free energy of a grain boundary, γ_{GB}, is less

than twice the surface free energy of a particle surface in contact with its vapor environment (air, nitrogen, argon, and so on), γ_{sv}. Thus when the individual particles consolidate into a densified and bonded ceramic, there is a net reduction in the free energy of the system. There is also a volume shrinkage (~ 50 percent) as the voids between the particles are eliminated. If heat (or heat and pressure) is maintained, the system will continue to lower its free energy by the process known as grain growth, which will continue until a metastable grain boundary array is obtained. Often, sintering, densification, and grain growth are accelerated by the use of additives, which may or may not produce a separate grain boundary phase. Even in relatively pure single-phase ceramics, grain boundaries tend to be favored sites for segregation of impurities, porosity, and entrapped gases. Thus grain boundary composition and structure and processes occurring at grain boundaries can dictate the properties of the ceramic piece.

Covalent compounds, especially those which dissociate rather than melt at normal pressures, such as SiC and Si_3N_4, are extremely difficult to sinter to full density even with the use of additives. At temperatures low enough to avoid dissociation atomic mobility is too low for densification to occur, and at temperatures where appreciable atomic mobility is possible the compounds start to decompose. Because both silicon carbide and silicon nitride are as hard as aluminum oxide (a standard abrasive and tool bit material) they are very difficult to machine. Fabrication methods that can reduce or eliminate machining are therefore essential to significantly increased utilization of these ceramics as engineering materials. With these basic considerations in mind, we will now take a look at various members of the silicon-based families of structural ceramics.

Silicon Carbide Ceramics

Silicon carbide can be produced with either a cubic, β, or a hexagonal, α, crystal structure. The silicon carbide–based ceramics include hot-pressed, sintered, reaction-sintered, chemically vapor-deposited (CVD), and fiber-silicon composites. Typical properties of each class of SiC material are presented in Table 2.

Hot-pressed silicon carbide can be fabricated to essentially full density and high strength by using additions of boron and carbon or of aluminum oxide to either α- or β-SiC starting powder (12, 13). Hot-pressing is typically accomplished at temperatures of ~ 1900° to 2000°C with pressures of ~ 35 megapascals (5000 pounds per square inch). The product is surpassed in strength only by hot-pressed silicon nitrides at low to moderate temperatures, and has a higher strength than any of the silicon nitride-based ceramics in the range 1300° to 1400°C (see Table 2). Given this outstanding behavior, one would think that considerable effort would have been expended on research to improve hot-pressed SiC during the past few years. This has not been the case for several reasons. Hot-pressed SiC has the major drawback of requiring expensive diamond machining to form shaped components. This has tended to focus industrial development on developing a sinterable SiC, which can be shaped into components in the "green" (unfired, soft and relatively easily shaped) state. Further, early failures of hot-pressed SiC turbine vanes in a test where hot-pressed Si_3N_4 vanes survived (14) were taken as an indication that hot-pressed SiC has a lower thermal shock resistance than hot-pressed Si_3N_4, and accordingly development emphasis shifted to the nitride. In addition, the lack of general availability of sintering-grade powders (powders with the correct particle size distribution and chemistry) and the general lack of facilities for hot-pressing at > 1900°C have served to restrict basic research on this form of SiC.

The recent successful development of sintered SiC is a major accomplishment in ceramic science and technology. Until 1973 it was widely believed that SiC could not be conventionally sintered to full density because its highly covalent bond precluded the degree of volume or grain boundary diffusion required for densification. However, in that year Prochazka at General Electric (15) demonstrated that submicrometer β-SiC powder, with a low oxygen content (< 0.2 percent oxygen) and small additions of boron and carbon (about 0.5 and 1 per-

Table 2. Typical properties of silicon carbide and silicon nitride ceramics.

Type	Bend strength (four-point) (MPa)			Young's modulus (GPa)	Coefficient of thermal expansion, α (10^{-6} °C^{-1})	Thermal conductivity, K (W m^{-1} °C^{-1})
	RT*	1000°C	1375°C			
Silicon carbides						
Hot-pressed (Al_2O_3 additive)	655	585	520	449	4.5	85–35
Sintered (α phase)	310	310	310	407	4.8	100–50
Reaction-sintered (20 percent free Si by volume)	380	415	275	345	4.4	100–50
SiC fiber–Si composite	275	275	275	340		70
CVD†	415	550	550	414		
Silicon nitrides						
Hot-pressed (MgO additive)	690	620	330	317	3.0	30–15
Sintered (Y_2O_3 additive)	655	585	275	276	3.2	28–12
Reaction-bonded (2.45 g/cm³)	210	345	380	165	2.8	6–3
β'-SiAlON (sintered)	485	485	275	297	3.2	22

*Room temperature.   †Chemically vapor-deposited.

cent, respectively), could be sintered to nearly full density without applying pressure at temperatures between 1950° and 2100°C under an inert gas or in a vacuum. The ultrafine powder provides a high thermodynamic driving force for densification as well as short diffusion distances, the boron appears to accelerate both volume and grain boundary diffusion, and the carbon removes the SiO_2 layer from the SiC powders. It is now widely thought that these three factors—ultrafine grain size, an additive to promote volume diffusion, and careful attention to the chemistry at the powder surface—are the key to producing other sintered, single-phase, covalent compounds such as Si_3N_4. Prochazka also demonstrated that sintered β-SiC could be formed into useful shapes by slip casting, die pressing, and extrusion. Nonuniform distribution of the carbon additive and exaggerated grain growth of α-SiC were found to interfere with densification. If one could use α starting powder rather than β, perhaps this problem could be overcome. In addition, there would be the advantage of using a less expensive, industrially available material. Coppola and McMurty (16) succeeded in developing such a sintered α-SiC, first reported in 1976. This material has so much promise that the Carborundum Company recently dedicated a multimillion-dollar facility to its commercialization, primarily in the area of automotive engine components. Supporting this optimism, sintered α-SiC has been successfully demonstrated in several piston engine applications.

Reaction-sintered SiC's cover a wide range of compositions and manufacturing processes. Although manufacturers use their own proprietary processes, in general a plastic body is formed of SiC powder, graphite, and a plasticizer. In some variants of the process, SiC powder plus a char-forming binder are used. The plastic body is pressed, extruded, injection-molded, or otherwise formed into a green body. The plasticizers are burned off or converted to a porous char by pyrolysis. Silicon metal as a liquid or vapor is infiltrated into the body and reacts with the graphite powder or char to form SiC in situ, which reaction-sinters the components. Excess silicon (typically 2 to 12 percent) is usually left to fill any voids, thus yielding a nonporous body. Such materials exhibit quite reasonable strengths to the melting point of silicon (~ 1400°C) or beyond, depending on the amount of free silicon retained. The presence of the free silicon is a problem in applications where temperatures above ~ 1300°C are likely to be encoun-

tered. The major advantage of these materials is that they maintain the geometry of the green preform after conversion to SiC. Thus, little machining is required and component cost is relatively low. A variety of successful experimental gas turbine components, such as combustors or stators, have been made by this materials-process route.

Reaction-formed SiC fiber–Si composites, developed by Hillig (17) at General Electric, are among the first engineered composite ceramic–ceramic structural materials. The process consists of starting with a graphite cloth, tow, felt, chopped fiber array, or any other possible precursor; forming a preform by any one of a variety of routes; and infiltrating the preform with liquid Si. The molten Si reacts with the filamentary graphite materials to form polycrystalline SiC fibers in a silicon metal matrix. The result is a fully dense oxidation-resistant body with about 30 to 50 percent Si (by volume) reinforced by 70 to 50 percent SiC fibers. (The high percentage of Si and the fibrous SiC morphology make this quite different from materials of the reaction-sintered SiC type.) Although this material shows promise of ease and versatility of fabrication, it is not yet clear what the production costs may be. The material permits the design of a composite component optimized for mechanical and thermal requirements. The presence of free silicon would be thought to limit use temperatures to ~ 1400°C. In spite of this, the material has been used in an experimental combustion liner appli-

cation at temperatures above 1425°C. Although the material is still in a very early stage of development, the fact that it is an "engineered" composite material makes it a very exciting new development.

Chemically vapor-deposited SiC is a fully dense material with no additives. However, the strength of CVD SiC, which on occasion can be very high, is quite variable. This large scatter in strength is due to the occurrence of large columnar grains and residual deposition stresses. The strength values shown in Table 2 are from the lower end of the scatter band. While CVD SiC has been produced in complex shapes, the technique may be costly for production. The material may find eventual use as a coating for high-temperature oxidation and erosion resistance on SiC ceramics formed by more conventional routes.

Silicon Nitride Ceramics

Silicon nitride exists in two phases, α and β, both of which have hexagonal crystal structures. The α phase has a unit cell approximately twice as large as that of the β phase. The silicon nitride–based ceramics include hot-pressed, reaction-bonded, and sintered materials and the β'-SiAlON's (a solid solution of Al_2O_3 and/or other metal oxides in the β-Si_3N_4 lattice). Typical properties are given in Table 2.

Hot-pressed Si_3N_4 is produced by either conventional uniaxial or hot isostatic pressing. One starts with α-Si_3N_4 pow-

39

der and adds a densification aid such as MgO, Y_2O_3, ZrO_2, or $SiBeN_2$. Under pressures of ~ 14 MPa and higher and temperatures of 1650° to 1750°C, some of the α-Si_3N_4 reacts with the additive and the thin layer of SiO_2 that coats each particle of Si_3N_4, producing a liquid silicate in which the remaining α-Si_3N_4 dissolves and from which it reprecipitates as elongated β-Si_3N_4 grains. On completion of the $\alpha \rightarrow \beta$ transformation, the elongated β grains are surrounded with a residual silicate or oxynitride grain boundary phase. The elongated nature of these grains, which are typically ~ 0.5 by ~ 4 micrometers, gives hot-pressed silicon nitride its high strength. Almost all hot-pressed silicon nitrides exhibit room-temperature flexural strengths of 690 MPa (100,000 psi) or higher and retain these strengths to at least 1000° to 1100°C, as well as exhibiting excellent thermal shock resistance and erosion and corrosion resistance.

Depending on the purity and phase composition of the starting Si_3N_4 powder, type and percentage of additive, milling and mixing procedures, and hot-pressing parameters (temperature, time, and pressure), one can obtain a wide variety of strength versus temperature, creep, or oxidation behaviors. Of all available high-temperature structural ceramics, hot-pressed Si_3N_4 with MgO as a densification aid (typified by the Norton Company's NC-132 material) most nearly approaches the reproducibility of mechanical properties expected of a true engineering material. However, the nature of the magnesium silicate grain boundary phase is such that these materials exhibit a rapid falloff in strength, creep, and oxidation resistance between 1200° and 1350°C.

Since it was apparent that the limitations of hot-pressed silicon nitride were due to the nature of the grain boundary phase and not intrinsic to the Si_3N_4 itself, Gazza and his co-workers in our laboratory [18] and Tsuge and his co-workers at Toshiba [19] focused attention on controlled modification of the grain boundary. This was an extremely difficult undertaking. There was no direct, but only inferential, evidence about the nature of the grain boundary chemistry and phase compositions. Most important, the grain boundary was thought to be only ~ 50 angstroms thick. [Subsequent direct observation of grain boundaries by Clarke and Thomas [20], using lattice imaging transmission electron microscopy (TEM), showed the boundaries to be only ~ 10 Å thick.] Since much of the grain boundary material is in the glassy state, even x-ray and lattice imaging TEM

techniques often yield little information. To distinguish this emphasis on the grain boundary from the more usual interest in total microstructure, a "grain boundary engineering" approach was adopted in our laboratory. Tsuge and co-workers focused on "grain boundary crystallization." This approach has been particularly successful in developing hot-pressed Si_3N_4 with Y_2O_3 as a densification aid. Gazza [21] and others [22] demonstrated that Y_2O_3 additions provide higher strength at both room and elevated temperatures (to 1400°C) as well as better creep and oxidation resistance than obtained with Si_3N_4 containing MgO. However, this material has been plagued with an intermediate-temperature (~ 1000°C) oxidation problem. Although it appears that postfabrication heat treatment [23] and proper attention to composition and phase equilibria, as discussed by Lange [22], can alleviate this problem, more work needs to be done before hot-pressed Si_3N_4 with Y_2O_3 will be as reliable as the MgO-containing varieties. Strengths as high as 965 MPa (140,000 psi) at 1200°C have been obtained with a Y_2O_3 + Al_2O_3 additive, using the grain boundary crystallization approach [24]. Through this focus on the grain boundary, the high-temperature strength of hot-pressed Si_3N_4 has been increased by an order of magnitude in less than 10 years. Nevertheless, hot-pressed Si_3N_4 is difficult to machine and components made from this material are relatively costly. While the high cost may be acceptable for some specialty applications, for more general use a lower-cost fabrication route must be developed.

Reaction-bonded Si_3N_4 has been developed largely to obtain a readily formable (little machining required) low-cost material [25, 26]. In contrast to hot-pressed Si_3N_4, reaction-bonded Si_3N_4 maintains its strength to temperatures beyond 1400°C and exhibits significantly lower creep rates. As a consequence of the reaction-bonding process, this material is of necessity at least 10 percent porous, which makes it less oxidation-resistant than hot-pressed Si_3N_4 at intermediate temperatures, limits its strength to less than 415 MPa (60 kilograms per square inch) or more typically to about 245 MPa (35 ksi), and considerably reduces its erosion resistance.

Fabrication of reaction-bonded Si_3N_4 components begins with a silicon metal preform made by slip casting, dry pressing, flame spraying, injection molding, or various other techniques. The preform is then nitrided in an atmosphere of pure N_2 or N_2 + H_2. The nitridation of such a

Si preform is a remarkable if still somewhat imperfectly understood phenomenon. For $3Si_{(s)} + 2N_{2(g)} \rightarrow Si_3N_{4(s)}$ there is a 23 percent expansion in the solid volume compared to Si, yet when this reaction is carried out on a preform there is essentially no change in dimensions (~ 0.1 percent). The reason for this appears to be that the first Si_3N_4 to form does so by a complex solid-liquid-vapor whisker growth into the void space of the Si preform, forming a skeleton-like structure, which in turn fixes the dimensions of the component. What appears to be a rather complex series of processing steps, and is a complex series of chemical reactions, yields a ceramic material with a unique property: it can be mass-produced to strict dimensional tolerances with little or no machining and at low cost.

Reaction-bonded Si_3N_4 has improved considerably over the past few years. Nevertheless, where high-strength, more oxidation-resistant material is required, it would be desirable to have a readily fabricable, fully dense Si_3N_4. This has been the impetus for the development of sintered Si_3N_4. However, reaction-bonded Si_3N_4 is still a viable candidate for many high-temperature applications and is currently under active development as an automotive turbocharger material in several laboratories.

Sintered silicon nitrides are a rather recent development. Although Si_3N_4 was sintered as early as 1973 [27], the balance between dissociation of the Si_3N_4 and densification during sintering was such that material of only about 90 percent theoretical density was obtainable. Indeed, the possibility of sintering fully dense Si_3N_4 was still an open question as late as 1976. Since then, several groups have succeeded in producing sintered Si_3N_4 of at least 95 percent density, and a few groups have obtained greater than 99 percent density. What occurred was that the progress being made by Prochazka and co-workers at General Electric in sintering SiC and other covalent compounds [15] was beginning to influence the thinking of the Si_3N_4 research community. Also, the concept of using a nitrogen overpressure and other techniques to suppress the density-limiting dissociation of Si_3N_4 at sintering temperatures was independently demonstrated by U.S. and Japanese investigators [28, 29]. Finally, the experiences gained in using the grain boundary engineering or crystallization approach on hot-pressed silicon nitride was seen to be directly applicable to sintered material. As shown in Table 2, commercially available sintered Si_3N_4 has strengths falling between those of the hot-pressed and reaction-bonded materials.

Sintered Si_3N_4 has been formed by injection molding and, provided isotropic shrinkage can be obtained, components require little machining. Recently, Giachello and Popper (30), in a joint program of the Fiat Research Center and the British Ceramic Research Association, demonstrated that it is possible to post-sinter a reaction-bonded silicon nitride preform to ~ 98 percent theoretical density, with increased strength and oxidation resistance. Mangels and Tennenhouse (31) at Ford Motor Company have independently followed a similar line of research and have, in fact, fabricated components of sintered reaction-bonded silicon nitride. With this development one could start with a sintering preform that would yield only 6 to 8 percent linear shrinkage, as opposed to 18 to 20 percent linear shrinkage for sintered components. It is also possible that sintered Si_3N_4 bodies of more than 95 percent theoretical density may be used as preforms for hot isostatic pressing. Such a development would be a major breakthrough toward attaining high-reliability, affordable, high-performance components such as turbocharger rotors and diesel pistons.

SiAlON's represent an important new class of ceramic materials that are solid solutions of metal oxides in the β-Si_3N_4 crystal structure. These solid solutions produce a distorted β-Si_3N_4 lattice; hence, they are referred to as β'-SiAlON's. SiAlON's were originally developed with Al_2O_3, but MgO, BeO, Y_2O_3, and others have all been found to yield β' solid solutions, as well as a variety of other phases. Although SiAlON's were intended for application in heat engines, these materials are still in early development and have not been used in engine demonstration programs to date. It is likely that they will play some role in future engine programs. Since SiAlON phases are present in the grain boundaries of most hot-pressed or sintered Si_3N_4, they control the high-temperature behavior of these materials. Therefore understanding the phase relationships in these systems is of major importance. The bulk of phase equilibrium studies in these systems have been performed by Jack and his students at the University of Newcastle upon Tyne (32) and Tien of the University of Michigan with Gauckler of the Max Plank Institute and their co-workers (33).

This brief review of the silicon carbides and nitrides has focused on processing and the development of high-temperature strength. There are, of course, many other properties that must be optimized for these materials to be utilized in high-temperature engineering systems. Creep and oxidation resistance have been briefly referred to above and progress has been achieved in improving these properties. Currently much attention is being directed to the time dependence of strength in these high-temperature structural ceramics. Like all materials at elevated temperatures, they exhibit a decreasing ability to carry a given load as time increases. Data on the time-dependence of high-temperature strength of these materials is presently being accumulated for the purpose of life prediction calculations. As data on the mechanisms of time-dependent failure become available, it can be anticipated that ceramists will develop SiC's and Si_3N_4's with improved time-dependent strength. Certainly, the sections above suggest that, in general, if one can define a specific materials shortcoming, the materials research community will be able to design a material to overcome the problem. Let us now turn from design of ceramic materials back to design with ceramic materials and see how they are being applied.

Applications

As previously stated, there is considerable incentive to use high-temperature structural ceramics in a wide variety of systems. However, the incentive has been counterbalanced by the serious engineering problem of how to deal with the brittle nature of ceramics. Any experienced designer knows that you do not subject brittle materials to appreciable tensile stresses, even if the materials are reputedly strong. Quite appropriately, even the most innovative designer is not willing to accept a high-risk design. A decade ago the concept of using ceramics as highly stressed components in any system was decidedly a high-risk undertaking. Recognizing both the potential and the risk, and the need to integrate materials and design innovation in one systems-oriented program that would be structured to allow for several iterations of the systems design, the U.S. Department of Defense's Advanced Research Projects Agency (DARPA) initiated a Brittle Materials Design Program in 1971. A contract was awarded to Ford Motor Company, with Westinghouse Electric Company as a subcontractor and the Army Materials and Mechanics Research Center as monitor, to develop a design capability for brittle materials. This capability was to be demonstrated through the successful development and application of ceramic hardware in gas turbine hot-flow-path components to operate at 1373°C, uncooled, for 200 hours in a vehicular engine and for 100 peaking cycles in an electrical utility size turbine test rig.

The major underlying goal of the DARPA Brittle Materials Design Program was to demonstrate and encourage the use of ceramics as engineering materials. When the program was initiated, it was debatable whether ceramics could survive the rigors of the gas turbine environment in any meaningful capacity. During the course of this pioneering program, which has just ended, both reaction-bonded Si_3N_4 and reaction-sintered SiC stators, reaction-bonded Si_3N_4 nose cones and shrouds, as well as reaction-sintered SiC combustors have been demonstrated by Ford over duty cycle conditions (up to 1373°C, uncooled) in test rigs for more than 200 hours. A ceramic rotor fabricated from a hot-pressed Si_3N_4 hub bonded to a reaction-bonded Si_3N_4 blade ring was also demonstrated for 200 hours at 50,000 revolutions per minute and a gas temperature equivalent to operation at a turbine inlet temperature of 1200°C and a rotor rim temperature of 1000°C—beyond the capability of uncooled superalloys. Most significantly, an engine was run with a ceramic nose cone, stator, and rotor at temperatures between 1225° and 1415°C for 36 hours (34). This demonstrated an operating advantage for ceramics of about 300°C over what is possible with uncooled superalloys. As the DARPA program was demonstrating that ceramic components could survive in a gas turbine environment, other programs both in the United States and abroad were initiated to demonstrate ceramics in a variety of engine types. Although there are active programs on ceramic engine technology in West Germany, Japan, Britain, and Sweden, as well as the United States (35), I will review progress only on several selected U.S. programs.

In the area of ceramics for gas turbine application, there have been three significant milestones within the past year. Under a joint DARPA-Navy program (36), Garrett Corporation has demonstrated the integration of more than 100 separate ceramic components into a modified, ceramic-configured (two stages of ceramic stator vanes and two stages of ceramic-bladed rotors) turboshaft engine, operation of the engine for several hours, and attainment of a more than 30 percent increase in horsepower. Although this engine still has many development problems, it has provided the first demonstra-

41

tion that, even with the design compromises required for their incorporation into an engine, ceramics can deliver the increased performance that their advocates have promised.

The second major event was the road and test track demonstration of a partially ceramic-configured Detroit Diesel Allison 404 gas turbine engine in a truck, under Department of Energy sponsorship. In this demonstration an engine with reaction-sintered SiC stator vanes and aluminosilicate ceramic regenerators was installed in the truck after $\sim$ 1800 hours of laboratory testing. The truck was then driven from Indianapolis to Detroit and subjected to about 25 hours of shock and vibration testing as well as about 100 hours of road tests (35). In the third achievement, the Solar Division of International Harvester, under contract from the U.S. Army Mobility Engineering Research and Development Command, conducted a successful, full-performance, 200-hour demonstration of an all-ceramic nozzle section (hot-pressed Si_3N_4 vanes with reaction-sintered SiC shrouds) in a 10-kilowatt turbogenerator. This is the first known instance of a ceramic-configured engine producing electrical power (35).

Progress has been made in the diesel engine area as well. Perhaps the most exciting advanced concept in diesel technology is the so-called adiabatic turbocompounded diesel. This engine would eliminate the water-cooling system, and the heat (energy) that would have been rejected to the cooling system is diverted to the exhaust, recovered via a turbine, and geared back into the engine output shaft. Fully developed, such an engine can reduce fuel consumption by as much as 25 percent, with a significant increase in specific power and a decrease in the maintenance problems associated with the water-cooling system. To attain such gains, ceramic pistons or piston caps, cylinder liners, exhaust gas manifolding, and valve train components are required. The U.S. Army Tank and Automotive Research and Development Command, jointly with Cummins Engine Company, is currently developing such an engine (9). The most highly stressed component in such an engine is the piston cap. An important milestone, therefore, is the demonstration of a ceramic piston cap in an engine. Such a demonstration was successfully completed in a 250-hour performance test in a single-cylinder engine during the past year, utilizing a hot-pressed-to-shape Si_3N_4 piston cap fabricated in our laboratory (35).

Ceramic components are under active development for other applications in piston engines, as turbochargers, diesel combustion prechambers, valve lifters, wrist pins, and push rod tips. The incentives for the use of ceramics in these applications are not high-temperature capability but rather lower cost, improved wear capability, increased stiffness, and reduced inertia for improved response and performance. Carborundum, AiResearch Casting Company, and Ford Motor Company among others, are actively pursuing development of Si_3N_4 and SiC turbochargers. Carborundum has sintered SiC valve lifters under test in racing cars, and thus far they are providing outstanding performance and durability.

In another area, Hague International is commercially marketing a high-temperature heat exchanger for use on slot forging furnaces (maximum temperature $\sim$ 1250°C). What is impressive in this application is that the heat exchange tubes are finned SiC elements about 50 inches long and that they operate successfully in the shock and vibration environment of a forge furnace (10). The corrosion and erosion resistance of SiC and Si_3N_4 ceramics, in particular, make them likely to be used as heat exchangers and combustors in the indirect firing of turbines with coal or heavy residual oil (3). A future application for Si_3N_4- or SiC-based heat exchangers would be in the solar energy cavity receiver for solar "power towers" operating hot-air turbines.

Aside from the high-temperature engineering applications for the SiC and Si_3N_4 materials, there are also potential applications in the metals processing, wear-resistant materials, and bearing areas. Hot-pressed Si_3N_4 in particular has shown significant potential for use in high-performance ball and roller bearings (37). Work on hot-pressed Si_3N_4 in the People's Republic of China (38) and on SiAlON in Britain (39) has shown that these materials give good performance as tool bits in machining selected materials.

The foregoing discussion clearly demonstrates that high-performance ceramics technology has come a long way in the past 10 years. We have gone from discussing whether ceramics can survive in an engine to actually having ceramics in engines on the highway, on the racetrack, and producing electrical power. Yet there is still a long way to go. The applications discussed above are for the most part one-of-a-kind demonstrations. For most applications ceramic components are not yet ready for introduction into commercial engines or other high-performance systems. What will it require for this technology to become commercialized?

Outlook

For high-performance ceramic materials to be successfully commercialized in mass-production applications such as automotive engines, several key problems must be solved. Primary among these is ceramic attachment. The section above described enough successes of ceramic application in engines to show that considerable progress has been made in the attachment area. What I did not mention above is that virtually every case of a ceramic component failure in an engine or a test rig, that I am aware of, has been the result of an attachment problem. Thus, advanced design and materials concepts for minimizing contact stresses in ceramic-to-ceramic and ceramic-to-metal attachments maintaining compliance and integrity at high temperatures and for long times constitute a crucial R & D area. The second most critical question is whether the relatively few components demonstrated to date can be scaled up to quantity production reliably, reproducibly, and affordably. This is a problem that the ceramics industry deals with each time a new technology using ceramics is introduced. Recent examples of reliable high-volume manufacturing of high-performance ceramics such as alumina substrates, envelopes for sodium vapor lamps, SiC igniters for the gas appliance industry, and ZrO_2 for automotive emissions sensors give reason for optimism that the levels of product uniformity required for industrial application can be attained. The third key area is that of nondestructive testing (NDT). Significant increases in NDT and other quality assurance methods will be required to ensure that only reliable components are put into use. At present, this is beyond the state of the art for the materials and applications discussed above. These current problems notwithstanding, it is clear that these materials are serious candidates for application in several areas of engine technology.

Applications such as valve lifters, exhaust port liners, and diesel precombustion chambers are fairly near-term ones that will not require major changes in production lines or engine technology. The same is true of the turbocharger applications, although here some material improvement and attachment work is required, so it is probably not quite so close. Ceramic heat exchangers for industrial waste heat recovery are already commercially available for some applications and will gradually spread into a variety of areas. The use of ceramics in large diesel engines through the adiabatic turbocompound engine is a good likelihood for commercialization by the 1990's. Ceramic components in truck and small industrial gas turbine engines can be phased in gradually and are also likely to be commercialized. The major question is whether the gas turbine, with ceramic components, will be adapted as an alternative engine for the automotive industry. The Department of Energy's Advanced Automotive Gas Turbine Program (*11*) is currently funding the development and evaluation of two such engines, one by AiResearch Company with Ford Motor Company as a subcontractor, and one by Detroit Diesel Allison in conjunction with Pontiac. These engines integrated into vehicles are due for demonstration around 1982 to 1984. Assuming technical success of these programs, there is still the issue of the investment costs for this new technology versus the costs of advanced piston engine technology. Conversion of the nation's automotive engine production lines is a multibillion-dollar undertaking. Whether ceramic gas turbine technology is fully utilized will depend on complex cost-benefit trade-offs between it and alternative technologies. What is important in our increasingly energy- and resource-scarce world is that such options are at least available. In any event, over the next decade we will begin to see

high-temperature ceramics utilized in applications that would have been unimagined a decade ago.

References and Notes

1. *Should We Have a New Engine?—An Automotive Power Systems Evaluation* (Rep. JPL-SP 43-17, Jet Propulsion Laboratory, Pasadena, Calif., 1975), vols. 1 and 2.
2. J. J. Burke, E. M. Lenoe, R. N. Katz, Eds., *Ceramics for High Performance Applications—II* (Brook Hill, Chestnut Hill, Mass., 1978).
3. *Proceedings of the Workshop on Ceramics for Advanced Heat Engines* (Rep. CONF 770110, ERDA, Washington, D.C., 1977).
4. *Proceedings of the Conference on Basic Research Directions for Advanced Automotive Technology* (Department of Transportation, Washington, D.C., April 1979).
5. R. W. Davidge, in *Nitrogen Ceramics*, F. Riley, Ed. (Noordhoff, Leyden, 1977), pp. 653–657.
6. J. L. Klann, *Advanced Gas Turbine Performance Analysis* (8th Summary Rep., Automotive Power Systems Contractor's Coordination Meeting, ERDA-64, May 1975), p. 173.
7. H. E. Helms and F. A. Rockwood, *Heavy Duty Gas Turbine Engine Program Progress Report* (Rep. DDAEDR 9346, contract NAS3-20064, NASA-Lewis Research Center, February 1978).
8. F. B. Wallace *et al.*, in (*2*), pp. 593–624.
9. R. Kamo, in (*2*), pp. 907–922.
10. R. A. Penty and J. W. Bjerklie, in *New Horizons Materials and Processes for the Eighties*, *Proceedings of the 11th National SAMPE Technical Conference* (SAMPE, Azusa, Calif., 1979); R. A. Penty, personal communications.
11. G. Thur, paper presented at the 17th Department of Energy Highway Vehicle Systems Contractors' Coordination Meeting, Dearborn, Mich., October 1979.
12. G. Q. Weaver and B. A. Olson, in *Silicon Carbide—1973*, R. C. Marshal, J. W. Faust, Jr., C. E. Ryan, Eds. (Univ. of South Carolina Press, Columbia, 1974), pp. 367–374.
13. S. Prochazka, in *ibid.*, pp. 391–402.
14. R. J. Bratton and D. G. Miller, in (*2*), p. 719.
15. S. Prochazka, in *Ceramics for High Performance Applications*, J. J. Burke, A. E. Gorum, R. N. Katz, Eds. (Brook Hill, Chestnut Hill, Mass., 1974), pp. 239–252.
16. J. A. Coppola and C. H. McMurty, in *National Symposium on Ceramics in the Service of Man* (Carnegie Institution of Washington, Washington, D.C., 1976).
17. W. B. Hillig, in (*2*), pp. 979–1000.
18. R. N. Katz and G. E. Gazza, in (*5*), pp. 417–431.
19. A. Tsuge, K. Nishida, M. Komatsu, *J. Am. Ceram. Soc.* **58**, 323 (1975).
20. D. R. Clarke and G. Thomas, *ibid.* **60**, 491 (1977).
21. G. E. Gazza, *Am. Ceram. Soc. Bull.* **54**, 778 (1975).
22. R. J. Bratton, C. A. Anderson, F. F. Lange, in (*2*), pp. 805–825.
23. G. E. Gazza, H. Knoch, G. D. Quinn, *Am. Ceram. Soc. Bull.* **57**, 1059 (1978).
24. A. Tsuge and K. Nishida, *ibid.*, p. 424.
25. D. J. Godfrey, *J. Br. Interplanet. Soc.* **22**, 353 (1969).
26. D. R. Messier and P. Wong, in (*15*), pp. 181–194; J. Mangels, in *ibid.*, pp. 195–206.
27. G. R. Terwiliger and F. F. Lange *J. Mater. Sci.* **10**, 1169 (1975).
28. H. F. Priest, G. L. Priest, G. E. Gazza, *J. Am. Ceram. Soc.* **60**, 81 (1977).
29. M. Mitomo, M. Tsutsumi, E. Bannai, T. Tanaka, *Am. Ceram. Soc. Bull.* **55**, 313 (1976).
30. A. Giachello and P. Popper, paper presented at the 4th International Meeting on Modern Ceramic Technologies, St. Vincent, Italy, 28 May to 1 June 1979.
31. J. A. Mangels and G. J. Tennenhouse, *Am. Ceram. Soc. Bull.* **58**, 884 (1979).
32. K. H. Jack, in (*5*), pp. 103–128.
33. L. J. Gauckler, S. Boskovic, G. Petzow, T. Y. Tien, in (*2*), pp. 559–572.
34. A. F. McLean and J. R. Secord, *U.S. Army Mater. Mech. Res. Cent. Rep. TR 79-12* (1979).
35. Presentations on these programs were made at the Conference on Ceramics for High Performance Applications III—Reliability, held at Orcas Island, Wash., July 1979.
36. W. L. Wallace, J. E. Harper, F. B. Wallace, "Ceramic Gas Turbine Engine Demonstration Program," Interim Rep. 13 on contract N00024-76-C-5352, May 1979.
37. C. F. Bersch, in (*2*), pp. 397–406.
38. H.-C. Miao, Y.-H. Liu, T.-C. Chiaug, paper presented at the 4th International Meeting on Modern Ceramic Technologies, St. Vincent, Italy, 28 May to 1 June 1979.
39. R. J. Lumby, B. North, A. J. Taylor, in (*2*), pp. 893–906.

Aircraft Gas Turbine Materials and Processes

B. H. Kear and E. R. Thompson

For more than three decades the development of the gas turbine engine has been paced by the availability of materials and the ability to process them into useful shapes. The most challenging materials problems have been encountered in aircraft gas turbines. This is because of the need to maintain high operating efficiencies without incurring unacceptable weight penalties. It is to the credit of materials technologists that these challenges have continued to be met, as engine designs have progressed to ever-increas-

ing levels of engineering sophistication and performance. As an indication of the remarkable progress that has been made over the years, it may be noted that, since the 1950's, thrust-to-weight ratios have tripled, fuel efficiencies have more than doubled, and the time between overhauls has increased from 100 to more than 10,000 hours.

The most significant developments in alloy design occurred in the early 1950's. Entirely new classes of heat-resistant nickel- and cobalt-base alloys were de-

veloped which became known later as the superalloys. At the same time, a new class of titanium alloys of high specific strength became available in usable structural forms. The superalloys proved to be of great utility in the hottest parts of the engine, such as burner and turbine sections, whereas the titanium alloys were ideal for the cooler compressor section of the engine. The effect of these developments was to cause a sharp increase in the use of superalloys and titanium alloys in engines, at the expense of conventional nickel- and iron-base alloys (Fig. 1).

Starting in the mid-1960's, the emphasis gradually shifted from alloy development toward process development. In the productive period that followed, several important advances were made in the materials processing area. Perhaps the most striking innovation was the introduction of directional solidification processing of turbine blades and vanes

B. H. Kear is a senior consulting scientist and E. R. Thompson is a group leader at United Technologies Research Center, East Hartford, Connecticut 06108.

43

Another development of equal significance was the adaptation of powder metallurgy processing techniques, such as hot isostatic pressing and superplastic forging, to the production of turbine disks. Along with these developments, which were focused primarily on the superalloys, important gains in processing capabilities were also achieved for titanium alloys. At the same time, advanced composites were developed, and these have found useful applications in the engine.

The situation today is that superalloys account for about 50 percent by weight

The effect is to drive the turbine, which in turn drives the compressor. The high-velocity gases expelled through the exhaust nozzle generate engine thrust. Additional thrust is derived from the relatively low-velocity bypass air driven by the fan and ducted outside the engine.

The indicated temperature and pressure variations correspond to maximum thrust developed on takeoff and represent important design parameters. From the materials standpoint, the crucial factor is the peak temperature developed at different locations in the engine, since

are attached to the compressor casing between adjacent rotor stages. The flow path, defined by successive stages of the compressor, decreases in the direction of air flow, corresponding to the reduction in volume as compression occurs from stage to stage (Fig. 2). The high-pressure compressor runs hotter and at higher speeds than the low-pressure compressor.

Blades and vanes must be capable of resisting aerodynamic loading. In addition, blades must be able to withstand high centrifugal loading and the effect of vibratory stresses caused by high-velocity air streaming from the spaces between blades. Disks must possess high load-bearing capacity, since they have to resist the centrifugal loading of the disk and blades. Some consideration must also be given to the effects of high strains developed at the critical points of attachment of the blades to the disk. Steels and titanium alloys have proved to be satisfactory materials for blades, vanes, and disks in the low-pressure compressor, but the more heat-resistant nickel-base alloys are preferred for the hotter sections of the high-pressure compressor.

Summary. Materials and processing innovations that have been incorporated into the manufacture of critical components for high-performance aircraft gas turbine engines are described. The materials of interest are the nickel- and cobalt-base superalloys for turbine and burner sections of the engine, and titanium alloys and composites for compressor and fan sections of the engine. Advanced processing methods considered include directional solidification, hot isostatic pressing, superplastic forging, directional recrystallization, and diffusion brazing. Future trends in gas turbine technology are discussed in terms of materials availability, substitution, and further advances in air-cooled hardware.

of advanced engines, with the balance roughly equally divided between titanium alloys, composites, and steels (Fig. 1). The trend appears to be in the direction of somewhat higher weight fractions of superalloys and composites, at the expense of titanium alloys and steels. In the superalloy area, the trend is toward increasing applications for directionally solidified (DS) and powder metallurgy (PM) products.

This article presents an overview of materials developments as they are related to specific components in the gas turbine engine, such as blades, vanes, disks, and combustors. A brief description will first be given of the design and operation of component parts in an engine to provide the reader with a better appreciation of the factors involved in materials selection.

Engine Components

Gas turbine engines such as the JT9D turbofan engine (Fig. 2) comprise three main sections: compressor (fan), burner, and turbine (*1*). The compressor raises the pressure and temperature of the incoming air and delivers it to the burner. In the combustion chamber, a fine spray of fuel is thoroughly mixed with the high-pressure air, and the mixture is ignited. The hot gases leaving the combustion chamber undergo rapid expansion in the turbine section, accompanied by a sharp drop in gas pressure and temperature.

this sets an upper limit on the required temperature capability of the material. Other materials design considerations of equal significance are (i) the magnitude of stresses developed by centrifugal forces, vibratory forces, and thermal gradients and (ii) the potential for oxidation (or hot corrosion) and erosion in the high-velocity gas stream.

Fan. The fan section of the engine is integral with the front part of the low-pressure compressor. This permits the fan blades to rotate at low tip speed, consistent with optimum fan efficiency. The fan blades are relatively long and thin components and are braced (shrouded) at the midspan for support and to prevent vibration. Air passing through the fan and exhausted through the ducts tends to carry away ingested foreign material, which otherwise might damage the engine core.

Materials for fan blades must be strong, elastically stiff, and resistant to damage by foreign objects. Experience has shown that titanium alloys satisfy these requirements. Composites with high specific strength, such as carbon-epoxy, have also been considered for this application, but at present lack sufficient resistance to foreign object damage.

Compressor. The compressor consists of a series of rotating blades and stationary vanes, which are arranged in stages concentric with the axis of rotation. The blades are fastened into slots around the periphery of individual disks; the vanes

Burner. The burner section is essentially an annular combustion chamber, in which several burner cans are arranged side by side around the inside of the chamber. Each burner can, which is a separate combustion chamber, is perforated with holes and slots to admit air for cooling. About one-third of the total volume of air entering the burner from the compressor discharge is mixed with the fuel to sustain combustion. The remainder of the air bypasses the fuel nozzles and is used downstream to cool the burner can surfaces and combustion products before the hot gases enter the turbine. The system is designed to maintain an even temperature distribution in the hot gases leaving the combustor and to ensure that the peak temperature does not exceed the allowable limit at the turbine inlet.

Burner materials must be formable, weldable, and resistant to corrosion, distortion, and thermal fatigue at high temperatures. The strength requirements are relatively low, but the strength must be maintained up to approximately 1100°C. A special class of heat-resistant sheet alloys have been developed for this purpose.

Turbine. The configuration of the main components of the turbine—the rotor and interstage guide vanes—is similar to that of the compressor, except that the gas path increases in the direction of flow to accommodate the expansion of gases between stages in the turbine. The blades are attached to the disks, using a characteristic fir-tree root design, which leaves space for expansion. Vanes are slotted into the turbine casing in stages between the rotors. To reduce vibrations and to increase turbine efficiency, the blade tips are sometimes shrouded. To permit a higher turbine inlet temperature, which is critical to improve operating efficiency, inlet guide vanes and first-stage blades are air-cooled. This is accomplished by passing compressor bleed air through longitudinal holes, tubes, or cavities in the airfoil sections (Fig. 3). The cooling air exits through small holes and slots at the leading and trailing edges of the airfoil. Air cooling is necessary only in the turbine inlet section, since the energy extracted from the hot gases by the first- and second-stage rotors reduces the temperature to a tolerable level. It is illustrative of the effectiveness of air cooling that even with a turbine inlet temperature as high as 1300°C, metal temperatures for inlet guide vanes and first-stage blades are easily maintained at 1100° and 950°C, respectively.

The materials requirements for turbine blades and vanes are similar, except for the reduced strength requirements in vanes. Stresses in vanes are less than 35 megapascals, whereas blades are subjected to longitudinal stresses up to 200 MPa in the midspan of the airfoil section. The blade root, which is attached to the disk, is outside the gas path and experiences a maximum temperature of approximately 750°C, but tensile stresses are much higher, in the range 275 to 550 MPa. The primary requirements are creep strength at high temperatures, thermal fatigue resistance, and resistance to environmental degradation by oxidation, corrosion, and erosion. Secondary requirements include castability, impact strength, and microstructural stability to ensure that properties are maintained for long periods of time. Some vane designs also require weldability. Turbine disks operate at temperatures that do not exceed approximately 750°C. The maximum temperature occurs at the outer edge or rim of the disk. The stresses developed by centrifugal loading are high at the rim and even higher in the bore. Primary materials requirements are for high burst strength at the operating temperature of the bore and good creep strength at rim operating conditions. The materials should also possess good fatigue strength, both low-cycle and high-cycle. A bewildering variety of superalloys, both nickel- and cobalt-base, have been designed for turbine applications.

Materials and Applications

Superalloys. The high-temperature alloys (2–5), known as the superalloys, occur in two broad classes: nickel-base and cobalt-base (Table 1). Nickel-base alloys are strengthened by precipitation of the $Ni_3(Al,X)$ γ' phase, where X is a solid solution hardening element, such as Ti, Nb, or Ta. In most advanced alloys, γ' precipitation hardening is supplemented by solid solution hardening of the γ matrix with refractory elements, such as W, Mo, or Re. The γ' particles precipitate on a fine scale (Fig. 4a), and both precipitate and matrix phase are coherent. This is a consequence of the close similarity in structure between the face-centered cubic γ and ordered face-centered cubic γ' phases. The cobalt-base alloys derive their strength from combined solid solution hardening and carbide dispersion strengthening. Fine networks of carbide phases (Fig. 4b) appear to be particularly effective in strengthening at high temperatures.

The nickel-base alloys have outstanding strength characteristics at temperatures in the range of 750° to 1000°C. The cobalt-base alloys are stronger and more corrosion-resistant at temperatures above about 1050°C. Accordingly, the nickel-base alloys are used for intermediate-temperature blade and disk ap-

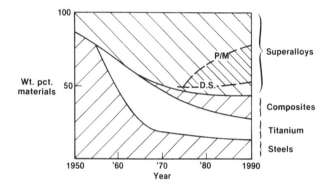

Fig. 1. Weight percent of materials used in most advanced aircraft gas turbine engines.

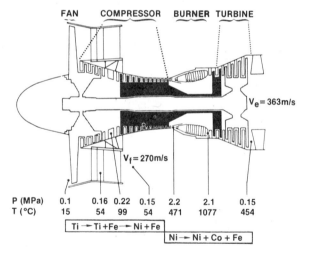

Fig. 2. Internal pressure (*P*) and temperature (*T*) variations in the Pratt & Whitney JT9D Turbofan engine at a sea-level take-off thrust of 19,500 kilograms and with a bypass ratio of 5:1.

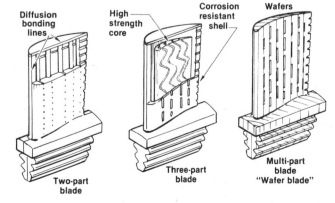

Fig. 3. Advanced air-cooled turbine blades.

Table 1. Nominal compositions of some nickel-base and cobalt-base alloys.

Alloy	Ni	Cr	Co	C	Ti	Al	Mo	W	Other
Blade alloys									
Waspaloy	Bal.*	19.5	13.5	0.08	3.0	1.4	4.0		0.08 Zr, 0.007 B
B-1900	Bal.	8.0	10	0.11	1.0	6.0	6.0		0.07 Zr, 0.015 B, 4.3 Ta
PWA 1422	Bal.	9.0	10	0.11	2.0	5.0		12.5	0.015 B, 1.0 Cb, 2.0 Hf
Disk alloys									
Incoloy 901	Bal.	12.5		0.10	2.6		6.0		34 Fe, 0.015 B
Waspaloy	Bal.	19.5	13.5	0.08	3.0	1.4	4.0		0.08 Zr, 0.007 B
IN-100	Bal.	12.4	18.5	0.07	4.3	5.0	3.2		0.06 Zr, 0.02 B, 0.8 V
Vane alloys									
X-40	10	25	Bal.	0.50				7.5	1.5 Fe
WI-52		21	Bal.	0.45				11	1.75 Fe, 2.0 Cb
MAR-M509	10	24	Bal.	0.60	0.2			7	0.5 Zr, 3.5 Ta
Burner alloys									
Hastelloy X	Bal.	22	1.5	0.10			9	0.6	18.5 Fe
Haynes 188	22	22	Bal.	0.10				14.5	0.08 La
Inconel 617	Bal.	22	12.5	0.07		1.0	9		

*Balance.

plications, whereas the cobalt-base alloys are generally preferred for high-temperature vane applications.

The essential requirement for blade alloys is adequate creep strength at elevated temperatures. Experience has shown that this requirement is best satisfied by exploiting the nickel-base alloys with a high γ' volume fraction. In the 1950's, alloys of this type, containing about 30 percent γ', were fabricated into blades by hot forging (Fig. 5). In the 1960's following the development of improved alloys containing up to 60 percent γ', forging was replaced by investment casting as the preferred method of blade fabrication. Later this was to evolve into directional solidification processing of blades, including both columnar-grained and single-crystal materials. Looking at the overall picture, it can be seen that such processing innovations, along with advances in alloy design, have been responsible for an increase of about 150°C in the permissible operating temperature of blades. Current interest in this area is focused on exploiting the higher strength capabilities of directionally solidified eutectics (6) and directionally recrystallized (7) conventional superalloys. The eutectic microstructure is composite in nature, consisting of a γ/γ' matrix and reinforcing fibers, or lamellae of a refractory phase. An example is the γ/γ'-α eutectic, which is reinforced with thin filaments of α-Mo (Fig. 4c). Such alloys promise to increase the temperature capability of blade alloys by a further 75°C in the 1980's (Fig. 5).

At one time, a clear distinction could be made between blade and disk alloys. Blade alloys were selected for intermediate-temperature creep strength, disk alloys for high strength at somewhat lower temperatures. As operating temperatures in the engine have increased, this distinction has become blurred because of the need to provide additional creep strength in the disk alloys. The solution to this problem has been to increase the γ' volume fraction of disk alloys at the expense of hot workability. As a consequence, conventional hot forging of disks, which is limited to workable low γ' alloys, has been gradually replaced by more flexible PM processing methods (Fig. 6). The new processes, which are known as superplastic forging and hot isostatic pressing, are applicable to all nickel-base alloys. Although appreciable gains in strength have been achieved with the new alloys and processing methods, this benefit is perhaps not as significant as the improvement in creep strength, because of the higher rim temperatures encountered in today's engines. For this reason, another likely development in the 1980's is the dual-property disk, in which the bore of the disk is optimized for load-bearing capacity and the rim for creep strength.

Cobalt-base alloys have found their widest application as vane materials. They are attractive for this application because they can be processed by relatively inexpensive investment casting techniques, without having to resort to vacuum melting. The early alloys, such as X-40 and WI-52 (Fig. 7), are examples of alloys in this air-melting category. For newer alloys, such as MAR-M509, which

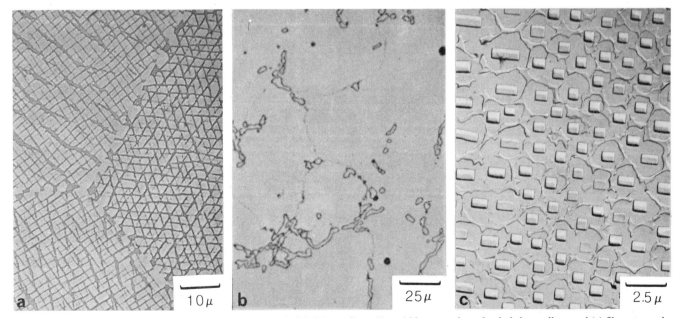

Fig. 4. Representative microstructures of (a) γ'-strengthened nickel-base alloy, (b) carbide-strengthened cobalt-base alloy, and (c) fiber-strengthened γ/γ'-α (Mo) eutectic alloy.

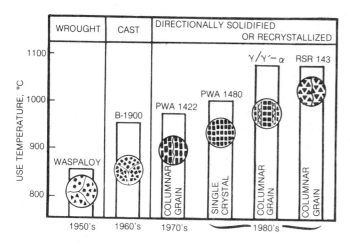

Fig. 5. Evolution of blade alloys.

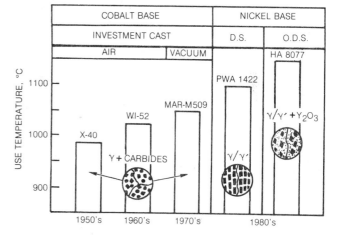

Fig. 6. Evolution of disk alloys.

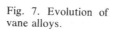

Fig. 7. Evolution of vane alloys.

contain more reactive elements, vacuum melting and casting is necessary. At present, the future of cobalt-base alloys in engines is uncertain, because of current and projected shortages of cobalt. It seems clear, however, that greater use will be made of directional solidification of vane alloys, because of the real improvements in thermal fatigue resistance exhibited by DS structures. It also seems likely that some use will be made of the new generation of oxide dispersion strengthened (ODS) alloys, which are now becoming available commercially. A lot will depend on the cost of fabricating these materials into vane shapes. These new alloys combine high-temperature strength with remarkable resistance to oxidation and hot corrosion.

Combustor components are fabricated from sheet alloys by a variety of conventional shaping and joining operations. High workability and good weldability are essential for burner alloys, which precludes the use of precipitation-hardenable systems. Experience has shown that the necessary strength at high temperatures can be achieved by refractory metal (W or Mo) solid solution strengthening of either nickel- or cobalt-base alloys. High Cr levels are necessary to maintain adequate resistance to oxidation and hot corrosion. Representative alloy compositions for service up to ~ 1100°C are given in Table 1.

Distortion of combustors due to thermal cycling and local hot spots remains a problem. Attempts are being made to improve the situation by exploiting the more heat-resistant ODS alloys, such as HA 8077 (NiCrAl with Y_2O_3 dispersion). The ODS alloys have the potential for use at an ~ 100°C higher temperature than conventional sheet alloys. Manufacture of ODS sheet products has been demonstrated, but costs are high. Improved manufacturing techniques, such

as mechanical alloying, promise substantial reduction in the cost of these materials (8, 9).

Titanium alloys. Titanium alloys are categorized by the microstructural phases that predominate near room temperature (10, 11). Titanium exhibits an allotropic phase transformation; β, the high-temperature phase, is body-centered cubic, and α, the low-temperature phase, is close-packed hexagonal. In pure titanium this transformation occurs at 882°C. Alloying elements may act to stabilize either the β or the α phase. Aluminum is the most important of the α-stabilizing elements. The β alloys have low creep strengths at elevated temperatures and are therefore limited to low-temperature applications. Table 2 lists selected properties of three conventional titanium alloys that are representative of current compressor disk and blade materials.

Materials used for moderate-temperature applications are often the heat-treatable $\alpha + \beta$ alloys, such as Ti-6Al-4V, with fine-grained microstructures of α and transformed β. The alloys selected for high-temperature compressor and

disk components may be near-α or α-lean β alloys. Examples of these respective alloys are Ti-8Al-1Mo-1V, which contains small amounts of the β-stabilizing elements Mo and V, and Ti-6Al-2Sn-4Mo-2Zr, where solid solution strengthening of the α phase by Al, Sn, and Zr produces improved creep resistance. The relatively good strength of Ti-8Al-1Mo-1V and its higher elastic modulus and lower density, as indicated in Table 2, favor its use for applications where material stiffness is important.

The application of titanium alloys to the blades and disks of the compressor section of the engine has been a major contributor to its evolution. This usage was prompted by the low density of titanium alloys, which leads to superior specific strength of these alloys at temperatures up to about 500°C. In rotating gas turbine components, tensile, creep, low- and high-cycle fatigue, fracture toughness, and erosion properties are the important selection criteria. Alloying and thermomechanical processing are used to achieve an appropriate balance of these properties and to improve the temperature capability of titanium alloys.

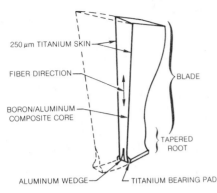

Fig. 8. Composite fan blade.

Despite the improvements that have been made, the upper use temperature for titanium alloys is disappointingly low relative to the melting point of titanium. Some promise for overcoming this temperature limitation is provided by the Ti_xAl_y-based intermetallics, TiAl and Ti_3Al. These intermetallics combine low densities with a high-temperature capability. They show loss of ductility above room temperature, and this problem is being addressed. Another approach, which may further enhance the specific properties of titanium alloys and their upper use temperature, involves the use of fibers, such as silicon carbide, for reinforcement. The eventual application of such composite systems awaits extensive materials development, component design, and evaluation efforts.

The initial developments in titanium alloys were not without significant problems. Some of these, such as alloy embrittlement in the presence of hydrogen concentrations higher than 150 parts per million and alloy element segregation and contamination, were solved by careful vacuum arc melting. Titanium alloys are now commonly produced by multiple consumable electrode arc melting under vacuum. This procedure degasses and homogenizes the cast structure.

Gas turbine components of titanium alloys are usually produced from hot-forged and heat-treated products. The properties of these alloys depend on microstructural morphology as well as composition. The morphology is subject to the thermomechanical processing history, and microstructural variations that occur in some products, in turn, result in variability of properties. This sensitivity is a consequence of effects such as that of cooling rate on the β-to-α transformation. Advances have been made in the fabrication of titanium alloys by PM techniques, and this approach promises to improve their compositional and microstructural homogeneity.

Composites. The high strengths, high stiffnesses, and low densities characteristic of high-performance fiber-reinforced materials make them attractive candidates for gas turbine structural applications (*12*) (Table 3). The most important current composite systems are graphite/epoxy, graphite/polyimide, and boron/aluminum. Graphite/epoxy is finding application as a material for fan exit guide vanes. This system has a use temperature up to 175°C. The graphite/polyimide system, with a use temperature to 325°C, is being evaluated for exhaust flaps in an advanced military engine. Another component subject to substantial development activity is the fan blade of boron/aluminum (Fig. 8). In comparison to the present titanium blade, the composite blade is capable of a higher tip speed and may, because of its higher modulus, be designed without the mid-span stiffener, which leads to improved aerodynamic efficiency. The 1980's should see increased applications of composites as engine components. Cost and reliability will be important considerations for these applications.

Coatings. The development of coatings to extend the life of a gas turbine airfoil began in about the mid-1950's. However, it was not until the late 1950's that the first practical application to cobalt-base alloy vanes was realized. Since then, successful applications of coatings have been made to blades, burner cans, and other critical components in the gas turbine engine (*13*).

The earliest coating methods involved some type of aluminizing treatment; that is, diffusion of aluminum into the surface layers of the alloy substrate. The effect of such a treatment on nickel- and cobalt-base alloys is to produce protective layers consisting mainly of the intermetallic compounds NiAl and CoAl (Fig. 9a). These compounds impart good oxidation and hot corrosion (sulfidation) resistance, because they form a continuous impervious scale of alumina on exposure to a high-temperature oxidizing environment. Two coating methods have been widely used for aluminizing superalloys: slurry fusion and pack cementation. In the slurry process, the part is sprayed with a suspension of aluminum or aluminum alloy and subjected to a high-temperature treatment to produce melting and interdiffusion between deposit and substrate. In the pack cementation process, the part is reacted with aluminum or aluminum alloy powder in the presence of an ammonium halide activator at an elevated temperature. The operation is carried out in an hermetically sealed container to maintain a controlled activity of the aluminum in the vapor phase. Typical coating thicknesses are in the

Table 3. Properties of reinforcing fibers.

Fiber	Strength (MPa)	Elastic modulus (GPa)	Density (kg/m³)
Graphite	2100–2400	200–400	1700–1900
Boron	3500	400	2500
Silicon carbide	3450	400	3000

Table 2. Selected properties of representative titanium alloys.

Alloy	Density (kg/m³)	Annealing/solution treatment temperatures (°C)	20°C elastic modulus (GPa)	Typical strength (MPa) 20°C	Typical strength (MPa) 538°C
Ti-6Al-4V	4428	732–760/899–968	11.3	1000	480
Ti-8Al-1Mo-1V	4373	760–788/non–heat-treatable	12.7	1000	630
Ti-6Al-2Sn-4Mo-2Zr	4539	704–843/829–913	11.3	1000	680

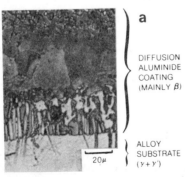

a

DIFFUSION ALUMINIDE COATING (MAINLY β)

ALLOY SUBSTRATE ($\gamma + \gamma'$)

20μ

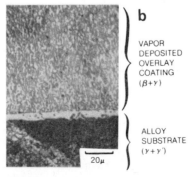

b

VAPOR DEPOSITED OVERLAY COATING ($\beta + \gamma$)

ALLOY SUBSTRATE ($\gamma + \gamma'$)

20μ

Fig. 9. Microstructures of coatings: (a) diffusion aluminide and (b) vapor-deposited MCrAlY overlay.

range 25 to 100 micrometers. Advanced coatings treatments include additions of Cr and Pt to further enhance the oxidation and hot corrosion resistance of the coating. Major improvements in hot corrosion resistance can be achieved by proper control of the Pt distribution in the aluminide layer. The Pt-rich coatings are synthesized by electroplating a thin (about 5 μm) layer of Pt before aluminizing.

A limitation of the diffusion aluminide coatings is that they seriously lack ductility at temperatures below about 750°C. Thus, they are highly susceptible to surface cracking under thermal cycling conditions. Another problem is the relatively poor adherence of the protective oxide scale to the coating alloy. Under thermal cycling conditions, it is not uncommon to encounter repetitive detachment or spallation of the alumina scale, leading to rapid degradation of the coating due to loss of Al. To resolve these problems, the emphasis in coating technology shifted in the mid-1960's to direct bonding or cladding of optimized coating compositions to the superalloy substrate. Ductility was improved by making adjustments in coating compositions so as to develop microstructures in which the brittle β-NiAl or β-CoAl is embedded in a ductile γ solid solution matrix. Oxide adherence was improved by making trace additions of rare earth elements, such as yttrium. The final result of this work was the introduction of the MCrAlY coatings, where M is Ni, Co, or Ni + Co (Fig. 9b). As in the case of diffusion coating, overlay coatings also benefit from additions of Pt.

Many methods of processing MCrAlY overlay coatings have been tested, including diffusion bonding, powder sintering, plasma spraying, vacuum evaporation, sputtering, and ion plating. Of these, electron beam evaporation has emerged as the preferred method of coating superalloy blades and vanes. In a typical arrangement, using continuous ingot feed, it is necessary to impart a complex motion to the part in order to achieve a uniform coating deposit.

An oxide scale on the alloy coating acts to some extent as a thermal barrier, because of its relatively poor heat transfer characteristics. Its effectiveness, however, is limited because its thickness is typically < 1 μm. To improve the situation, the concept of applying thermal barrier coatings, on top of existing coatings, has emerged (14). So far the method has been applied successfully to sheet metal components, such as burner cans and exhaust liners. The coatings are based on yttria-stabilized zirconia, and

are applied by plasma-spraying layers 125 to 500 μm thick. To minimize the effects of stresses due to thermal expansion mismatch between alloy substrate and zirconia deposit, coatings compositions are normally graded—that is, the composition gradually changes from base metal to pure zirconia over the coating thickness. Thermal barrier coatings are now being developed for turbine airfoils, to reduce metal temperatures and thereby reduce cooling-air requirements. Some of the best results have been obtained with a 250-μm plasma-sprayed coating of yttria-stabilized zirconia over a 100-μm layer of NiCrAl coating (Fig. 10). Metal temperature reductions of ~ 100°C have been achieved in an engine test of experimental air-cooled vanes. Increased applications of ceramic coatings for thermal insulation and as abradable seals are expected in the 1980's.

Advanced Processes

Directional solidification. In 1960 it was demonstrated that the creep properties of superalloys can be improved markedly by directional solidification to align all the grain boundaries parallel to the direction of the applied tensile stress (15). The effect was attributed to the elimination of grain boundaries transverse to the applied stress, since such boundaries are highly susceptible to cav-

itation and cracking under creep conditions. In 1967 it was shown that further improvements in properties could be achieved by eliminating the grain boundaries altogether; that is, by directional solidification to produce single crystals (16). Following these leads, in the late 1960's, Pratt & Whitney Aircraft developed various commercial processes for the directional solidification of bars, ingots, slabs, and more complex shapes. The most notable achievement was the directional solidification of cast-to-size turbine blades (17) (Fig. 11).

The DS process evolved from conventional investment casting practice. In conventional casting, the melt is poured into a ceramic shell mold, and solidification is allowed to occur in a relatively uncontrolled manner by radiation to the walls of the vacuum chamber (Fig. 12a). Since the mold preheat is usually about half the melt temperature, the melt experiences a chilling effect in the mold, which results in fairly rapid solidification and the formation of a fine polycrystalline structure. In directional solidification the situation is similar, except for a higher mold preheat temperature and some provision for controlling temperature gradients in the mold. This is accomplished by attaching the ceramic mold, open at its base, to a water-cooled copper chill plate. Following the introduction of the melt into the mold and the commencement of solidification in the chill zone, the temperature gradient

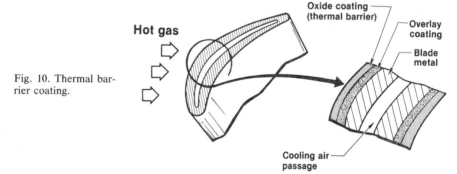

Fig. 10. Thermal barrier coating.

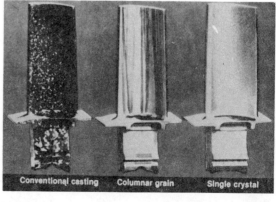

Fig. 11. Effect of processing on grain structure of cast-to-size turbine blades.

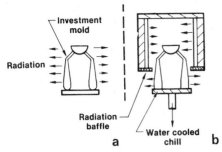

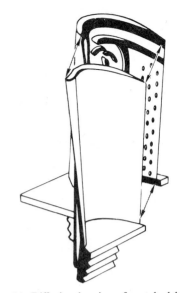

Fig. 12. Comparison of (a) conventional casting and (b) directional solidification.

Fig. 14. Diffusion brazing of matched blade halves.

in the mold is adjusted to promote the advance of the solid-liquid interface in a direction normal to the chill surface. Thus, a characteristic directionally aligned or columnar-grained structure is developed in the casting. A useful consequence of the steep temperature gradients developed in the chill zone is the formation of a strong <100> or cube texture, which is incorporated into the columnar grain structure. Current practice favors the arrangement depicted in Fig. 12b. The desired temperature gradient is maintained by gradually lowering the mold from the hot zone, through an array of heat shields, into the vacuum chamber, where radiation losses can occur freely.

The procedure for processing single-crystal superalloys emerged quite naturally from the basic DS process. A crystal selector is incorporated into the mold just above the chill zone. After developing the desired <100> texture in the chill zone, the crystal selector is used to exclude all but one grain, which then expands to fill the mold. To obtain single crystals in any desired orientation, conventional seeding techniques must be used in conjunction with the DS process. The oriented seed crystal is attached to the base of the mold in contact with the chill plate. The maximum allowable growth rate for the DS processing of single crystals is 500 centimeters per

hour, while columnar-grained material can be produced at rates up to 40 cm/hour.

Nickel-base superalloys respond well to DS processing. This applies to both conventional γ' precipitation-hardened alloys and eutectic superalloys, such as γ/γ'-α (Mo) and γ/γ'-TaC. Eutectic superalloys, however, require steeper temperature gradients and much lower growth rates (typically ~ 2 cm/hour) to obtain optimally aligned structures. Significant improvements in creep rupture properties by DS processing have been obtained for both types of superalloys (Fig. 5). Conventional superalloys also exhibit improved thermal fatigue properties in the <100> orientation. These property advantages have been exploited in DS superalloy castings of first- and second-stage turbine blades and inlet guide vanes. At present, superalloy castings of the columnar grain type are the most widely used in commercial engines. However, the situation is changing, and it appears that before long single-crystal castings will displace them for the most critical applications in the engine. This is because of the higher temperature capabilities that have been achieved in a new class of superalloys, designed specifically to exploit the single-crystal character of the material. An example of an alloy in this category is PWA 1480, which offers a 25°C advantage over its columnar-grained predecessor, PWA 1422 (Fig. 5). It remains to be seen whether the demonstrated advantages in creep strength of certain eutectic superalloys, such as γ/γ'-α, will offset the higher manufacturing costs due to the lower growth rates.

In contrast to these developments, the situation with respect to cobalt-base superalloys for vane applications has remained relatively static. As indicated in Fig. 7, conventional investment casting has maintained a dominant position in the processing of cobalt-base superalloys since the 1950's, although the temper-

ature capabilities of the alloys have continued to improve. The picture may be changing because of continuing shortages in the supply of cobalt. It now seems likely that the 1980's will see the gradual adoption of DS nickel-based superalloys, such as PWA 1422, for vane applications, along with ODS nickel-base alloys, such as HA 8077.

Superplastic forging. In 1963 it was found that fine-grained superalloys exhibit high ductility or superplastic behavior when deformed at high temperatures under appropriately low strain rate conditions (18). This paved the way for the subsequent development of a new process for the hot deformation processing of superalloys, which has become known as the Gatorizing forging process.

The desired fine-grained microstructure can be obtained most conveniently by hot extrusion. Typically, the material is subjected to high deformation rates at temperatures just below the γ' solution temperature. This is a crucial aspect of the process, since the high deformation rate causes spontaneous recrystallization on a very fine scale and the presence of the γ' particles serves to stabilize the fine grain structure. The resulting grain size is in the range 1 to 5 μm, depending on the extrusion ratio and temperature. When the optimum fine-grained structure has been achieved, the material may be worked into shape by exploiting its superplastic character. This is accomplished by controlled strain rate, isothermal forging—that is, superplastic forging. After forging, the creep strength of the material is recovered by a solution heat treatment to coarsen the grain structure, followed by quenching and aging to obtain the opti-

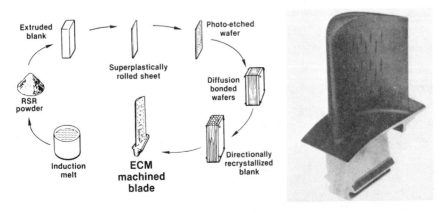

Fig. 13. Sequence of steps involved in fabrication of a wafer blade.

mum distribution of the γ' hardening phase.

Superplastic forging has been used to fabricate superalloy turbine disks and even complete rotors starting from pre-alloyed powder. A particular advantage of the process, compared with conventional hot forging, is its ability to generate a forged profile that more closely matches the final shape of the article. This reduces machining costs, because less material has to be removed in the finish machining operation.

A characteristic feature of the fine-grained superplastic material, at least when produced from prealloyed powder, is its remarkable chemical and micro-structural uniformity. Another favorable aspect is its ability to form a directionally aligned or columnar-grained structure when subjected to directional recrystallization (DR) under the influence of a steep thermal gradient (19). The resulting grain structure bears a superficial resemblance to that obtained by directional solidification. However, there are important distinctions. The DR grain structure is finer in scale and much more homogeneous than the DS structure. Furthermore, the DR structure exhibits not one, but several grain textures, including, <100>, <110>, and <111>, depending on processing parameters and alloy composition.

This ability to develop textured columnar-grained structures entirely by solid-state processing has stimulated interest in new methods for the fabrication of air-cooled blades and vanes of advanced design, starting from superplastically formed sheet stock (20, 21). One such processing scheme is depicted in Fig. 13. The critical step in the process is the diffusion bonding of thin photoetched wafers to produce the desired configuration of internal cooling passages. Following directional recrystallization of the bonded structure, the actual blade profile is formed by electrochemical machining. An experimental air-cooled wafer blade, similar to that shown in Fig. 13, has already withstood higher turbine inlet temperatures than even the most sophisticated one-piece or two-piece blade castings (Fig. 3). This is because of the much more efficient air-cooling schemes that are attainable with the multiple-wafer fabrication technique.

Hot isostatic pressing. In 1955, a process was invented for the gas-pressure diffusion bonding of materials, which later became known as hot isostatic pressing (HIP) (22, 23). After 20 years, the process has found its greatest commercial significance in the production of fully dense, shaped parts from prealloyed powders. It has also been used

to some extent for the healing of casting defects.

The superalloy powder is packed inside a thin-walled collapsible container, which is a geometrically expanded version of the final shape. After vacuum degassing at an elevated temperature, the container is sealed, pressure checked, and subjected to the simultaneous application of pressure and temperature for a time sufficient to cause full densification of the powder. Finally, the shaped part is obtained by stripping off the container.

At this time, processing of superalloy disks, near-final shape, has become a firmly established technique. The development phase of the process, however, is far from over. Remaining to be formulated are procedures for more effectively controlling the size and distribution of fatigue-limiting flaws, such as ceramic inclusions, which are invariably found mixed in with the alloy powder. The most promising immediate solution to this problem is to screen out the finer size fractions (44 μm) of alloy powder for fabrication purposes. In this way, an upper limit can be set for the maximum flaw (inclusion) size, which controls fatigue crack initiation. This is an important design consideration in highly stressed parts, such as disks.

Hot isostatic pressing has also found useful application in the healing of defects in conventionally cast blades and vanes, such as small shrinkage pores. When such pores are not surface-connected, HIP may be used to close them. Surface-connected pores can be healed only if a coating is applied before HIP. This treatment results in a significant improvement in rupture life.

In addition to these current applications, HIP is also being considered for the rejuvenation of parts, such as blades, vanes, and disks, that have sustained creep or fatigue damage in service. Blade and vane materials that have been extended well into steady-state creep can be restored to their original condition by HIP and reheat treatment. The HIP cycle evidently heals the micropores that develop during creep. Fatigue-induced cracks in disk materials have also been healed by HIP. However, it is not yet clear how best to handle the surface-connected cracks, which tend to be oxidized or otherwise degraded. Another area of interest is the fabrication of laminated composite structures, or graded monolithic structures. A good example of the latter is the dual-property disk concept, which envisages a bore that possesses high load-bearing capacity and a rim that has good creep strength. Conceivably, this structure could be made by HIP,

utilizing two different powder compositions.

Diffusion brazing. Conventional brazing is widely used in the fabrication of gas turbine components and parts, but not usually for joining critical superalloy components that are exposed to high temperatures and corrosive environments. This is because it is difficult to design a filler material that satisfies all property requirements, both physical and mechanical, including complete compatibility with the workpiece. In recognition of these shortcomings in conventional brazing, a new process called diffusion brazing was developed in 1974 (24).

Diffusion brazing, as applied to the joining of superalloy components, is a vacuum brazing operation. The process makes use of a filler, or interlayer material, that closely matches the composition of the workpiece except for the addition of an appropriate melt depressant, such as boron, to form a eutectic with a low melting point. The interlayer is placed between the mating surfaces of the workpiece, and a slight normal pressure is exerted to maintain physical contact. The temperature is increased to the predetermined brazing temperature, where the eutectic melts and alloys with the workpiece. Under isothermal conditions, the melting point of the eutectic gradually rises as the boron diffuses away into the workpiece. The process is considered to be complete when no eutectic remains in the joint. A subsequent heat treatment is employed to erase all traces of the original interface.

Diffusion brazing produces good joints in a variety of nickel-base superalloys, including dissimilar alloy combinations. Furthermore, bond strengths comparable with base-metal strength have been achieved, even in high-temperature creep-rupture tests. Successful applications for diffusion brazing have included joining of vane clusters, attachment of hardened tips to blades, and bonding of two-part blades (25) (Fig. 14). With respect to the two-part blade application, good joints have been obtained irrespective of whether the matching blade halves have DS columnar-grained or single-crystal structures.

Future Perspective

Among the factors that will influence the application of materials to future aircraft gas turbines will be the cost and availability of certain strategic elements, such as Cr, Co, Ta, and Pt. This may dictate the limitation of specific alloys to the most critical engine components. Such constraints will promote the development of new material systems and processes. As design schemes for gas turbine components increase in sophistication, it seems clear that manufacturing innovations will play an increasingly important role.

Some anticipated changes are outlined below:

1) Low- to moderate-temperature static and rotating components will be manufactured from high-performance composite materials.

2) Near-final shape disks will be produced that are graded in composition, microstructure, and anisotropy to optimize mechanical properties.

3) New burner configurations will permit the use of new, improved sheet materials.

4) Multipiece construction of vanes and blades to provide highly efficient cooling will become common practice.

The emphasis in military engines will be on improving system reliability and performance. Higher turbine inlet temperatures will continue to be the primary goal, since this is the most effective way to increase power output. Commercial engines will be designed primarily for fuel efficiency, possibly even at the expense of some sacrifice in component durability. An important consideration will be the attainment of higher metal temperatures for turbine blades, to reduce cooling-air requirements. The achievement of higher design strengths for disks and higher resistance to environmental degradation for vanes will also be important goals, because of the benefits of reduced weight, lower engine cost, and increased operating life.

References and Notes

1. *The Aircraft Gas Turbine Engine and Its Operation* (Pratt & Whitney Aircraft Manual No. 200, United Technologies Corp., Hartford, Conn., 1970).
2. C. T. Sims and W. C. Hagel, Eds., *The Superalloys* (Wiley, New York, 1972).
3. P. R. Sahm and M. O. Speidel, Eds., *High Temperature Materials in Gas Turbines* (Elsevier, Amsterdam, 1976).
4. B. H. Kear, D. R. Muzyka, J. K. Tien, S. T. Wlodek, Eds., *Superalloys: Metallurgy and Manufacture* (Claitor's, Baton Rouge, La., 1976).
5. E. F. Bradley, Ed., *Source Book on Materials for Elevated Temperature Applications* (American Society for Metals, Metals Park, Ohio, 1979).
6. J. L. Walter, M. F. Gigliotti, B. F. Oliver, H. Bibring, Eds., *Proceedings of the Third Conference on In-Situ Composites* (Ginn, Lexington, Mass., 1979).
7. A. R. Cox, J. B. Moore, E. C. VanReuth, *AGARD Conf. Proc. No. 256* (1978).
8. J. S. Benjamin, *Met. Trans.* 1, 2943 (1970).
9. _____, *Sci. Am.* 234, 40 (May 1976).
10. C. Hammond and J. Nutting, *Met. Sci.* 11, 474 (1977).
11. S. R. Seagle and L. J. Bartlo, *Met. Eng. Q.* 8, 1 (1968).
12. G. M. Ault and J. C. Freche, *J. Astronaut. Aeronaut.* 17 (No. 10), 48 (October 1979).
13. Z. A. Foroulis and F. S. Pettit, Eds., *Proceedings of Symposium on Properties of High Temperature Alloys with Emphasis on Environmental Effects* (Electrochemical Society, Princeton, N.J., 1976).
14. C. H. Liebert and F. S. Stepka, *NASA Tech. Memo. TM X-3352* (1976).
15. F. L. VerSnyder and R. W. Guard, *Trans. Am. Soc. Met.* 52, 485 (1960).
16. B. H. Kear and B. J. Piearcey, *Trans. AIME* 238, 1209 (1967).
17. F. L. VerSnyder and M. E. Shank, *Mater. Sci. Eng.* 6, 213 (1970).
18. J. B. Moore and R. L. Athey, U.S. Patent 3,519,503 (1970).
19. M. M. Allen, V. E. Woodings, J. A. Miller, U.S. Patent 3,975,219 (1976).
20. W. H. Brown and D. B. Brown, U.S. Patent 3,872,563 (1975).
21. A. R. Cox *et al.*, Pratt & Whitney Aircraft report on DARPA contract F33615-76-C-5136 (1979).
22. H. A. Saller, S. J. Paprocki, R. W. Dayton, E. S. Hodge, U.S. Patent 687,842 (1966).
23. H. D. Hanes, D. A. Seifert, C. R. Watts, *Hot Isostatic Pressing* (Metals and Ceramics Information Center Rep. MCIC-77-34, Battelle Columbus Laboratories, Columbus, Ohio, 1977).
24. D. S. Duvall, W. A. Owczarski, D. F. Paulonis, *Weld. J.* 54, 203 (1974).
25. J. Mayfield, *Aviat. Week Space Technol.* 111, 41 (3 December 1979).
26. We thank C. P. Sullivan, M. J. Donachie, Jr., G. W. Goward, and M. L. Gell for their constructive criticism of the manuscript and for valuable information incorporated in several figures. In addition, we appreciate the helpful comments made by C. E. Sohl and R. G. Bourdeau.

Metallic Glasses

John J. Gilman

During almost all of the 8000 years that metals have been used by humans, their structures have consisted of aggregates of crystals. Therefore, the discovery by Klement *et al.* (*1*) that selected metal alloys can be quenched fast enough to circumvent crystallization caused considerable excitement among metallurgists. They showed that very fast cooling ($\sim 10^6$ °C per second) can yield metallic materials that are rigid and have liquid-like molecular structures.

By analogy with other supercooled liquids, quenched metallic alloys are called glasses. Since they are derived from liquids rather than gases or plasmas, they do not necessarily have the same structures as other noncrystalline metals. Also, since associations of atoms often exist in liquids, they are not necessarily "amorphous," but instead may contain well-defined short-range order.

Perhaps the prime virtue of metallic glasses is that they can be produced in useful forms economically. As a result, on a comparative cost basis, they are potentially the strongest, toughest, most corrosion-resistant, and most easily magnetized materials known to man. Their costs are very low because they are formed directly from the liquid without passing through the many steps that are used in conventional metallurgy. They can be made from the least expensive of all metallic raw materials, iron.

To make a thin strip of steel in the conventional way, an ingot is first cast; this is hot-rolled to form a billet, the billet is flattened by further rolling into a narrow plate, and then a series of cold-rolling and annealing steps is used to reduce the plate to thin strip stock. In all, six or eight steps are needed.

In contrast, thin strips of metallic glass are cast in one step. An entire spool can be produced in a matter of a few minutes. It is estimated that about four to five times less energy is consumed in going from a liquid metal to a thin, metallic glass strip than would be consumed by

Summary. The novel internal structures of metallic glasses lead to exceptional strength, corrosion resistance, and ease of magnetization. Combined with low manufacturing costs, these properties make glassy ribbons attractive for many applications. These materials also have scientific fascination because their compositions, structures, and properties have unexpected features.

John J. Gilman is director of the Corporate Development Center, Allied Chemical Corporation, Morristown, New Jersey 07960.

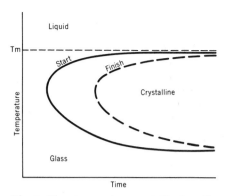

Fig. 1. Time-temperature-crystallization diagram. For a glass to be obtained, the cooling curve from the liquid state must avoid intersecting the crystallization "start" line.

conventional metallurgical processing.

Supercooled liquid alloys have remarkable combinations of properties as a result of their unusual molecular structures. Of all their physical properties, perhaps the most remarkable is the ductility of many of them (2). This is the basis of their interest for engineering uses because it makes them tough and easy to handle and increases their reliability markedly. It is exemplified by one of the first metallic glass products, nickel-base brazing foils (3). These were introduced to the marketplace in the spring of 1978, and their use for joining together parts of aircraft engines and other devices has grown since then. Their utility comes from the fact that nickel-base brazing alloys are brittle when they are crystalline, so the conventional form is powder; but when these alloys are rapidly quenched to obtain glassy ribbons, they are ductile. Such ribbons can be more easily handled than conventional powders.

The ductilities of glasses pose an interesting scientific question: For the same alloy composition, why is the crystalline form brittle but the glass ductile? The answer lies in subtleties of the atomic arrangements and the nature of chemical bonding in metals.

Because iron is the least expensive metal, ferrous glasses have the greatest commercial value, and they are emphasized in this article. However, metallic glasses have been made in many alloy systems, using elements from all parts of the periodic table including the precious metals (for example, $Pd_{80}Si_{20}$), transition metal pairs ($Cu_{60}Zr_{40}$), low-density metals ($Ti_{60}Be_{40}$), metal-metalloids ($Fe_{80}B_{20}$), alkali metals ($Rb_{85}O_{15}$), alkaline earth metals ($Ca_{70}Mg_{30}$), rare earth metals ($La_{76}Au_{24}$), and actinides ($U_{70}Cr_{30}$). These many alloys and their properties have been reviewed by Waseda and Toguri (4) and by Chen (5).

Formation

In principle, any liquid can be quenched so rapidly that it does not have enough time to crystallize before its atoms (or molecules) begin to move so sluggishly that the structure within it becomes "frozen." This may be understood by considering Fig. 1, a time-temperature-crystallization diagram. Here the C-shaped curves indicate the times at which crystallization will start and finish if a liquid is quickly cooled below its melting point to a particular temperature level. The "nose" of the C-curve indicates the minimum time for crystallization to start. The higher the temperature is above the nose, the smaller the supercooling, and therefore the longer it takes to start crystallization. The lower the temperature is below the nose, the higher the viscosity of the liquid, and therefore the longer it takes for crystallization to start.

To avoid crystallization and thereby obtain a glass, the time taken to cool below the C-curve's nose must be less than the time at which the nose is positioned. For silicates and many organic polymers, the position of the nose may lie at hours or days. This makes it easy to bypass crystallization in these materials. For pure metals, because of their simple atomic structures, it may take less than a microsecond for crystallization to begin at the nose temperature. However, for the selected alloys that form interesting

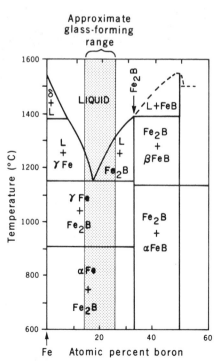

Fig. 2. Portion of the iron-boron phase diagram. The glass-forming region spanning the deep eutectic composition at 17 atomic percent boron is stippled.

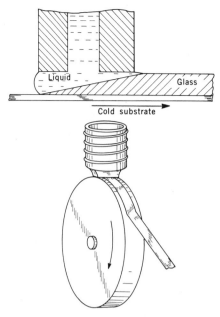

Fig. 3. Schematic diagram of the planar-flow-casting process, in which a slotted nozzle is brought very close to the surface of a rapidly moving cold substrate.

metallic glasses, the nose lies at a few milliseconds. Thus cooling rates of 10^5° to 10^6°C per second are adequate.

The alloys that can most easily be obtained in glassy form are usually eutectic compositions (6). The phase diagram of Fig. 2 illustrates this for the practically important Fe-B system. At a eutectic composition, the melting temperature is a minimum, but this is not the most important feature of a eutectic liquid for easy glass formation. What appears to be most important is the large structural change that must occur between the liquid just above and the two solid phases just below the eutectic temperature. If the atoms in the liquid are associated into quasi-molecules (as they often are), this increases the amount of change that must occur.

Complex structural changes during crystallization take time because much diffusion must occur and often the atomic movements must occur in specific sequences. Thus the nose of the C-curve is moved toward longer times.

In order to form useful materials, selected alloys must be quenched not only rapidly but also in a continuous way, so that useful filaments can be obtained. This can be done by "jet-casting" or by "planar-flow-casting." In jet-casting a thin stream of liquid metal is ejected from a laminar flow nozzle and then quenched. A cold liquid such as brine that flows cocurrently with the liquid metal may be used to quench the stream into a wire [the Kavesh process (7)], or the stream may be impinged onto a revolving piece of cold metal such as copper, thereby producing a flat ribbon [the Pond process (8)].

If wide strips are desired, the metal can be passed through a nozzle with a thin slit in it that lies in close proximity to a cold rotating metal wheel (Fig. 3). As the metal flows from the slit it is very quickly quenched into a glass [the Narasimhan process (9)]. Since the geometry of this process is planar, it can produce strips of any desired width (ignoring practical problems that intensify as the width increases).

From a practical viewpoint, the production of metallic glass filaments has several advantages when compared with conventional metallurgy. Being a direct casting process, it eliminates a number of forging, rolling, annealing, and drawing steps. Since the alloys are eutectics, the maximum process temperatures are lower than those for pure metals or dilute alloys. Also, the shaping of the metal occurs in the liquid state so the forces needed for shaping are very small, leading to lightweight equipment. Finally, the processing is intrinsically fast (up to 2 kilometers per minute, 6000 feet per minute) because of the rapid quenching involved.

Some disadvantages are that heat must be extracted from the system very rapidly in order to achieve continuous operation. And heat must flow out of the mate-

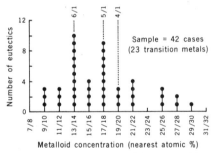

Fig. 4. Distribution of eutectic compositions for pairs consisting of a transition metal plus B, C, or P. All known cases are shown.

Table 1. Yield strengths of iron-based glasses compared with some conventional polycrystalline steels.

Material	Yield strength	
	kpsi	MPa
Glasses		
$Fe_{40}Ni_{40}P_{14}B_6$	350	2400
$Fe_{80}B_{20}$	525	3600
$Fe_{60}Cr_6Mo_6B_{28}$	650	4500
Steels		
Type 302 (cold-worked)	200	1400
AISI 4340	240	1600
Ausformed	290	2000
Tire cord (0.8 percent C)	400	2800

rial in a short time, so at least one dimension of the glassy product must be small. This limits feasible shapes to powders, wires, ribbons, and thin shells. Typical ribbons are 0.05 millimeter thick and 50 millimeters wide.

Another disadvantage of metallic glasses is that they have limited thermal stability. As they are heated, atomic motion occurs in them at temperatures of a few hundred degrees Celsius. At slightly higher temperatures they devitrify into one or more crystalline phases.

Compositions and Atomic Structures

The search for alloys that readily yield metallic glasses is aided by the fact that eutectic compositions are favored (10). This also provides clues about the structures of glasses because the structure of the eutectic liquid must anticipate the structure of the glass. When a pure metal cools through its melting temperature, it crystallizes into a solid of the same composition. But a eutectic alloy is converted into two solids just below its melting point, time permitting. If time does not permit, it is supercooled and eventually becomes a glass. Thus it is apparent that solids of eutectic compositions have markedly different structures depending on the cooling rates at which they are formed. This is a key to the unusual properties of metallic glasses, whose compositions are often ones that metallurgists have discarded as worthless for centuries. These compositions are brittle in their conventional polycrystalline forms, but not when they are glassy.

A second important clue to the structures of metallic glasses is provided by the observation that eutectic compositions tend to lie at simple atomic ratios (11). Many important glass-forming alloys consist of transition metals plus metalloids (B, C, N, Si, or P). The frequencies with which eutectics of transition metals plus B, C, or P lie at particular compositions are plotted in Fig. 4. If arbitrary compositions yielded eutectics, the distribution would be flat. Instead, it has two strong peaks: one at the ratio 5/1, the other at 6/1.

The presence of specific compositional ratios suggests specific chemical bonding. This is consistent with the strong interactions between transition metal atoms and metalloids which lead to refractory borides, carbides, silicides, and so on. However, these compounds are brittle, whereas the glasses tend to be ductile. In order to account for these and other facts, it may be postulated that the eutectic liquids consist of random-

ly packed clusters or quasi-molecules, rather than randomly packed atoms. Each cluster tends to be made up of a metalloid atom surrounded by five (or six) transition metal atoms. The clusters are bound internally primarily by stereoregular d-orbital bonds, while the clusters are held together by a Fermi gas of s electrons. In a glass derived from such a liquid, ductility is possible through gliding of dislocations between the clusters without a need for disrupting d bonds. On the other hand, the local short-range order within the clusters allows d-orbital symmetry to be partially preserved.

Clear evidence of the existence of short-range order within metallic glasses has been provided by anelastic relaxation studies (12), magnetic textures and annealing phenomena (13), and Mössbauer spectroscopy (14). Other types of studies have also provided evidence of short-range order, but not necessarily of the molecular clusters hypothesized above. Absence of long-range order has been amply confirmed by x-ray, electron, and neutron spectroscopy (15).

Properties

Some of the properties of metallic glasses are outstanding, such as the ease with which they can be magnetized and their mechanical toughness. Selected alloys also have very small temperature coefficients of electrical resistivity (16) and of lineal expansion (17). And acoustic waves can propagate through them for remarkably long distances (18). At low temperatures, some become superconductors (19), and other interesting electronic phenonomena occur in them such as the propagation of spin waves (20). There is too little space here to discuss the many properties that have been measured. Only the strengths, corrosion resistances, and magnetizabilities will be described. The reviews already men-

Table 2. Toughnesses (tearing energies) of some metallic glasses compared with other materials. Note that about 10,000 times more energy is needed to tear the iron-based glass than ordinary silicate glass.

Material	Tearing energy (J/cm^2)
$Fe_{80}P_{13}C_7$ glass	11
$Pd_{80}Si_{20}$ glass	4
$Cu_{57}Zr_{43}$ glass	6
Steel	2
Aluminum alloy	1
Elastomeric glass	< 0.1
Polymeric glass	< 0.1
Silicate glass	< 0.001

tioned contain a wealth of further information.

Strength. Metallic glasses have good elastic stiffness, and some of them resist plastic deformation better than the strongest steels, such as those used in aircraft landing gear. They also resist cracking very effectively, which is especially interesting when their high strengths are considered, because crack resistance tends to decrease markedly as strength increases.

No other ferrous materials have as high yield stresses as the best metallic glasses, although some steels have comparable breaking strengths (Table 1). Ductile iron-based compositions were first discovered by Chen and Polk (21). An example of one of their compositions is the multicomponent alloy $Ni_{39}Fe_{38}P_{14}B_6Al_3$. A much simpler composition is the binary alloy $Fe_{80}B_{20}$ discovered by Ray and Kavesh (22), which has a yield strength of 525,000 pounds per square inch (3600 megapascals). It has a correspondingly high hardness [1100 DPH (diamond pyramid hardness)] and a good abrasion resistance. The strongest composition to date is one found by Ray (23), $Fe_{60}Cr_6Mo_6B_{28}$, which has a yield strength of 650,000 psi (4500 MPa).

After plastic flow begins, glasses show little strain-hardening. Therefore, they are plastically unstable in tension, but they can sustain very large plastic strains when they are loaded in bending, torsion, and compression. Their yield stresses are essentially independent of temperature below about 0.7 of their glass transition temperatures. This is reflected in the small dependence of their yield stresses on applied strain rates.

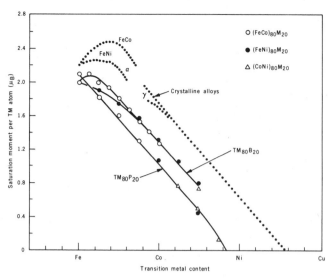

Fig. 6. Dependence of magnetic moments for glasses on alloy composition. Note that phosphorus depresses the magnetic moment more than does boron, Abbreviations: *TM*, transition metal; *M*, metalloid.

Since they yield abruptly when an applied shear stress reaches a critical value, and the yield stress depends little on temperature, it is apparent that metallic glasses do not exhibit Newtonian viscosity. Instead, they approximate an ideal elastic-plastic material. In several ways their plastic behavior resembles that of a crystalline metal that has been heavily strain-hardened.

Plastic flow in metallic glasses occurs very heterogeneously at low temperatures and high strain rates. This is readily confirmed by polishing a flat surface on a specimen and then observing it with a microscope after the specimen has been deformed. It will show a broad spectrum of steps on the surface that have resulted from localized plastic shearing; that is, from the motion of dislocations through the structure.

In pure crystalline metals the dislocations that create plastic flow are very mobile because of the periodic atomic structure. But in glasses they have very low mobilities because the structure is not periodic. Low dislocation mobility requires a high driving stress for plastic flow; hence the yield stress is high. Its magnitude can be accounted for quantitatively by means of a very simple theory (24).

One consequence of plastic flow is good resistance to crack propagation. This makes a material resistant to impact and insensitive to surface defects such as scratches. It is measured by a fracture toughness parameter. This parameter is one feature that differentiates metallic glasses from silicate glasses; the latter can be quite strong, but they have very little resistance to cracking. Metallic glasses have fracture toughness parameters much larger than those of silicate glasses (25). Their resistance to tearing is especially high. Measurements by Ki-

mura and Masumoto (26) are shown in Table 2.

A unique mechanical feature of metallic glass strips is that they are not only strong in the longitudinal direction, but just as strong in the transverse direction. This biaxiality of strength makes such strips attractive for reinforcing compliant materials such as rubber and for constructing composites. More discussion is given elsewhere (27).

Corrosion resistance. All metals oxidize readily in aqueous electrolytes. But some become "passivated" by the formation of a protective oxide film on their surfaces. An example is the case of iron, which rusts readily because simple iron oxide is porous and hence not protective. If several percent chromium is added to the iron, a protective chromate film forms on the surface and stainless steel results.

On the surfaces of iron-chromium glasses extremely protective films form

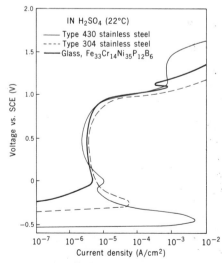

Fig. 5. Potentiostatic corrosion curves for two stainless steels and a chromium-bearing glass. The glass passivates at comparatively low potentials. Abbreviation: *SCE*, standard calomel electrode.

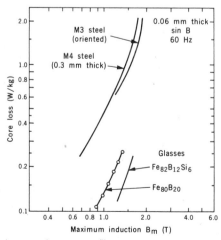

Fig. 7. Transformer core losses at 60 hertz for two glass compositions compared with conventional magnetic steel. Note the loss reduction of nearly an order of magnitude.

Table 3. Characteristics of two 15-kVA transformers. The total loss for the conventional transformer is 322 watts, compared with 180 watts for the transformer having a Metglas magnetic ribbon core. Note also the large differences in operating temperature and exciting current.

Operating parameter	Core material	
	Silicon steel	Metallic glass
Exciting current, amperes	2.5	0.12
Core loss, watts	112	14
Copper loss (15-kVA), watts	210	166
Total loss (15-kVA), watts	322	180
Energy saving, kilowatt-hours per year	0	1250
Temperature, °C	100	70

that consist of hydrated chromium oxyhydroxides, and "super–stainless steels" result (28). Such glasses are particularly resistant to chlorides (seawater) and sulfate environments. For example, in a standard ferric chloride solution conventional stainless steels are severely attacked, whereas an appropriate iron-chromium glass is barely touched.

A more general indication of the superior corrosion resistance of metallic glasses is provided by polarization curves (Fig. 5)—plots of the currents flowing through specimens at various constant applied potentials. The current is proportional to the rate of corrosive attack. Note that the metallic glass specimen is passivated at much lower potentials than either of the stainless steels. Also, its maximum corrosion rate is 1000 times smaller than that of type 430 and 10 times smaller than that of type 304 stainless steel.

Not just the transition metal content but also the metalloid content is important in determining the corrosion resistance of a glass. Naka et al. (28) have shown that ferrous glasses containing 5 to 10 percent chromium are most corrosion-resistant if they contain several percent phosphorus in addition to other metalloids. Silicon is least effective, while boron and carbon lie between phosphorus and silicon in effectiveness.

Ease of magnetization. Strong steels are usually relatively difficult to magnetize, but ferrous glasses are among the most easily magnetized of all ferromagnetic materials. Some can be magnetized by the application of fields in the millioersted range; that is, fields 100 times smaller than the earth's field. Previously, only nickel-iron alloys such as Permalloy could be so easily magnetized. But conventional, easily magnetized materials differ from metallic glasses in being annealed, and therefore mechanically soft, whereas the glasses are mechanically hard. Also, they are more difficult to manufacture.

Despite their irregular atomic structures, ferrous glasses exhibit high saturation magnetizations. This is illustrated by Fig. 6, which shows how the saturation magnetizations vary with composition in the series Fe, Co, and Ni. The magnetic moment per transition metal atom is decreased in the glasses by the irregular structure and by the presence of the metalloid atoms. The latter effect is clearly demonstrated by the fact that boron shifts the curve less than phosphorus does.

The easy magnetization in glasses implies that magnetic domain walls move through them easily. This was confirmed by O'Handley (29), who made direct measurements of domain wall velocities as a function of an applied magnetic field. Such measurements yield damping constants (viscous drag) for domain wall motion. Comparative data show that domain walls do indeed move through metallic glasses with exceptional ease. Some reasons for this are that obstacles such as grain boundaries are absent; glasses have high electrical resistivity, which damps eddy currents; and crystal anisotropy is absent (although some short-range-order anisotropy is present).

Domain wall mobility translates into excellent macroscopic magnetic properties. In particular, the loss factor is very low in comparison with conventional materials (Fig. 7). In addition, some glasses exhibit substantial changes in magnetization when stresses are applied to them (magnetostriction), and elastic waves propagate through them with exceptionally little attenuation because they have such high yield stresses. This combination gives them interesting magnetoacoustic properties. Reviews of the magnetic behavior of metallic glasses have been written by Luborsky et al. (30) and by Hilzinger et al. (31).

Role in Engineering

As a result of the invention of metallic glasses, metallurgical engineering has a new branch. Unlike science, engineering demands that economic considerations be applied to any material that attempts to compete with other materials. Thus the fundamental features of metallic glasses that become important are the ease and speed with which they can be manufactured, the facility with which they can be fabricated into product components, and their ability to reduce operating costs of devices in which they are incorporated.

To construct components, metallic glass ribbons can be woven to make fabrics, braided to make tubes, helically

Fig. 8. Photograph of a 15-kVA transformer being tested. This transformer has a toroidal metallic glass core.

wrapped to make cylinders, or laminated to make plates. Such components may be used in structures for their strength, in hostile environments because of their corrosion resistance, or in magnetic circuits to save energy and improve performance. Under some conditions, shock compression can be used to make objects from glassy powders (32).

Structural applications may include high-strength control cables, sheathing on electrical and optical cables, pressure vessels, flywheels for storing energy and power, mechanical transmission belts, torque transmission tubes, rocket casings, reinforcing belts in rubber tires, and more.

Because chromium-bearing glasses resist general corrosion and pitting so well in chlorides and sulfates, they may be attractive for marine and biomedical uses. Products include naval aircraft cables, torpedo tubes, chemical filters and reaction vessels, electrodes, razor blades, scalpels, suture clips, and others.

Ease of magnetization combined with hardness makes metallic glasses very attractive for carrying flux in a variety of magnetic devices, including motors, generators, transformers, amplifiers, switches, memories, recording heads, delay lines, transducers, and shielding. Some of these applications take advantage of the mechanical hardnesses of glasses to minimize wear, or they maximize acoustic wave propagation.

Because of their widespread use, transformers are particularly attractive. The core losses in transformers that one finds in every residential neighborhood for reducing high transmission voltages to lower household voltages waste approximately $500 million worth of electricity per year in the United States. Metallic glass transformer cores can potentially cut this waste in half. Much of the waste occurs because transformers are often operated continuously, whether any load is being placed on them or not. An extreme example is the doorbell transformer. The only time this transformer is used is when someone presses the doorbell button, but the transformer

dissipates energy continuously. Transformers in the electricity distribution system are also not used continuously. Late at night, for example, the loads on them are small. On the other hand, they tend to be overloaded in the early evenings of hot summer days.

Some advantages of putting metallic glass cores into power transformers are illustrated by the data of Table 3, which compares some of the operating characteristics of two 15-kilovolt-ampere transformers: one with a conventional magnetic steel core, the other with a metallic glass core (Fig. 8). In the transformer with the metallic glass core the exciting current is reduced by a factor of 21 and the core loss by a factor of 8. The low core loss also allowed design improvements to be made which reduced the copper loss (Joule heating) and gave an overall loss improvement by a factor of 1.8 at full load. For electricity costing $0.05 per kilowatt-hour, the energy savings for this transformer correspond to about $63 a year. The total electric generating capacity of the United States is about 0.5 terawatt. If the rate of savings that these data indicate could be scaled up to include the whole system of generators, substations, and distribution transformers, the annual savings would be $2 billion to $3 billion.

Even larger energy savings can be realized if motors can be developed that use metallic glasses effectively.

Role in Science

Studies of metallic glasses are improving our understanding of the metallic state, both liquid and solid. More detailed knowledge of the short-range atomic structures of metallic glasses will lead to further insight, because the availability of glassy specimens allows measurements to be made of materials in composition regimes that were not accessible previously. These composition ranges together with novel atomic patterns lead to electronic structures that have not been studied in the past.

Structural defects such as vacancies, impurities, dislocations, and Bloch walls take on new meanings in these materials, as do excitations such as phonons and spin waves. Surfaces and their defect structures in metallic glasses need a great deal of definition and reference to surfaces on crystals.

As scientific knowledge of these glasses grows, there is every reason to expect that it will reinforce and accelerate their effect on technology. Just as recent technological advances have provided rich scientific opportunities in this field, so will scientific studies enrich the technology.

References and Notes

1. W. Klement, R. H. Willens, P. Duwez, *Nature (London)* **187**, 869 (1960).
2. J. J. Gilman, *Phys. Today* **28**, 46 (1975).
3. N. DeCristofaro and C. Henschel, *Weld. J. (Miami, Fla.)* **57**, 33 (1978).
4. Y. Waseda and J. M. Toguri, *Metall. Soc. Can. Inst. Min. Metall.* **16**, 133 (1977).
5. H. S. Chen, *Rep. Prog. Phys.*, in press.
6. M. H. Cohen and D. Turnbull, *Nature (London)* **189**, 131 (1961).
7. S. Kavesh, U.S. Patent 3,845,805 (1974); in *Metallic Glasses* (American Society for Metals, Metals Park, Ohio, 1978), p. 36.
8. R. B. Pond, U.S. Patent 2,879,566 (1959).
9. D. Narasimhan, U.S. Patent 4,142,571 (1979).
10. H. J. V. Nielsen, *Z. Metallkd.* **70**, 180 (1979).
11. J. J. Gilman, *Philos. Mag.* **37B**, 577 (1978).
12. B. S. Berry, in *Metallic Glasses* (American Society for Metals, Metals Park, Ohio, 1978), p. 161.
13. E. M. Gyorgy, in *ibid.*, p. 275.
14. C. L. Chien and R. Hasegawa, *J. Appl. Phys.* **47**, 2234 (1976).
15. G. S. Cargill, *Solid State Phys.* **30**, 227 (1975).
16. R. Hasegawa and L. E. Tanner, *J. Appl. Phys.* **49**, 1196 (1978).
17. K. Fukamichi, H. Hiroyoshi, M. Kikuchi, T. Masumoto, *J. Magn. Magn. Mater.* **10**, 294 (1979).
18. M. Dutoit, *Phys. Lett. A* **50**, 221 (1974).
19. W. L. Johnson, *J. Appl. Phys.* **50**, 1557 (1979).
20. S. J. Pickart, J. J. Rhyne, H. A. Alperin, *Phys. Rev. Lett.* **33**, 424 (1974).
21. H. S. Chen and D. E. Polk, U.S. Patent 3,856,513 (1974).
22. R. Ray and S. Kavesh, U.S. Patent 4,036,638 (1977).
23. R. Ray, U.S. Patent 4,140,525 (1979).
24. J. J. Gilman, *J. Appl. Phys.* **46**, 1625 (1975).
25. L. A. Davis, *Metall. Trans.* **10A**, 235 (1979).
26. H. Kimura and T. Masumoto, *Scr. Metall.* **9**, 211 (1975).
27. Y. T. Yeow, *J. Compos. Mater.*, in press; *Compos. Techn. Rev.*, in press; see also J. J. Gilman, *Met. Prog.* **116**, 42 (July 1979).
28. M. Naka, K. Hashimoto, T. Masumoto, *Corrosion* **32**, 146 (1976); *J. Non-Cryst. Solids* **28**, 403 (1978).
29. R. C. O'Handley, *J. Appl. Phys.* **46**, 4996 (1975).
30. F. E. Luborsky, P. G. Frischmann, L. A. Johnson, *J. Magn. Magn. Mater.* **8**, 318 (1978).
31. H. R. Hilzinger, A. Mager, H. Warlimont, *ibid.* **9**, 191 (1978).
32. C. F. Cline and R. W. Hopper, *Scr. Metall.* **11**, 1137 (1977).

High-Strength, Low-Alloy Steels

M. S. Rashid

High-strength, low-alloy (HSLA) steels represent a new class of materials which evolved from plain carbon or mild steel. They are the precursors of dual-phase steels. HSLA steels have a balance of mechanical properties and provide suitable replacements for other steels in many applications. Their development was triggered by economic considerations and societal need, but is based on solid technical foundations (1).

and less than 1 percent other elements such as manganese and silicon, with the balance being iron. Their typical mechanical properties are yield and tensile strength in the ranges 150 to 200 and 280 to 350 megapascals, respectively, and total elongation or ductility of 30 to 40 percent in a 50.8-mm gage length. Yield strength (YS) is defined as the load needed per unit cross-sectional area of the steel to produce plastic flow and is

Summary. High-strength, low-alloy (HSLA) steels have nearly the same composition as plain carbon steels. However, they are up to twice as strong and their greater load-bearing capacity allows engineering use in lighter sections. Their high strength is derived from a combination of grain refinement; precipitation strengthening due to minor additions of vanadium, niobium, or titanium; and modifications of manufacturing processes, such as controlled rolling and controlled cooling of otherwise essentially plain carbon steel. HSLA steels are less formable than lower strength steels, but dual-phase steels, which evolved from HSLA steels, have ferrite-martensite microstructures and better formability than HSLA steels of similar strength. This improved formability has substantially increased the utilization potential of high-strength steels in the manufacture of complex components. This article reviews the development of HSLA and dual-phase steels and discusses the effects of variations in microstructure and chemistry on their mechanical properties.

At the turn of the century, when the first horseless carriages were appearing on the scene, plain carbon sheet steel was not a significant product of the steel industry. At that time steel was being used mainly for its strength, rigidity, and ductility in applications such as railroad equipment, machinery, and weapons. But the growth of consumer goods industries, especially automobiles, changed this. These industries found that sheet steel not only provided strength and rigidity but also could be formed easily into the intricate shapes desired for consumer products. Also, sheet steel had an attractive surface after forming, was easily joined to other parts, and was available at reasonable cost.

Metallurgically speaking, plain carbon sheet steels include all flat-rolled products less than 5.84 millimeters thick that contain less than 0.15 percent carbon

usually measured at 0.2 percent strain. Ultimate tensile strength (UTS) is the maximum load-carrying ability of the steel per unit cross-sectional area. The strain at UTS is referred to as the uniform elongation (e_u); the strain to failure is called total elongation (e_T). Both e_u and e_T are measures of steel ductility or formability, the two latter terms being used synonymously.

The microstructure of plain carbon steel consists of large ferrite crystals or grains, with grain boundary iron carbides and a small volume fraction of pearlite (Fig. 1). Ferrite is almost pure iron, is also known as α-iron, and has a low solubility for carbon. The grain boundary carbide composition is iron carbide (Fe_3C), which is also known as cementite. Pearlite is characterized by a lamellar structure consisting of alternate layers of ferrite and cementite and forms in the steel under certain cooling conditions.

As sheet steel usage grew, stronger

steels with different properties were needed for new applications or to improve durability or manufacturability in existing applications. In the early 1900's high strength was developed in hot-rolled steels by alloying with carbon, sometimes as much as 0.3 percent. However, mechanical properties such as fracture toughness, weldability, and formability suffer with increasing carbon content, and other ways of strengthening low-carbon steel had to be developed. Several research breakthroughs in the 1950's and 1960's triggered the development of present-day HSLA steels (2). These advances were a result of an improved understanding of the effect of interactions between alloying elements, such as vanadium, niobium, and titanium, and processing variables on grain refinement, steel strength, and ductility. Vanadium, niobium, and titanium are also referred to as microalloying elements in HSLA steels because of the small quantity of addition required. Methods for controlling the shape of nonmetallic sulfide inclusions in the steel also evolved. All steels contain nonmetallic inclusions such as oxides and sulfides as a consequence of the steelmaking process. Steel mechanical properties can be improved by controlling the shape of these inclusions.

Like plain carbon steel, HSLA steels contain less than 0.15 percent carbon. In addition, they typically contain about 1.0 percent manganese, about 0.1 percent microalloying elements, less than 0.6 percent silicon, and about 0.1 percent alloying elements for inclusion shape control, the balance being iron. The microstructural constituents of HSLA steels are similar to those of plain carbon steel, but the ferrite grain size is far smaller (Fig. 2) because of the microalloying addition and special processing. High strength is developed in these steels by grain refinement, substitutional strengthening due to manganese and silicon, and precipitation strengthening by micro-alloy precipitates.

HSLA steels are currently produced with the following typical mechanical properties: yield strengths, 350 to 700 MPa; ultimate tensile strength, 450 to 850 MPa; and total elongation, 14 to 27 percent. They are high-strength only in comparison with plain carbon steel, since quenched and tempered high-carbon steels may have tensile strengths as high as 1100 MPa. The reason for this limitation in strength is that other properties, such as fracture toughness, formability, weldability, and fatigue resistance, and processing considerations must also be weighed in the balance. Of-

The author is a senior research engineer in the Metallurgy Department, General Motors Research Laboratories, Warren, Michigan 48090.

ten one property is enhanced at the expense of another. Consequently, no single mechanical property is maximized but many properties are optimized so that the steel is rendered suitable for various applications.

Although HSLA steels have an excellent balance of mechanical properties, their formability or ductility was not enough for the manufacture of many press-formed components for automotive applications. For these parts, involving such modes of forming as stretching and drawing, the reduced formability of HSLA steels was a serious stumbling block. To make HSLA steels more widely acceptable to the automobile industry, it was imperative to improve their strength-formability balance. This problem was resolved in the mid-1970's by the development of dual-phase steels (3, 4).

Steels described thus far have microstructures consisting of ferrite, pearlite, and iron carbides. The temperature range of stability of these phases is shown in the iron-carbon equilibrium phase diagram (Fig. 3). Ferrite or ferrite and cementite coexist in steel at temperatures below 723°C, the eutectoid or critical temperature (Fig. 3). For carbon contents less that 0.8 percent, cementite is not stable above the critical temperature, but austenite, or γ-iron, is stable to almost 1500°C.

When austenite is cooled slowly under equilibrium conditions, phase transformations occur at the temperatures indicated in Fig. 3. When cooled rapidly under nonequilibrium conditions, the austenite transforms to metastable phases, termed martensite and bainite, which cannot be shown on the equilibrium phase diagram. Martensite forms very rapidly by a shear mechanism when the steel is cooled quickly below a characteristic martensite start temperature. Martensite is far stronger than the other phases. Low-carbon martensite is needle- or lath-shaped. High-carbon martensite is usually twinned; the atomic arrangements of portions of the martensite grains are oriented in mirror-image fashion to accommodate strains introduced by the austenite-to-martensite transformation. Bainite is essentially an aggregate of ferrite grains with finely divided carbide particles and forms when the austenite is cooled and held just above the martensite start temperature.

Dual-phase steels are characterized by a ferrite microstructure with a uniform distribution of about 20 percent martensite by volume (Fig. 4). Their stress-strain curve is characteristically different (Fig. 5) from that of plain carbon or

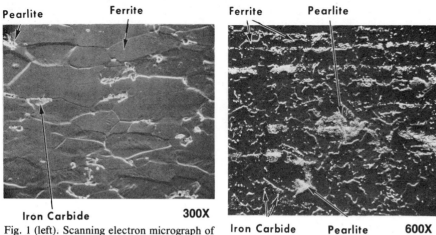

Pearlite   Ferrite   Ferrite   Pearlite

Iron Carbide   300X   Iron Carbide   Pearlite   600X

Fig. 1 (left). Scanning electron micrograph of plain carbon steel. The microstructure consists of a ferrite matrix with grain boundary iron carbides and a small volume fraction of pearlite.   Fig. 2 (right). Scanning electron micrograph of a high-strength, low-alloy steel. The microstructure consists of a fine-grained ferrite matrix, grain boundary iron carbides, and islands of pearlite.

HSLA steels. All metals exhibit elastic behavior (a linear stress-strain relationship) at low strains and plastic behavior (a nonlinear stress-strain relationship) at higher strains. The transition from elastic to plastic behavior is continuous in dual-phase steels. Other steels exhibit yield point elongation or plastic deformation at a constant load before the steel work-hardens and strength increases with increasing strain (Fig. 5). Yield strength is low in dual-phase steels compared to tensile strength, and strength increases rapidly as a function of strain as a result of a very high initial strain hardening rate. The relation of strength to uniform elongation (ductility) for plain carbon and HSLA steels can be represented by the same curve (Fig. 6). Similar data for dual-phase steels fall on a separate curve; these steels have higher ductility and are more formable than HSLA steels of equivalent tensile strength (Fig. 6).

Currently, dual-phase steels are produced with tensile strengths in the range 485 to 850 MPa. They have the same nominal chemistry range as HSLA steels and are produced by proper annealing of certain HSLA steels to convert their ferrite-pearlite microstructure to a ferrite-martensite one. Dual-phase steels are generally regarded as an extension of the broad family of HSLA steels, although dual-phase microstructures can also be produced using steel with no micro-alloying additions. HSLA and dual-phase steels are now produced and used commercially. Their balance of mechanical properties combined with the superior formability of dual-phase steels has substantially increased the utilization potential of high-strength steels in consumer products.

In this article I discuss some steel-strengthening mechanisms and present a brief treatise on how small amounts of alloying elements and controlled process-

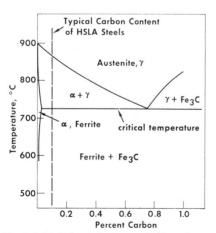

Typical Carbon Content of HSLA Steels

Austenite, γ

α + γ

α, Ferrite

critical temperature

γ + Fe₃C

Ferrite + Fe₃C

Temperature, °C

900
800
700
600
500

0.2 0.4 0.6 0.8 1.0
Percent Carbon

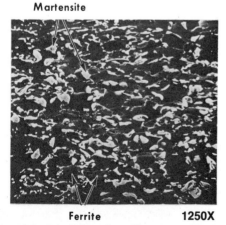

Martensite

Ferrite   1250X

Fig. 3 (left). Schematic representation of a portion of the iron-carbon phase diagram.   Fig. 4 (right). Scanning electron micrograph of a dual-phase steel. The microstructure consists of a fine-grained ferrite matrix with a uniform distribution of about 20 percent martensite by volume.

Table 1. Qualitative representation of the effects of various microstructural phenomena on the mechanical properties of HSLA steels. Beneficial effects are indicated by +, detrimental effects by −, and small or insignificant effects by 0 (2).

| Property | Effect | | | | |
	Fine grain	Pearl-ite	Solid solu-tion	Precipi-tation	Inclusion shape control
Strength	+	+	+	+	0
Formability	+/0	−	−/0	−	+
Toughness	+	−	−	−	+
Fatigue resistance	+	+	+	+	+/0
Weldability	0	−	−	−/0	+/0
Cost	0	0	−	−/0	−

ing can substantially improve mechanical properties of steels similar in chemistry to plain carbon steel. The contributions of the various strengthening mechanisms in ferrite-pearlite and ferrite-martensite high-strength steels are contrasted, and our present understanding of the mechanisms involved in transforming the former to the latter microstructure is discussed.

Strengthening Mechanisms

Plastic deformation occurs in polycrystalline metals by movement of dislocations on certain crystallographic slip planes within each crystal or grain. A dislocation is an extra half-plane of atoms in an otherwise periodic lattice; it is referred to as a linear defect and can move in the lattice under an applied stress. Strength can be increased by retarding dislocation motion. This is usually done by increasing the number of obstacles—precipitates, substitutional and interstitial solute atoms, grain boundaries, or other dislocations—in the path of the moving dislocations (5).

A uniform distribution of very small precipitates such as carbides, nitrides, or borides in a crystal lattice effectively retards dislocation motion. This contribution is enhanced when there is geometric matching, or coherence, between atoms of the precipitate and the matrix lattices. The volume, size, distribution, and coherence of precipitates can be controlled by steel chemistry and processing parameters. Moving dislocations are stopped when they run into precipitates, and the dislocation lines usually bend to avoid the precipitate and attempt to continue motion by a process called cross-slip (5). This involves slip on another set of crystallographic planes. Only a limited number of such planes are available, however, and dislocation motion is finally stopped; a much higher stress is required for further motion, resulting in stronger steel.

Strengthening by solute atoms is the result of elastic strains due to the difference in size between solute and solvent atoms and to solute-dislocation interactions. Solute atoms that are similar in size to atoms of the solvent metal—for example, manganese or silicon in an iron lattice—occupy positions otherwise occupied by the solvent metal atoms in the crystal lattice and are called substitutional solutes. Smaller atoms such as carbon or nitrogen occupy spaces between the solvent metal atoms and are called interstitial solutes. Because of their very small size, interstitial solutes also form clusters or "atmospheres" around dislocations and effectively retard their motion (5).

The most common strengthening mechanism in metals is the interaction of dislocations with each other. A well-annealed lattice has of the order of 10^6 dislocations per square centimeter. Their number increases with increasing strain, as do the interactions between them (5).

However, because of the large number of dislocations, it is difficult to specify group behavior in a simple mathematical way or to calculate the exact strengthening contribution of the interactions.

Size of the crystal grains in the polycrystalline metal also has a significant effect on mechanical properties. Steel strength can be increased by decreasing grain size and thereby decreasing the distance available for free dislocation travel. The boundaries between grains are regions of lattice irregularity; dislocations can move relatively freely through the crystal but are impeded at the boundaries. Therefore, the smaller the effective size of the grains, the shorter the distance available for free dislocation travel.

Yield strength is more dependent on grain size than tensile strength. The beneficial effect of ferrite grain refinement on yield strength and on the temperature at which steels undergo the ductile-to-brittle transition was demonstrated in the early 1950's. Hall (6) and Petch (7) reported independently that a simple empirical relationship existed between yield strength and grain size, namely

$$\sigma_y = \sigma_i + \frac{k}{\sqrt{d}} \qquad (1)$$

where σ_y is yield strength; σ_i is friction stress, a measure of the difficulty of moving dislocations through the lattice; k is the strength coefficient, a measure of the extent to which dislocations are piled up at obstacles; and d is ferrite grain size, in terms of equivalent diameter. This was one of the first demonstrations of a quan-

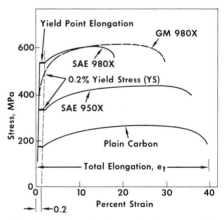

Fig. 5. Schematic stress-strain curves for plain carbon, HSLA, and dual-phase steels. SAE 950X and 980X are Society of Automotive Engineers designations for HSLA steels of different strength levels. GM 980X is a dual-phase steel developed by General Motors. GM 980X is more ductile than SAE 980X, although both steels have similar tensile strengths.

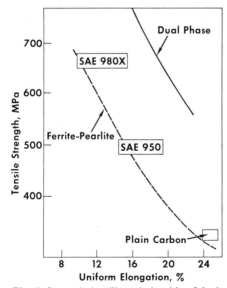

Fig. 6. Strength-ductility relationship of dual-phase steels compared with that for plain carbon and HSLA steels. The dual-phase steel curve is far above that for ferrite-pearlite steels (13).

titative relationship between microstructure and mechanical properties.

Later, researchers used this relationship to explain and analyze quantitatively the effects of other structural factors involved in strengthening steels (2). This was done by expanding the friction stress to take into account solid solution strengthening by elements such as silicon and manganese and to allow for the volume fraction of pearlite in the microstructure and the effect of carbide and nitride precipitates, namely

$$\sigma_i = \text{function of } A + B\,(\%\text{ pearlite}) + C\,(\%\text{ Mn}) + D\,(\%\text{ Si}) + \ldots + \Delta Y$$

where A is the friction stress in an unalloyed and unstrengthened ferrite matrix; B, C, D, . . . are empirical constants; and ΔY, the difference between σ_y measured and σ_y calculated, is assigned to the contribution of precipitates.

Examples of this empirical quantitative approach to structure-property relationships are shown in Fig. 7, where σ_y is plotted as a function of grain size. Each percent of manganese increases the strength about 48 MPa, each percent of silicon increases the strength about 150 MPa or more, and these contributions are additive. Individual contributions of various amounts of manganese, silicon, or pearlite to σ_y, and similar contributions due to precipitation strengthening related to vanadium and niobium additions, are also shown in Fig. 7. The various contributions to yield strength are superimposed schematically in Fig. 7 for a typical HSLA steel of 550-MPa yield strength and 650-MPa tensile strength. A typical stress-strain curve for one such steel (SAE 98OX) is shown in Fig. 5.

Similar relationships between other mechanical properties and various structural strengthening elements have been developed for HSLA steels. These effects are summarized qualitatively in Table 1. Grain refinement appears to be the most desirable method for improving several mechanical properties simultaneously. However, strengthening elements sometimes enhance one mechanical property at the expense of another. For example, microalloy precipitates increase steel strength but decrease formability and toughness. Such interactive relationships must therefore be considered in evaluating the contributions of various structural strengthening elements to each mechanical property. The relative contribution of each strengthening mechanism can be controlled through proper adjustment of chemical composition and processing variables.

Processing for Microstructural Control

Very briefly, the manufacture of steel consists of melting the steel and then pouring the molten metal into molds to solidify. These ingots are heated at elevated temperatures that are in the austenite region of the iron-carbon diagram (Fig. 3) to homogenize the steel, and then hot-rolled down in several stages to slabs 50 to 100 mm thick from an ingot 500 to 600 mm thick. The slabs are reheated as before and rolled down to a sheet of the desired thickness. As the sheet emerges from the rolling mill it is air- or force-cooled from the rolling temperature to a predetermined temperature and then coiled.

Effect of processing parameters. Steel microstructure and mechanical properties can be adjusted by controlling processing parameters during slab rolling. Some important parameters are shown qualitatively in Fig. 8; namely (i) rate and

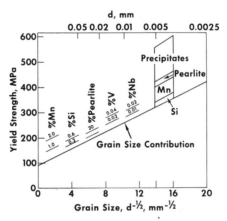

Fig. 7. Effects of grain size, solid solution strengtheners manganese and silicon, pearlite content, and precipitation strengthening elements vanadium and niobium on steel yield strength. A schematic representation of the origin of various strengthening contributions to yield strength of a typical HSLA steel with a tensile strength about 650 MPa is also shown (2).

amount of deformation, particularly in the final rolls, (ii) finishing temperature, or the temperature at which hot-rolling is completed, (iii) time lapse between deformation and the beginning of the austenite-to-ferrite transformation, (iv) rate of cooling through the transformation range, and (v) coiling temperature. A portion of a schematic continuous-cooling diagram is also shown in Fig. 8. The continuous-cooling diagram defines contour lines for the start and finish of each phase transformation as a function of cooling time and gives an indication of the cooling rate required to develop the desired microstructure.

Work done during hot-rolling deforms the austenite grains and increases the strain energy in the lattice. With enough time and a sufficiently high temperature, the system tends to approach a low-energy equilibrium state by decreasing its high strain energy. This is manifest by recrystallization or nucleation of new grains having lower energy polygonal configurations, at the expense of deformed grains. The amount of deformation in the last rolling pass controls the number of lattice defects or dislocations present at the finishing temperature. A large number of defects provides increased nucleation sites for recrystallization of austenite grains, which in turn results in a finer ferrite grain size on transformation of the austenite to ferrite.

Ferrite grain size can also be reduced by reducing the finishing temperature, because the two are directly related. In plain carbon steel, recrystallization occurs very rapidly during hot-rolling above 800°C and the recrystallized austenite grains grow to a fairly large size. The ferrite grain size that results on cooling through the austenite-ferrite transformation is then correspondingly large. On finish rolling below 800°C, recrystallization takes place more slowly and the austenite may only partially recrys-

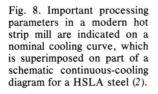

Fig. 8. Important processing parameters in a modern hot strip mill are indicated on a nominal cooling curve, which is superimposed on part of a schematic continuous-cooling diagram for a HSLA steel (2).

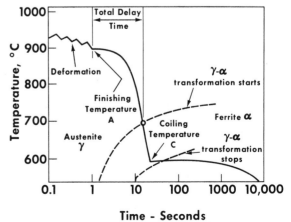

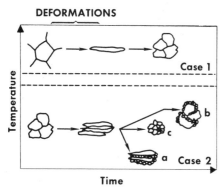

DEFORMATIONS

Fig. 9. Schematic representation of changes in austenite grains during controlled hot rolling (10).

tallize. More ferrite nucleation sites are available in the partially recrystallized austenite, resulting in a finer ferrite grain size.

The cooling rate after hot-rolling governs the time available for transformation and grain growth. With higher cooling rates, less time is available for austenite grain growth, and this results in a smaller ferrite grain size on transformation. Furthermore, rapid cooling through the transformation range limits ferrite grain growth. Coiling temperature governs the amount of ferrite grain growth in the coil and is maintained as low as practicable to minimize such growth.

Effect of microalloying elements. The addition of a microalloying element such as vanadium, niobium, or titanium retards recrystallization and growth kinetics in the austenite and ferrite by several orders of magnitude compared to plain carbon steel, resulting in a finer ferrite grain size (2, 8, 9). The precise nature of this effect on recrystallization is still unresolved. Evidence has been presented both for and against the idea that the microalloyed elements must precipitate to inhibit recrystallization of the hot-rolled austenite (1).

The microalloy elements exist as carbide and nitride precipitates in the ingot and slabs. However, all but the largest precipitates dissolve in the steel during heating before hot rolling. During hot rolling, some strain-induced precipitation occurs at lattice defects and prior grain boundaries. These precipitates retard austenite recrystallization by pinning grain boundaries and subboundaries and limiting their motion (8). Also, dissolved microalloy solutes restrain grain boundaries and slow down their migration rate, retarding recrystallization (9). Similar mechanisms retard austenite grain growth following recrystallization. The high density of lattice defects and precipitates in the austenite provides an increased number of ferrite nucleation sites on cooling through the transformation, resulting in substantial grain refinement.

Higher finishing temperatures are generally necessary in microalloyed steels compared to plain carbon steel. When the finishing temperature is above 1000°C, austenite recrystallizes (10) and grains grow in size (Fig. 9, case 1), but are still smaller than would be expected in steels without microalloying additions. At lower finishing temperatures, say 1000°C, various morphologies may be developed, depending on steel composition and extent of prior hot-rolling history. Complete recrystallization may occur (Fig. 9, case 2c), generating a finer grain size than before (case 1). Partial recrystallization may occur at the grain boundaries of deformed austenite grains (case 2a) or grain growth may occur following complete recrystallization (case 2b). On transformation to ferrite, the austenite morphology developed by case 2c will yield the most uniform fine-grained ferrite.

At still lower finishing temperatures (below 950°C), the deformed austenite

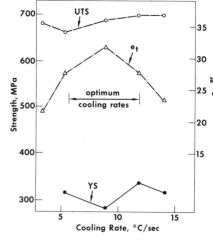

Fig. 11. Effect of cooling rate on mechanical properties of a dual-phase steel. Yield strength (*YS*) and ultimate tensile strength (*UTS*) are affected minimally in the cooling rate range evaluated, but ductility or total elongation (*e*$_T$) is substantially altered.

will not have time to recrystallize and ferrite will nucleate from unrecrystallized austenite. The resulting ferrite grains are polygonal despite the distortion of the unrecrystallized grains, and the ferrite grain size becomes even finer than in case 2c.

As with plain carbon steel, rapid cooling between the finishing and coiling temperatures (points *A* and *C* in Fig. 8) inhibits austenite grain growth before transformation to ferrite and also lowers the transformation temperature, with further benefit to ferrite grain size. It also controls the intensity of precipitation strengthening and brings about a slight dislocation strengthening.

Controlled cooling. Rapid cooling is usually attained by pouring large volumes of cooling water over red-hot steel. On initial contact, the large difference between the steel and water temperatures promotes rapid heat transfer. However, an insulating layer of steam forms rapidly above the steel, lowering the heat transfer rate. Gradually, the steam jacket is penetrated and the steel and water again come in contact, resulting once again in a very rapid heat transfer. Efficient heat transfer can be attained by breaking down the insulating steam jacket as soon as it is formed.

Research has shown (11) that streams of water falling at low pressure on hot steel strip in a system called laminar flow cooling produce a rodlike flow of cooling water, which falls 1 to 2 meters before breaking into large droplets. Although to the eye this rodlike stream looks little different from any other low-pressure stream, it has sufficient energy to penetrate most effectively the pro-

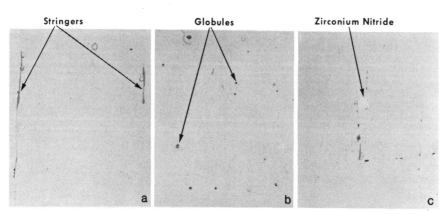

Fig. 10. Effect of inclusion shape control elements on shape of manganese sulfide inclusions. (a) Elongated stringers result from inadequate shape control. (b) Inclusions are globular, indicating adequate shape control. (c) Zirconium combines with nitrogen, forming zirconium nitride particles, and manganese sulfides are elongated (×200).

tective steam blanket formed on the hot strip, achieve the required heat transfer and cooling rate, and develop the required microstructure and mechanical properties in HSLA steels. As discussed before, coiling temperature governs the amount of ferrite grain growth in the coil and is maintained as low as practicable to minimize such growth.

An added consequence of the hot-rolling process is its effect on the shape of nonmetallic inclusions in the steel. Since the latter influences steel strength and ductility, methods have evolved for controlling the shape of such inclusions by addition of suitable elements.

Inclusion shape control. Most steels, including plain carbon steel and HSLA steels, contain nonmetallic inclusions from the steelmaking process. These inclusions are globular in the ingot and consist primarily of manganese sulfide. At hot-rolling temperatures, the manganese sulfide inclusions are very plastic and are elongated into stringers in the direction of rolling (Fig. 10a). Consequently, steel ductility is inferior in the direction transverse to the rolling direction. To maintain their globular shape and prevent excessive stretching out of the sulfides during hot rolling it is necessary to alter their composition.

Zirconium or rare earth elements, such as cerium and lanthanum, form high-melting-point sulfides that are not readily deformable at hot-rolling temperatures. When these elements are added to molten steel, sulfur combines with them instead of manganese, forming rigid sulfides (*12*). After hot rolling, rare earth or zirconium sulfides remain globular (Fig. 10b). Most conventional HSLA steels are therefore treated with rare earth elements or zirconium for inclusion shape control to minimize anisotropy in mechanical properties. The use of zirconium is restricted, however, to steels that do not rely on nitride precipitation strengthening, because nitrogen competes with sulfur for the zirconium, forming zirconium nitrides (Fig. 10c). Besides tying up nitrogen instead of sulfur and failing to control the shape of inclusions and improve transverse ductility, zirconium nitride particles could contribute to overall reduction of ductility.

Dual-Phase Steels

Dual-phase steels are generally produced by continuous annealing (*4, 13*), but can also be produced as-rolled or directly off the hot mill (*14*). Continuous annealing involves moving a previously

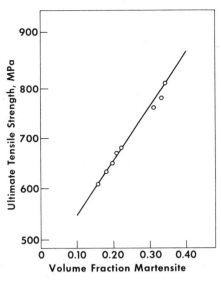

Fig. 12. Effect of volume fraction of martensite on tensile strength of a dual-phase steel. Tensile strength increases linearly with volume fraction of martensite (*15*).

rolled strip of steel through a long furnace that is maintained at the desired annealing temperature. The required time at that temperature is ensured by controlling the rate at which the strip is fed through the furnace. In as-rolled dual-phase steels the desired microstructure is developed immediately after hot rolling by control of composition and cooling conditions. The manufacture of as-rolled dual-phase steels is similar to that of other HSLA steels, but the alloy composition is chosen so that 80 to 90 percent of the steel transforms to ferrite between the finish rolling and coiling temperatures and the cooling rate is controlled so that the remaining 10 to 20 percent austenite transforms to martensite in the coil.

Continuous-annealing parameters. Continuously annealed dual-phase steels are produced by heating HSLA steels of certain compositions intercritically, namely at temperatures in the ferrite plus austenite region of the iron-carbon phase diagram (Fig. 3), or supercritically, namely in the austenite region, and then cooling to room temperature. Processing parameters and other factors discussed with regard to HSLA steels do influence the mechanical properties of continuously annealed dual-phase steels. However, for each steel composition, the final microstructure and mechanical properties are governed principally by the continuous-annealing time and temperature, cooling rate after annealing, and coiling temperature.

For continuously annealed dual-phase steel, the composition is designed to have enough hardenability to produce about 20 percent martensite by volume

at cooling rates possible on available processing equipment. Hardenability is defined as the ability of a steel to harden on rapid cooling. This occurs by the formation of martensite rather than ferrite and pearlite and can be enhanced by alloying with certain elements—for example, carbon or manganese. Steels with lower hardenability require rapid cooling, such as a water quench, whereas those with higher hardenability develop the required microstructure during air cooling. Lower cooling rates are generally preferred, because a larger number of lattice defects are quenched into the lattice at higher cooling rates and this can adversely affect ductility. Furthermore, optimum cooling rates exist for each steel composition for each heating time-temperature combination. For example, an e_T of more than 27 percent was reported for the steel in Fig. 11 for cooling rates between 5° and 12°C per second, e_T being less at other cooling rates. However, yield and tensile strength were minimally affected.

The continuous-annealing temperature determines the carbon content and strength of the martensite and hence influences the mechanical properties of the steel. In steels heated to lower intercritical annealing temperatures—for example, just above the critical temperature—the proportion of austenite formed is at a minimum and it has a high carbon content, because of the greater solubility of carbon in austenite than in ferrite (see Fig. 3). This austenite will transform to high-carbon twinned martensite on cooling. At higher annealing temperatures, the volume fraction of austenite is larger and it has a lower carbon content and hence lower hardenability. On cooling, some of the austenite will transform back to ferrite and the rest to low-carbon lath martensite. High-carbon martensite is stronger than low-carbon martensite, and hence their contributions to steel strength and ductility are different. Steels heated supercritically transform entirely to austenite. The volume fraction and carbon content of the martensite in these steels are different, and so are their contributions to mechanical properties.

The volume percent of martensite influences both strength and ductility of the steel. Tensile strength increases linearly (*15*) with volume percent of martensite (Fig. 12); ductility is inversely related to tensile strength (Fig. 6) and hence to volume percent of martensite.

After continuous annealing, the steel is cooled to below its martensite start temperature to ensure occurrence of the transformation, and is then coiled. Ferrite grain growth in the coil is not of much concern here; however, some tempering or softening of the martensite occurs as the coil cools to room temperature. Since martensite is a metastable phase it decomposes on reheating or tempering into more stable aggregations of ferrite and carbide, which have somewhat lower strength but improved ductility. Proper control of cooling parameters in the coil affords an additional means for controlling steel ductility.

The structure-property relationships. These relationships are quite complex in dual-phase steels. They are not fully understood but can be described in qualitative terms (3, 4, 15). Mechanical properties are influenced by steel composition and microstructure prior to continuous annealing. The fine grain size of the HSLA steel is inherited by the dual-phase steel, and Eq. 1 can be used to estimate the grain size contribution to yield strength. Ferrite grains do not appear to grow at continuous-annealing temperatures. Instead, some grain refinement seems to occur due to the martensite transformation. The strengthening contribution of precipitates in dual-phase steels may not be the same as those observed in HSLA steel because precipitate morphology is modified after continuous annealing. The relationship implied in Fig. 7 is altered, and the exact contribution of precipitates to the strength and ductility of dual-phase steel has yet to be isolated. However, substitutional strengthening contributions (Fig. 7) in the starting material appear to carry over into dual-phase steels.

Transformation mechanism. Before continuous annealing the steel microstructure consists of a parent ferrite phase with grain boundary iron carbides, small volume fractions of pearlite (Fig. 2), and a fine distribution of strengthening precipitates. On heating above the critical temperature the iron carbides go into solution, forming carbon-rich pools of austenite. On cooling back to room temperature part of the austenite transforms to martensite, part transforms back to ferrite, and a small fraction remains as austenite. Some precipitates are dissolved and large volumes of the ferrite are relatively free of them. For steels heated intercritically, some ferrite does not transform but the morphology of precipitates within this untransformed ferrite is usually altered (4).

These microstructural changes produce a stress-strain behavior (Fig. 5) that is quite different from that of plain carbon or HSLA steels. Dual-phase steels have no yield point elongation, a low yield strength, a high work hardening

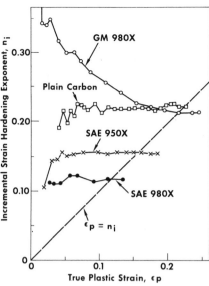

Fig. 13. Variation of incremental strain hardening rate, n_i, with increasing true plastic strain in various steels; n_i is relatively constant in the ferrite-pearlite steels but not in the dual-phase steel.

rate, and better ductility than HSLA steels of similar tensile strength. The low yield strength and absence of yield point elongation are attributed to the presence of numerous mobile dislocations in the ferrite generated by stresses imposed by the martensite transformation (3). Interactions between these dislocations and strain-induced transformation of the retained austenite to martensite at low strains contribute to the high work hardening rate. The high tensile strength is directly related to the volume percent of martensite (15). Mechanisms for the high total elongation and improved formability are quite complex; they are not fully understood but are generally explained as a consequence of the high work hardening rate.

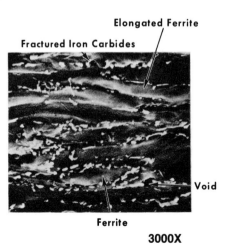

Elongated Ferrite

Fractured Iron Carbides

Void

Ferrite

3000X

Fig. 14. Scanning electron micrograph of a ferrite-pearlite HSLA steel at fracture initiation. Ferrite grains are elongated, voids are created at ferrite–iron carbide interfaces, and some iron carbides are fractured.

Deformation mechanism. The deformation behavior of most metals, especially plain carbon steel, may be described (16) in terms of the simple empirical relationship between stress and strain, namely

$$\sigma = k\epsilon_p{}^n \qquad (2)$$

or

$$\log \sigma = \log k + n \log \epsilon_p \qquad (3)$$

where σ is the true stress, defined as the load at any instant divided by cross-sectional area of the specimen at that instant; k is the strength constant, or σ at $\epsilon_p = 1.0$; ϵ_p is the true plastic strain, defined as the change in length referred to the instantaneous length of the specimen; and n is the strain hardening exponent, a measure of the ability of the metal to distribute strain. When Eq. 2 is satisfied, a plot of $\log \sigma$ against $\log \epsilon_p$ will be a straight line with slope n. Furthermore, it can be shown (16) that the true strain at maximum load, ϵ_u, also called the true uniform strain, will numerically equal n when Eq. 2 is satisfied.

Equation 2 is sometimes not satisfied, in which case a plot of Eq. 3 is a curved line (17). An incremental value of n, n_i, must then be calculated for each segment of this curve for small increments of strain, where

$$n_i = \frac{\log \sigma_j - \log \sigma_{j-1}}{\log \epsilon_{p(j)} - \log \epsilon_{p(j-1)}}$$

for $j \to 1$ to m, and m is the number of small segments on the stress-strain curve. The true uniform strain, ϵ_u, will now equal n_i at maximum load, and not the n value obtained for the full true stress-true strain curve (18).

A plot of n_i against ϵ_p will give an indication of the variation of the strain hardening behavior with increasing deformation in either type of steel. Equation 2 is satisfied in plain carbon and ferrite-pearlite HSLA steels, and a plot of n_i against ϵ_p is a straight line. This is not so for dual-phase steels (19). In the latter steels, n_i is very large at low strain but decreases progressively with increasing strain and is almost equal to that for plain carbon steels at maximum load (Fig. 13). The different variation of n_i with ϵ_p in the two steels gives an indication of the differences in their microstructural deformation characteristics.

The microstructure of dual-phase steels prior to continuous annealing consists of a parent ferrite phase with grain boundary iron carbides and some pearlite. When they are strained, plastic deformation takes place by strain hardening of the ferrite with practically no deformation of the iron carbides. Since deformation occurs predominantly in a single phase, n_i is constant and represents the strain hardening behavior of this phase alone (19).

Dual-phase steels consist of transformed and untransformed ferrite, martensite, and some retained austenite, all of which have different strengths. When dual-phase steel is strained, slip leading to deformation occurs first in the constituent with the lowest yield strength. When this constituent work hardens to the yield strength of the second constituent, slip begins to occur in it. This continues until all constituents are involved in the deformation process. The observed n_i at any particular strain is therefore the work hardening rate of one or several different constituents that may be undergoing deformation at that strain. Since deformation is shared consecutively by various constituents, strain is distributed more uniformly and failure is delayed. As a result, dual-phase steels have better formability than their parent HSLA steel (19).

After the ferrite is highly strained in the ferrite-pearlite steel, voids form preferentially at the ferrite–iron carbide interfaces (Fig. 14) and failure is initiated. In dual-phase steels, the martensite does not deform until the retained austenite transforms to martensite, and one or both ferrites, when present, are highly strained (19). Therefore voids leading to failure do not form until more extensive deformation has occurred and the martensite is also highly strained (Fig. 15). Hence failure occurs at much higher strains in dual-phase steels than in ferrite-pearlite steels of the same strength.

Attempts have been made to treat dual-phase steels as two-component composites, and their mechanical properties have been predicted with reasonable success (20, 21). However, further refinement of these models is necessary

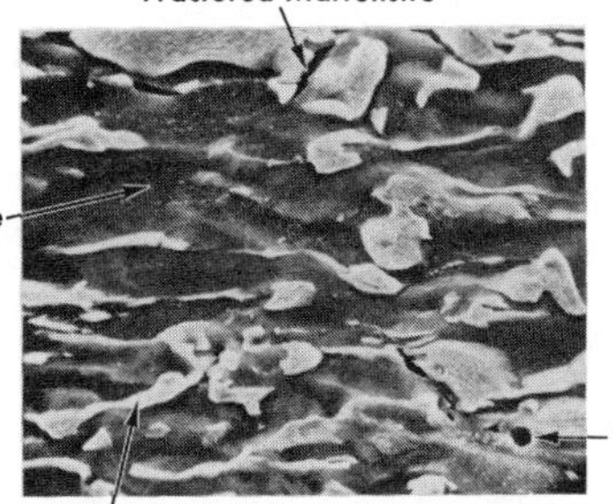

Fig. 15. Scanning electron micrograph of a dual-phase steel at fracture initiation. Both ferrite and martensite grains are elongated. Some martensite islands are fractured and voids are formed at the ferrite-martensite interfaces.

so that they may be used effectively in designing new steels with improved combinations of strength and ductility. Commercial utilization of all HSLA steels is growing rapidly and limitations in their mechanical properties are now being discovered. This has provided a new impetus to develop steels with modified plain carbon steel chemistry but with strength-ductility combinations that far surpass those of the available HSLA steels.

Conclusion

Slight modification of the plain carbon steel composition and close control of processing variables produce desirable properties in HSLA and dual-phase steels. Although an extension of the broad family of HSLA steels, dual-phase steels have different microstructures and mechanical properties. Their strength-ductility relationships are such that high-strength steel usage may be extended; greater weight reduction in steel products with complex shapes is now possible. The HSLA steels produced thus far have an optimized balance of mechanical properties that makes them suitable for various commercial products. However, in order to efficiently design components and conserve material and energy, the steel should be matched to the application. To accomplish this a more complete understanding of HSLA steels at the basic level is needed, and research in this field will undoubtedly lead to further steel developments.

References and Notes

1. M. Korchynsky, Ed., *Microalloying 75, Proceedings of the Conference, Washington, D.C., October 1975* (Union Carbide Corp., New York, 1975).
2. _____ and H. Stuart, in *Proceedings of the Symposium on Low Alloy High Strength Steels, Nuremberg, West Germany, May 1970* (Union Carbide Corp., New York, 1970), pp. 17–27.
3. S. Hayami and T. Furukawa, in (*1*), pp. 78–87.
4. M. S. Rashid, *SAE Trans.* **85** (No. 2), 938 (1976); *ibid.* **86** (No. 2), 935 (1977).
5. J. P. Hirth and J. Lothe, *Theory of Dislocations* (McGraw-Hill, New York, 1968).
6. E. O. Hall, *Proc. Phys. Soc. London Sect. B* **64** 747 (1951).
7. N. J. Petch, *J. Iron Steel Inst. London* **174**, 25 (1953).
8. J. D. Jones and A. B. Rothwell, *Iron Steel Inst. London Pub.* **108**, 78 (1968).
9. J. N. Cordea and R. E. Hook, *Metall. Trans.* **1**, 111 (1970).
10. A. B. LeBon and L. N. de Saint-Martin, in (*1*), pp. 90–99.
11. E. R. Morgan, T. E. Dancy, M. Korchynsky, *Met. Prog.* **89** (No. 1), 125 (1966).
12. L. Luyckx, J. R. Bell, A. McLean, M. Korchynsky, *Metall. Trans.* **1**, 3341 (1970).
13. E. G. Hamburg and J. H. Bucher, *SAE Trans.* **86** (No. 1), 730 (1977).
14. A. P. Coldren and G. Tither, *J. Met.* **30** (No. 4), 6 (1978).
15. J. F. Butler and J. H. Bucher, in *Proceedings of the VANITEC Conference on the Production and Properties of Dual Phase Steels and Cold Pressing Steels, Berlin, October 1978* (Vanadium International Technical Committee, London, 1978), pp. 3–12.
16. G. E. Dieter, *Mechanical Metallurgy* (McGraw-Hill, New York, 1961), p. 56.
17. S. P. Keeler, *Machinery (N.Y.)* **74** (No. 2), 88 (1968); *ibid.* (No. 3), p. 94; *ibid.* (No. 4), p. 94; *ibid.* (No. 5), p. 92; *ibid.* (No. 6), p. 98; *ibid.* (No. 7), p. 78.
18. L. A. Erasmus, *Metall. Met. Form.* **42**, 94 (April 1975).
19. M. S. Rashid and E. R. Cprek, in *Formability Topics—Metallic Materials* (ASTM STP 647, American Society for Testing and Materials, Philadelphia, 1978), pp. 174–190.
20. R. G. Davies, *Metall. Trans.* **9A**, 41 (1978).
21. R. Lagneborg, in *Proceedings of the VANITEC Conference on the Production and Properties of Dual Phase Steels and Cold Pressing Steels* (Berlin, October 1978), pp. 43–52.

Sintered Superhard Materials

R. H. Wentorf, R. C. DeVries, F. P. Bundy

In the hierarchy of hardness one may recognize three broad classes of materials:

1) The common materials of construction, such as organic materials, some ceramic materials, and metals, including most steels.

2) The harder materials, such as aluminum oxide and silicon carbide, which are used mostly as abrasives, and the cemented carbides of tungsten and titanium, which are used mostly as cutting tools. These harder materials are used to shape the common materials of construction into useful articles. Much of our material wealth and productivity depends on the effective use of these harder materials.

Summary. Diamond or cubic boron nitride particles can be sintered into strong masses at high temperatures and very high pressures at which these crystalline forms are stable. Most of the desirable physical properties of the sintered masses, such as hardness and thermal conductivity, approach those of large single crystals; their resistance to wear and catastrophic splitting is superior. The sintered masses are produced on a commercial scale and are increasingly used as cutting tools on hard or abrasive materials, as wire-drawing dies, in rock drills, and in special high-pressure apparatus.

3) The superhard materials, diamond and cubic boron nitride (CBN). The best ways to shape or sharpen the harder materials widely used as tools in modern industry generally involve the use of diamond as a component of an abrasive wheel or a fixed tool. For example, approximately a quarter carat (0.05 gram) of diamond is used up in shaping the tools used to manufacture a typical automobile.

Diamond is also widely used in tools for the cutting and shaping of the harder materials of construction such as stone, concrete, glass, ceramics, and other hard or abrasive materials which chemically react only slightly with carbon at the cutting edge. Cubic boron nitride, which is somewhat softer than diamond, finds use in shaping hard iron or nickel-containing metals that chemically attack diamond

rapidly at the hot cutting edge and thereby make a diamond tool deteriorate at uneconomic rates.

Both diamond and CBN are thermodynamically stable only at very high pressures, as indicated in Fig. 1. Diamond is found naturally, presumably brought to the earth's surface after being formed at great depths. Most of the industrial diamond used today is synthesized at high pressures from graphite (*1*); CBN is formed under similar conditions of temperature and pressure from graphitic hexagonal boron nitride (*2*). One of the factors affecting the extent of use of these materials is the high cost of the apparatus required to synthesize them. Other factors are their size, shape, and coherence. Until recently, most of the synthetic materials came in pieces whose maximum size was less than about 1 millimeter. This was suitable for sawing and grinding purposes but the great need for larger pieces to be used as cutting tools or wire-drawing dies remained unsatisfied by the synthesized material. This need was met as much as possible by natural single crystals of diamond despite their limitations of costliness, variability, susceptibility to massive fracture along easy cleavage planes, and uneven wear.

These limitations arise from the crystal structure of diamond, which is such that each carbon atom has four nearest neighbors at the corners of an imaginary tetrahedron. They are stacked together so that the entire array has the same symmetry properties as a cube. The carbon atoms are small and the carbon-carbon bonds are stiff and highly directional, so that the bond energy per unit volume of diamond exceeds that of any other substance, hence the surpassing hardness. However, if one makes imaginary slices through the crystal on various planes, one finds that there are four different planes across which the number of bonds per unit area is the least. Furthermore, impurities are most easily accommodated on these planes, because most impurity atoms are larger than carbon atoms and so prefer positions where the misfit strain is the smallest. Thus the crystal tends to separate most easily across these less strongly bonded planes. These planes are the (111) or octahedral face planes. They pass through three (but not four) corners of a cube.

The so-called "grain" as well as many of the wear and fracture properties of diamond crystals may be explained by invoking the lower resistance of (111) planes to tensile or cleavage failure (*3*). In the gem trade, the "easy" polishing directions on different crystal faces are determined mainly by the ease with which microscopic octahedral-faced fragments can be pulled out of the surface being polished. In single-point cutting tools or wire-drawing dies, attention is paid to the orientation of the crystal to minimize the possibilities of catastrophic cleavage and rapid or uneven wear. For example, the hole in a single-crystal wire-drawing die wears to an out-of-round shape owing to the differing resistance to abrasion along different crystal directions as the wire rubs the diamond. Usually, a single diamond crystal bears some obvious natural crystal faces to permit easy identification of the weak planes; otherwise an x-ray diffraction pattern is needed to indicate them because the crystal is optically isotropic, except for residual strains.

Several kinds of natural polycrystalline diamond masses are also available and find some use despite their relative scarcity and the difficulties of shaping and holding them. For example, framesite is weak, composed of relatively large crystals held together more by impurities than by diamond-diamond bonding, and is not widely used. A more durable category, called carbonado, has extensive diamond-diamond bonding despite mineral oxide impurities. Ballas contains very little impurity, is generally

The authors are members of the technical staff at the General Electric Company, Research and Development Center, Schenectady, New York 12301.

←Oil well drilling bit armed with sintered diamond teeth. [*R. H. Wentorf, General Electric Co., Research and Development Center, Schenectady, New York*]

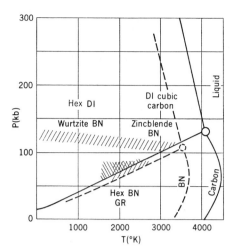

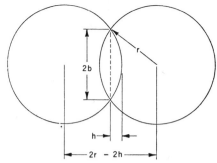

Fig. 1 (left). Pressure-temperature phase diagram for carbon and boron nitride (*BN*). (*DI*, diamond; *GR*, graphite.)   Fig. 2 (right). Overlapping spheres.

white instead of dark in color, and is extremely strong (*4*). The natural formation processes for these materials remain largely unknown.

The advantages of a strong, polycrystalline mass are its greater "toughness" (the easy cleavage planes of its component crystals lie randomly toward one another, so that propagation of a crack throughout the mass is difficult) and its more uniform wear, independent of direction. Such masses lack the aesthetic appearance of single crystals but their greater utility makes efforts to prepare them worthwhile.

Aspects of Sintering Superhard Materials

By sintering one means the bonding together of a mass of small particles into a larger, coherent piece. During this process the contact patches between particles grow and the total volume of the mass shrinks. Eventually the mass becomes sufficiently dense and strong even though some voids may remain.

As a simple model of a sintered mass, consider a face-centered cubic array of close-packed spheres, each of radius r, which have common circular contact patches of radius b, and center-center distances of $2r - 2h$, as shown in Fig. 2. Then a cap of volume $(3r - h)\pi h^2/3$ is missing from each sphere. A unit cell, of edge length $a = 2\sqrt{2}\,(r - h)$, contains four spheres, each with 12 nearest neighbors. Then F, the actual volume fraction of the cell occupied by spheres will be

$$F = 4[4\pi r^3/3 - 12(3r - h)\pi h^2/3]/16\sqrt{2}(r - h)^3 \quad (1)$$

When the spheres are flattened so that the contact patches from six coplanar spheres touch, the voids between spheres will no longer be connected. This occurs for $2(r - h) = \sqrt{3}\,r$, or $h/r = 0.134$. The total contact patch area, C, per unit cell, arises from 12 patches on each of four spheres, or 24 net patches, each of area πb^2, so that

$$C = 24\,\pi b^2 = 24\,\pi h(2r - h) \quad (2)$$

For mechanical strength and transmission of heat, one considers the ratio $C/a^2 = M$. Table 1 shows that M increases rapidly at the higher values of F. A practical lower limit for carrying most of the

strength of the grains might be $M = 2$, where F is about 0.92. Foreign material among the spheres will be trapped when F exceeds 0.962.

The driving forces for sintering are the decreases in free energy accompanying reductions of stresses and surface area. Pure materials usually sinter reasonably rapidly at absolute temperatures about 0.6 to 0.7 that of melting. Solvent impurities may assist sintering but they may also affect the sintered mass undesirably.

The embedding of diamond particles in some kind of a tough matrix (usually metal) to form a hard composite mass is satisfactory for many uses of diamond, but it is not a substitute for direct sintering between diamond grains when the properties necessary for a cutting tool, such as high hot strength and good thermal conductivity, are desired. The conductivity, strength, stiffness, hot hardness, and thermal expansion of possible matrix materials are so inferior to those of diamond that the properties of the composite tend more to those of the matrix than of diamond (see Table 2).

The melting point of diamond is about 4000 K (see Fig. 1), so that a suitable sintering temperature would be about 2400°C, about the same temperature as that used in the manufacture of graphite from coke. The required pressure for diamond stability then would be about 70 kilobars.

However, the hardness of diamond is a practical obstacle to achieving this pressure over the entire surfaces of all the consolidating particles. A vessel into which diamond grains have been poured may contain about 30 percent voids, depending on grain shape and size distribution. Tamping or squeezing the vessel scarcely affects the void fraction because the hard particles do not crush much; instead they indent the walls of the vessel. Where the diamond particles touch each other, the local pressures are very high, in the range 300 to 1000 kbar. In other places (most places) the local pressure is that of the voids. Thus, at an average pressure of, say, 70 kbar on the vessel, one has heavily indented vessel walls, a network of local high-pressure contact patches, and a large area at a relatively low pressure. As the sintering temperature is reached, some reduction in strength and hence local pressure will occur, and the contact patch area will increase slightly. But at 2400 K the transformation of diamond into graphite is rapid at low pressures. One volume of diamond expands to about 1.6 volumes of graphite, and so the low-pressure voids become filled with graphite until

Table 1. Values of F and M for various values of h/r up to 0.134, above which the analysis is irrelevant.

	Values of h/r						
	0	0.04	0.06	0.08	0.10	0.12	0.134
F	0.704	0.826	0.863	0.895	0.927	0.953	0.962
M	0	0.801	1.242	1.715	2.21	2.75	3.14

Table 2. Properties of diamond and steel.

Property	Diamond	Steel
Melting point (K)	4000	1800
Hardness (kg/mm²)		
At 25°C	7000+	1000
At 1300°C	1500	10
Youngs' modulus (dyne/cm² at 25°C)	8 to 12 × 10¹²	2 × 10¹⁰
Thermal conductivity (watts cm⁻¹ K⁻¹ at 25°C)	7 to 20	0.7
Linear thermal expansion coefficient 10⁻⁵ K⁻¹ at 25°C	0.1	1

the local graphite pressure is high enough to repress further graphite formation. However, this pressure is not necessarily high enough to force the graphite back into diamond. Thus the 70 kbar of external pressure is borne by a mixture of diamond and graphite, and in order to increase the fraction of diamond, an increase in pressure will be necessary so that the diamond grains yield and the graphite pressure increases enough to form some diamond. From separate experiments on the simple conversion of graphite into diamond, it appears that pressures on the order of 100 to 130 kbar are required (5). Such pressures are difficult to achieve on a practical commercial scale.

The presence of solvent catalyst metals, such as manganese, iron, or nickel facilitates the transformation between diamond and graphite (1). These molten metals are also excellent solvents for most other metals and their carbides and hence might be expected to attack the walls of the vessel containing the mass to be sintered. Too much metal among the diamond grains reduces the stress gradients that can drive the consolidation process.

Another difficulty that occurs with a mixture of catalyst metal and diamond is the formation of stable arches or shells before the entire mass is consolidated. Such a shell is essentially all diamond in perfect mechanical contact and hence changes shape very slowly. Inside it the pressure falls to that of the diamond-graphite equilibrium; no consolidation occurs and graphite may form from diamond there. Suitable vessel design can frequently eliminate this problem, thereby reducing the overall pressure required for consolidation.

Finally, let us consider a mass of sin-tered diamond embedded in a pressure-transmitting medium at high pressure. This medium is necessarily a solid at the low temperatures (20° to 200°C) prevailing during the reduction of pressure to 1 atmosphere. Since nothing expands so little as diamond during pressure reduction, the outer surfaces of the sintered diamond mass are exposed to strong tensile stresses as the pressure-transmitting medium tries to expand more rapidly. These stresses can be high enough to tear the sintered mass apart if it is not strong enough or convex enough.

From the foregoing comments one may conclude that the production of sintered superhard materials is beset with several peculiar and difficult problems. Indeed, they blocked the development of practical production of sintered diamond bodies for many years despite numerous attempts and much experimentation in many laboratories.

Progress Toward Practical Solutions

The availability of apparatus capable of maintaining temperatures of 1500 to 2000 K together with pressures of 50 to 70 kbar, where diamond is stable, fostered new work on sintering diamond.

Katzman and Libby (6) described the formation of sintered diamond compacts with a cobalt binder. They found the best conditions to be about 20 minutes at 1600°C and 62 kbar using a mixture of cobalt and diamond powders (20:80, by volume) of particle sizes 1 to 5 micrometers held in a tantalum vessel. The sintered masses were about 6 millimeters in diameter and 3 to 8 mm long. There appeared to be some grain growth or diamond-diamond bonding. The masses had about half the life of a single-crystal diamond when used to dress abrasive wheels. The process never reached commercial production.

Stromberg and Stevens (7) reported on the sintering of diamond at 1800° to 1900°C and 60 to 65 kbar. They loaded cleaned natural diamond powder of sizes 0.1 to 10 μm into tantalum containers. The vessels, about 3 mm in size, were heated in vacuum to remove gases and then sealed by electron beam welds before being loaded into a high-pressure apparatus and exposed to high pressure and temperature for about an hour. In some experiments about 1 percent of boron, silicon, or beryllium powder was added to react with residual gases and form strong carbides for bonding. An indium sheath surrounding the vessel helped prevent tearing apart of the mass as the pressure was reduced. The diamond masses obtained had densities of up to about 99 percent and microhardnesses of 75 to 90 percent that of solid diamond. Efficient mixing of diamond particles in a range of sizes suitable for high packing density produced the highest final densities. No tests of strength or wear resistance were reported. The process was never practiced on a commercial scale.

Hall (8) found that strong diamond compacts could be made at about 2000 to 2500 K and 65 kbar using short sintering times of 20 to 200 seconds. The diamond powder is contained in graphite tubes and no special cleaning procedures are mentioned. The products are usually black cylinders with densities of about 3.2 grams per cubic centimeter. Extra bonding agents (borides, nitrides, carbides, and oxides) may be added to the diamond before sintering. Such diamond compacts are available and have found

Table 3. Mechanical properties of sintered superhard materials [from (11, 16–19)].

| Type of compact | GPa | | | | | | Sound velocity (V_1, m/sec) | Poisson's ratio, ν | Relative wear resistance |
	Compressive strength	Modulus of elasticity, E	Transverse rupture strength	Simple compressive stress	Simple shear stress	Knoop hardness			
Diamond									
Regular (two-phase)	6.9 (11)	890 (11)	1.35 (11)			63.7 to 78.4 (19)	16330 (18)		200 (19)
"Treated" Anvils	—	840 (11)	0.83 (11)	35 (16)	20 (16)				
Cubic boron nitride									
Single-phase		890 (18)				44	16365 (18)	0.138 (18)	20 (19)
Comparative materials									
Cemented tungsten carbide	3.7 to 5.5	460 to 675 (18)	1.7 to 2.7	7 (16)	4 (16)	18 to 22	6400 to 7000 (18)	0.20 to 0.24 (18)	2 (19)
Single-crystal diamond	8.69 (17)	950 (17) to 1100 (11)	1.05 (11)			≤ 88	17780 (<100>) 18620 (<111>) (18)		95 to 245 (19)

Table 4. Machining performance of sintered diamond and CBN.

Material	Total number of pieces cut per tool
Machining silicon-aluminum SAE 332 (33)	
Compax® diamond tool	412,000
Cemented carbide	2,400
Turning rubber filled with nickel and aluminum powder (34)	
Compax tool	6,000
Cemented carbide	140
Machining flame-sprayed alloys (Colmonoy No. 43) (35)	
CBN tool	3,000
Cemented carbide	9
Milling of type 380 aluminum line bore crankcase (36)	
Compax tool	12,000 to 14,000
Cemented carbide	3,000
Machining glass filled polypropylene (facing operation) (37)*	
Compax tool	7,000
Cemented carbide	400

*Cutting speed for Compax is approximately two times that of carbide. ®Registered trademark of General Electric Co., USA

uses as cutting and dressing tools, and as gauge blocks.

Hibbs and Wentorf (9) described commercially available diamond and CBN compacts and some of their uses in cutting and shaping hard materials. These sintered masses are characterized by extensive direct bonding between the hard particles and a minimum of second phase. Temperatures in the range 1500° to 2000°C are suitable for their formation and these temperatures require pressures of the order of 50 to 70 kbar so that the cubic crystals are thermodynamically stable during sintering. The grain size in the diamond and CBN compacts can range from a few to several hundred micrometers in size. Smaller grain sizes permit finer cutting edges on the finished

tool. Usually the hard materials are consolidated as a layer about 0.5 mm thick on the circular face of, or as a core within, a cylinder of cobalt-cemented tungsten carbide which provides strength and convenience in use (10). Cutting tool blanks are commercially available in disks up to about 14 mm in diameter and wire-drawing die blanks of comparable sizes are also available. The General Electric Company has been manufacturing sintered diamond materials since 1972, and these materials enjoy wide utility and acceptance for many industrial purposes, where they replace not only large single crystals but also often other hard sintered materials such as tungsten carbide and aluminum oxide.

Relation of Microstructure to Property

Although sintered superhard materials are relatively new, some understanding of the microstructural features and their relation to macroscopic properties has been developed. The stronger and more successful man-made sintered diamond compacts are brittle, two-phase materials—predominantly randomly oriented diamond ($\geq$ 90 percent by volume) with the remainder a metallic phase (Fig. 3). The former is continuous and the latter essentially continuous as determined by its etching behavior (11). The modulus of elasticity of the two-phase composite was found to be 890 gigapascals (11), which approaches that of single-crystal diamond.

When the metal phase was dissolved away, the modulus of the polycrystalline diamond, corrected for porosity, was 840 GPa (11). (The pores were uniformly distributed and were less than 1 μm in diameter.) This value is only about 5 percent lower than that for the original material containing the metallic phase. This result strongly suggests that there is diamond-

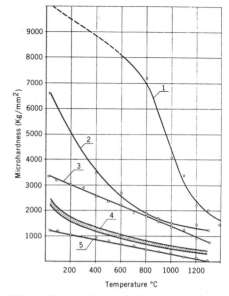

Fig. 4. Hot hardness of sintered superhard materials. Curve 1, diamond; curve 2, CBN; curve 3, silicon carbide; curve 4, different kinds of electrocorundum; curve 5, carbide VK-8 (92 percent tungsten carbide, 8 percent cobalt).

diamond bonding in the composite and that the metallic phase is a minor contributor to the strength of the material, acting only as a mechanical filler. Besides this indirect evidence, there is convincing direct support for diamond-diamond bonding from microscopic observations of grain boundaries.

Another prominent microstructural feature of some sintered diamond is deformation twinning, which is seen as groups of lines within grains (Fig. 3). When diamond is deformed under conditions that inhibit brittle fracture, it can deform by twinning on (111) planes. This has been observed in the naturally occurring polycrystalline aggregate called framesite (12) and has purposely been introduced into single crystals (13). Similar structures have been reported in single-crystal diamond indentors (14) and in anvils subjected to high pressures (15). The deformation twins within grains in Fig. 3 are visible because of relief polishing which emphasizes the greater wear resistance of the twinned region compared to the rest of the crystal. If the pressure-temperature conditions are adequate, this type of deformation structure can be expected when the diamond grains are large enough to sustain such deformation, and this microstructural feature may be significant in grains that are less than about 1 μm in size. In addition, under semihydrostatic conditions, as might exist with the presence of a softer phase, the deformation structure will probably not be generated everywhere in the compacted aggregate.

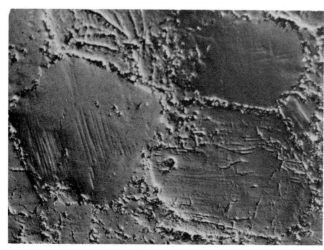

Fig. 3. Photomicrograph of polished surface of man-made sintered diamond mass.

Mechanical Properties

A summary of the mechanical properties of sintered diamond with some comparisons with single-crystal diamond and with cemented tungsten carbide is shown in Table 3. These data, which are from several sources (11, 16–19), are primarily for the same sintered material (20–25).

The effect of grain size on mechanical properties has been reported (21). In common with other brittle polycrystalline materials, an inverse relation exists between grain size and ultimate tensile strength.

A direct relation between remanent magnetism of sintered diamond and hardness in material sintered with cobalt has been reported (25). The intensity of the remanent magnetization is thought to be related to the residual graphite. As the amount of decomposed diamond (that is, as graphite in the pore regions) increases, the hardness and magnetic intensity decrease, unless the graphite reprecipitates as diamond from the cobalt-carbon liquid.

Dunn and Lee studied the fracture and fatigue of sintered diamond compacts with a cemented carbide substrate (26). The conclusions from their study are: (i) that the failure load falls from 1900 pounds at 5000 cycles to 1200 pounds at 200,000 cycles; (ii) fracture stress is reduced under cyclic loading; (iii) failure is by shear mode brittle fracture; (iv) there is no appreciable difference in results at 20°C and at 300°C; (v) the failure load corresponding to an infinite number of cycles is far above reported values of force for cutting rocks under similar geometric conditions.

Thermal Properties

Thermal degradation of some sintered diamond compacts is a well-documented phenomenon. Lee and Hibbs (27) and Dennis and Christopher (28) cited a 700°C temperature limit, and it is recommended that this temperature not be exceeded in brazing other sintered diamond compacts (22). Thermal cracking is related to the thermal stresses set up by the differential thermal expansion of diamond and the metallic second phase.

Hot hardness data on sintered superhard materials (29) are shown in Fig. 4. On the basis of comparative hot strength data, Hibbs and Wentorf (9) cited the advantages of sintered diamond and CBN tools over tungsten carbide–based materials with respect to machining hard alloys. Machining of these materials at temperatures of their yield strengths is clearly possible with sintered CBN. Cubic boron nitride has the distinct added advantage over diamond of greatly reduced reactivity with tool steels and superalloys (30).

There are claims of increased fracture toughness and improved strength of CBN ceramics if they contain some of the wurtzite form of boron nitride (31). This form is as hard as the cubic form but appears to be less stable than CBN with respect to conversion back to the soft graphite form. The toughening mechanism suggested is by the stress-induced phase transformation of the metastable wurtzite boron nitride to the graphitic form of boron nitride. The high internal stresses thus created control the extent of further transformation of the constrained metastable form. In grinding ap-

plications, the tensile stresses release the compressive stress around the potential graphitic boron nitride. This explanation may be valid, but it ignores grain size effects and the fact that the cleavage systems of the two types of hard boron nitride are quite different. It may also be that the longer life of the cutting tools containing wurtzitic boron nitride is simply related to the lubrication provided by the soft form of boron nitride as it forms from the less stable wurtzite form. There's room for additional study of this and other physical "metallurgy" phenomena in sintered superhard materials.

Sintered diamond and CBN have been proposed as heat sink materials because of the intrinsic high thermal conductivity of single crystals of these materials. Values as high as 9 watts per centimeter per degree Celsius have been reported for sintered CBN (32), but there is little evidence for application of these materials because of cost considerations.

Applications of Sintered Superhard Materials

In the short time since their introduction to technological applications, sintered superhard materials have successfully established places as both cutting tools and wire-drawing dies, and encouraging results have been achieved in some rock-drilling applications. Examples of a layer of sintered diamond on cemented carbide are shown in Fig. 5. Wire-drawing dies with sintered diamond cores are shown in Fig. 6. Single-crystal diamond is being replaced in many cutting and dressing operations in-

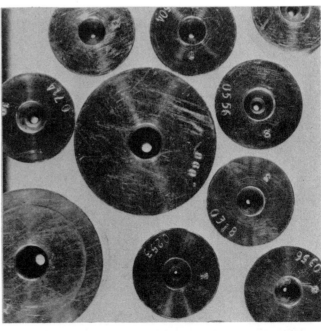

Fig. 5 (left). Sintered diamond layers on cemented tungsten carbide substrates.   Fig. 6 (right). Sintered diamond wire-drawing dies.

71

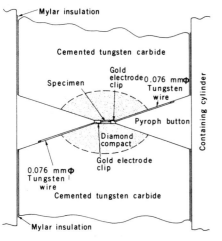

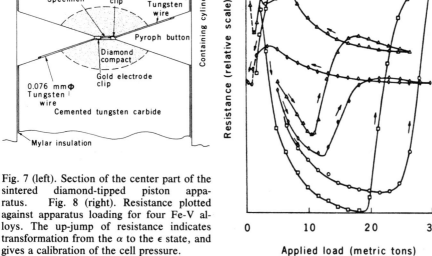

Fe 6V
Fe 12V
Fe 16V
Fe 20V

Resistance (relative scale)

Applied load (metric tons)

Fig. 7 (left). Section of the center part of the sintered diamond-tipped piston apparatus. Fig. 8 (right). Resistance plotted against apparatus loading for four Fe-V alloys. The up-jump of resistance indicates transformation from the α to the ϵ state, and gives a calibration of the cell pressure.

volving ceramics, glass, concrete, plastics, fiberglass composites, presintered and sintered tungsten carbide, and nonferrous metals such as copper, or aluminum-silicon alloys that contain hard particles. Sintered diamond is practically as hard as single-crystal diamond and in addition is not plagued by catastrophic single-crystal cleavage and anisotropic wear. The availability of large-size pieces of different shapes is another asset. The most severe restriction for diamond is still chemical reactivity under

cutting and grinding conditions with iron-, cobalt-, and nickel-based alloys. For these materials sintered CBN may be used. As a general rule, for finer finishes a finer grain size compact should be used in both cutting and wire-drawing operations. There is as yet no published data on turning optical finishes with sintered materials, but the prospects look attractive. Table 4 gives some typical examples of the performance of sintered diamond in cutting tools (33–37).

Perhaps the most spectacular com-

parison of single-crystal diamond with the sintered aggregate is as wire-drawing dies. Here the isotropic wear characteristic and the lack of extensive easy cleavage planes of the sintered material results in outstanding performance. Enough copper wire to circle the earth at least seven times has been drawn through one 0.203-cm die without change in roundness or dimension. With the recent availability of larger size die blanks, cemented carbide is also being replaced in this application. For a summary of the performance of these materials, see Table 5 (38).

Sintered Diamond for
Ultrahigh-Pressure Apparatus

To generate the highest possible static pressures in specimens for scientific experiments the specimen material must be confined between the faces of pistons of very hard, strong material and compressed to smaller volume. In the past, the most useful ultrahigh-pressure apparatuses have included pistons made of cemented tungsten carbide (39–41) or of single-crystal diamond (42), and these materials are still being used extensively today. Even in the best designs, in respect to keeping shear stress as low as possible, it has been determined that some plastic flow begins in carbide apparatus at about 52 kbar (43, 44), and in single-crystal diamond at 700 to 1500 kbar (45, 46). Cemented tungsten carbide apparatus, properly used, can be made to operate up to 200 to 230 kbar, but loading beyond that results only in extensive plastic distortion without increase of cell pressure.

In the case of single-crystal diamond pistons the operating limits are usually set by catastrophic cleavage fracture. Also, because of the limited size and high cost of nearly perfect single-crystal diamond for pistons, the physical dimensions of the specimens that can be pressurized in this manner are very small. In most cases the specimens are monitored by optical, spectroscopic, or x-ray diffraction methods.

In recent years the availability of sintered diamond in special geometric forms has made it possible to fabricate ultrahigh-pressure apparatus of relatively large size (compared to single-crystal diamond apparatus) in which the electrical behavior of specimens can be studied up to pressures of about 600 kbar over the temperature range of about 400 to 2.6 K. We will illustrate this type of equipment and its use by an apparatus developed at the General Electric Research

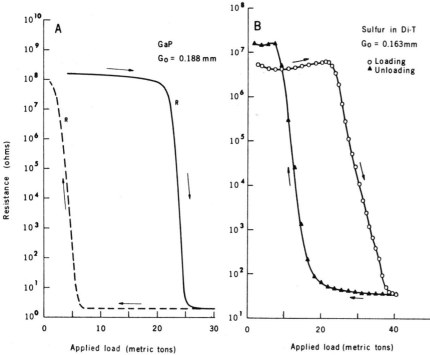

Fig. 9. (A) Insulator-to-metal transition in GaP at room temperature at about 220 kbar. (B) Resistance plotted against pressure behavior of sulfur.

72

and Development Center in recent years (*47*). Other experimenters have developed versions that utilize the basic principles, but differ in detail.

Figure 7 shows the details of the center part of the apparatus, particularly the specimen and electrode arrangements in the high-pressure zone between the flat faces of the pistons. In this case the pistons were about 13.7 mm in diameter and the flat-pressure face about 1.25 mm in diameter. In most cases the specimen was in the form of a thin rectangular flake less than 0.025 mm thick and 0.60 mm long, resting in the plane of the equator sandwiched between two pyrophyllite stone disks, and with each end in contact with a tiny gold foil electrode. The gold electrodes in turn were connected by very thin tungsten wires to the metallic part of the piston because the sintered diamond is essentially an electric insulator. One electrode made contact with the top piston and the other with the bottom piston. The pistons themselves were electrically insulated from the cylinder wall by 0.025-mm Mylar film so that they could serve as part of the electrical monitoring circuit. The pyrophyllite stone gasket between the adjacent ends of the pistons served as a gasket and as pressure gradient medium, as well as an electric insulator.

For experimentation to very low temperatures this apparatus was fitted with a "squirrel-cage" clamping arrangement (*48*) so that it could be loaded at room temperature to a desired pressure in a hydraulic press, clamped at that loading, removed from the hydraulic press, and inserted in a cryogenic container and cooled to temperatures as low as 2.6 K with liquid helium. This arrangement was used particularly to study the behavior of "high-pressure metals" derived from insulator or semiconducting substances (such as silicon, tellurium, or selenium), especially in respect to electrical superconduction.

Some examples of very-high-pressure phenomena that have been observed with this type of apparatus are as follows. Some ordinary metals change to different metallic structures when subjected to high pressures. For example, iron and many of its alloys take the α (body-centered cubic) structure under ordinary conditions, but upon application of high pressure, transform to a more dense ϵ (hexagonal close-packed) structure with higher electrical resistivity. When iron is alloyed with cobalt or with vanadium, the pressure of the transition is increased by an amount roughly proportional to the fraction of the alloying element. Shock compression tests

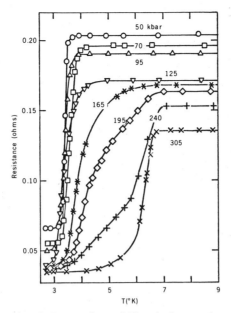

Fig. 10. Recent General Electric data on the pressure dependence of the T_c of tellurium, presented as $R(T)$ for various pressures. Note the large increase of T_c in the 110 to 190 kbar pressure region.

of such alloys at Los Alamos Scientific Laboratory by Loree *et al.* in 1966 (*49*) showed that the transition in Fe20V is at about 500 kbar.

Figure 8 shows the $\alpha \rightarrow \epsilon$ transition as observed by a resistance-jump in a sintered diamond-tipped apparatus for four different Fe-V alloys. This is the first time these high-pressure transitions had been observed in a static high-pressure apparatus.

An example of the transformation of an insulator to a metal at room temperature is gallium phosphide, shown in Fig. 9A. In this experiment the resistance remained at about 10^8 ohms, the insulation resistance of the apparatus, up to a loading of about 22 tons, beyond which it dropped quite abruptly by eight orders of magnitude to about 2 ohms. This is the result of a change of structure from a covalent-bonded diamond cubic form to a tinlike metallic form. Upon unloading, the GaP reverts to an insulating, covalently bonded structure.

Figure 9B shows the electrical behavior of sulfur when compressed at room temperature to over 500 kbar pressure (*50*). As in the case of GaP the indicated resistance remained constant at the apparatus insulation value until, at about 25 tons of loading, the resistance of the sulfur specimen started decreasing below that value. Beyond a loading of about 37 tons the resistance leveled out at about 30 ohms with the specimen in a semimetallic state, having a positive temperature coefficient of resistance. Temperature cycling experiments at pressures in the range of 250 to 450 kbar showed that in this interval sulfur acts as a semiconductor with the band gap, or activation energy of conduction, decreasing rapidly with pressure. This band gap went to zero between 450 and 500 kbar, and for greater pressures the sulfur acted as a semimetal.

Some of the "high-pressure-induced metals" become superconducting at low temperatures. One of the most interesting examples is tellurium, which goes from a semiconductor form to a metallic form at about 40 kbar. Soviet studies (*51*) of the superconduction critical temperatures, T_c, of this material up to about 150 kbar indicate three different metallic forms. More extensive studies of the su-

Table 5. Compax® diamond die blank performance (*38*). These results were obtained under production conditions in different wire manufacturer's plants and are believed to be representative of the performance that can be expected by others. However, variables such as rod quality, type of wire machinery, lubrication systems, die fabrication techniques, die practices, and drawing speed can alter performance.

Application	Wire sizes (mm)	Relative performance
Aluminum	0.64 to 2.6	3 times over natural diamond
Copper	1.84 to 4.6	200 times over carbide
Copper	0.40 to 2.05	10 times over natural diamond
Tin-plated copper	0.50 to 1.45	8 times over natural diamond
Nickel-200	0.33 to 1.45	10 times over natural diamond
Aluminum (5056)	3.05 to 4.76	150 times over carbide
Tungsten	0.18 to 0.62	4 times over natural diamond
Molybdenum	0.38 to 1.02	70 times over carbide
Molybdenum	0.18 to 1.02	5 times over natural diamond
Brass-plated steel (tire cord)	0.17 to 0.96	4 times over natural diamond
Brass-plated steel (tire cord)	0.17 to 0.96	20 times over carbide
Galvanized high carbon steel	0.17 to 1.05	36 times over carbide
Stainless steel (304 and 316)	0.41 to 1.6	6 times over natural diamond
Stainless steel (302)	0.36 to 0.71	3 times over natural diamond
Stainless steel (302)	1.1 to 1.6	10 times over carbide
Nickel-chrome-iron alloy (60-15-25)	0.23 to 0.91	5 times over natural diamond
Low carbon steel	1.55 to 3.2	40 times over carbide

® Registered trademark of General Electric Co., USA

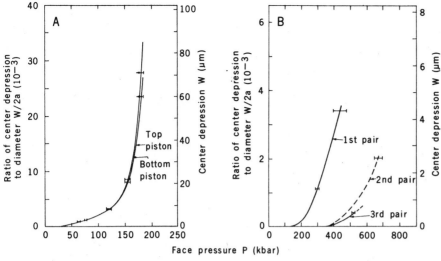

Fig. 11. Face center deformation plotted against face loading pressure for (A) 3 percent cobalt cemented carbide pistons, and (B) sintered diamond-tipped pistons. Note that the yield stress in (B) varies significantly with the quality of sintering of the diamond in the three different cases.

perconduction temperature of high-pressure tellurium metal, done recently with the sintered diamond-tipped apparatus to pressures over 300 kbar, indicate that there is an additional phase change in the 160- to 190-kbar pressure region, as shown in Fig. 10 (52). Note from the R(T) curves shown that the resistance jump associated with the specimen going superconducting goes from about 3.5 K at 125 kbar to about 6.5 K at 195 kbar. Such a large change in T_c almost certainly means a significant change in the structure of the metallic phase.

Recently, quantitative studies of the permanent deformation of piston faces subjected to known pressures have yielded information on yield strength of the piston materials. The results for 3 percent cobalt-cemented tungsten carbide are shown in Fig. 11A (44). Taking the definition of the plastic yield stress as that at which there is 0.2 percent permanent deformation, it is found that the yield stress of this hard grade of carbide is about 50 ± 5 kbar. Figure 11A indicates that plastic yielding limits the face pressure of Bridgman-type or Drickamer-type opposed piston apparatus to about 200 kbar.

The same kind of tests applied to pairs of sintered diamond-tipped pistons gave the results shown in Fig. 11B. Such tests show the difference in the completeness and quality of the sintering of the diamond particles of these development pieces. The first pair showed a nominal yield strength of only a little over 200 kbar, while the third pair was much better at about 500 kbar. Thus a well sintered set of such pistons, properly used, is capable of a few experiments up to the 600 to 800 kbar range.

Somewhat higher pressures are attainable by going to much smaller pressure face diameters at the expense of reducing the specimen size by a large factor, and hence also at the expense of the possibility of getting good electrical measurements. This size effect on the maximum attainable pressures applies to single-crystal diamond apparatus as well as to the sintered diamond kind. To date the highest reliably reported static pressure experiments extend up to about 700 kbar with sintered diamond apparatus and electrical monitoring, and to 1700 kbar with gasketed opposed single-crystal pistons with optical and x-ray monitoring. Since diamond is the hardest and strongest known material, both theoretically and practically, the attainment of higher static pressures can result only from improved geometrical stress designs. The pressure gains by this means probably will be only incremental.

References and Notes

1. F. P. Bundy, H. M. Strong, R. H. Wentorf, Jr., in *Chemistry and Physics of Carbon*, P. L. Walker, Jr., and P. A. Thrower, Eds. (Dekker, New York, 1973), vol. 10, pp. 213–263; R. H. Wentorf, Jr., *Ber. Bunsenges. Phys. Chem.* **70**, 975 (1966); H. M. Strong and R. H. Wentorf, Jr., *Naturwissenschaften* **59**, 1 (1972).
2. R. H. Wentorf, Jr., *J. Chem. Phys.* **34**, 809 (1961); F. P. Bundy and R. H. Wentorf, Jr., *ibid.* **38**, 1144 (1963).
3. E. M. Wilkes and J. Wilkes, *Philos. Mag.* **4**, 158 (1958); R. H. Wentorf, Jr., *J. Appl. Phys.* **30**, 1765 (1959); F. P. Bundy, *Sci. Am.* **231**, 62 (August 1974).
4. F. L. Trueb and W. C. Butterman, *Am. Mineral.* **54**, 412 (1969).
5. F. P. Bundy, *J. Chem. Phys.* **38**, 631 (1963); K. Ned. Akad. Wet.-Amsterdam Proc. Ser. B **72** (No. 5), 302 (1969).
6. H. Katzman and W. F. Libby, *Science* **172**, 1132 (1971).
7. H. D. Stromberg and D. R. Stevens, *Ceram. Bull.* **49**, 1030 (1970); U.S. Patent 3,574,580, 13 April 1971.
8. H. T. Hall, *Science* **169**, 868 (1970).
9. L. E. Hibbs, Jr., and R. H. Wentorf, Jr., *High Temp.–High Pressures* **6**, 409 (1974).
10. R. H. Wentorf, Jr., and W. A. Rocco, U.S. Patents 3,745,623, 17 July 1973, and 3,767,371, 23 October 1973, assigned to General Electric Company.
11. P. Gigl, in *High Pressure Science and Technology, 6th AIRAPT Conference, July, 1977, Boulder, Colorado*, K. D. Timmerhaus and M. S. Barber, Eds. (Plenum, New York, 1979), vol. 1, pp. 914–922.
12. R. C. DeVries, *Mater. Res. Bull.* **8**, 733 (1973).
13. _____, *ibid.* **10**, 1193 (1975).
14. C. Phaal, *Philos. Mag.* **10**, 887 (1964).
15. H. K. Mao *et al.*, *Rev. Sci. Instrum.* **50** (No. 8), 1002 (1979).
16. F. P. Bundy and K. J. Dunn, in *High Pressure Science and Technology, 6th AIRAPT Conference, July 1977, Boulder, Colorado*, K. D. Timmerhaus and M. S. Barber, Eds. (Plenum, New York, 1979), vol. 1, pp. 931–939.
17. R. Chrenko and H. M. Strong, *Physical Properties of Diamond* (Rep. 75CRD089, General Electric Company, Schenectady, N.Y., 1975).
18. R. H. Gilmore, private communication.
19. J. F. Spanitz, paper presented at *Diamond, Partner in Productivity*, symposium of Industrial Diamond Association of America, November 1979.
20. Because of the inherent variability of the product from different processes and the difficulty of comparing results from different testing procedures, we have chosen essentially "in-house" data. For procedures and property data on other sintered products see (21–25).
21. D. C. Roberts, *Ind. Diam. Rev.* (July 1979), pp. 237–245; P. A. Bex and D. C. Roberts, *ibid.* (January 1979), pp. 1–7.
22. P. A. Bex, *ibid.* (August 1979), pp. 277–283.
23. M. D. Horton, in *High Pressure Science and Technology, 6th AIRAPT Conference, July 1977, Boulder, Colorado*, K. D. Timmerhaus and M. S. Barber, Eds. (Plenum, New York, 1979), vol. 1, pp. 923–930.
24. Y. Notsu, T. Nakojima, N. Kawai, *Mater. Res. Bull.* **12**, 1079 (1977).
25. _____, *ibid.* **14**, 1065 (1979).
26. K. J. Dunn and M. Lee, *J. Mater. Sci.* **14**, 882 (1979).
27. M. Lee and L. E. Hibbs, Jr., in *Wear of Materials—1979*, K. C. Ludema, W. A. Glaeser, S. K. Rhee, Eds. (American Society of Mechanical Engineers, New York, 1979), pp. 485–491.
28. M. D. Dennis and J. D. Christopher, Technical Paper MR75-986, 17th Annual Seminar of Society of Manufacturing Engineers, Kansas City, Mo., 17 May 1975.
29. T. N. Loladze and G. V. Bockuchava, in *Proceedings of the International Grinding Conference, 18–20 April*, M. C. Shaw, Ed. (Carnegie-Mellon Univ. Press, Pittsburgh, 1972).
30. A. Furst, *Am. Met. Mark. Metalwork. News Ed.* **87** (No. 249), (24 December 1979).
31. A. N. Pilyankevich and N. Claussen, *Mater. Res. Bull.* **13**, 413 (1978).
32. F. R. Corrigan, in *High Pressure Science and Technology, 6th AIRAPT Conference, July 1977, Boulder, Colorado*, K. D. Timmerhaus and M. S. Barber, Eds. (Plenum, New York, 1979), vol. 1, pp. 994–999.
33. F. W. Krumrei, Industrial Diamond Association of Japan, 30th Anniversary Meeting and Seminar, Tokyo, 17 May 1978.
34. Specialty Materials Department, General Electric Company, Worthington, Ohio 43085, Case History 107.
35. _____, Case History 202.
36. T. Ford and E. Krumrei, *Am. Mach.* (July 1976).
37. Specialty Materials Department, General Electric Company, Worthington, Ohio 43085, Case History 113.
38. Product data sheet SMD 85-324, General Electric Company, December 1979.
39. P. W. Bridgman, *J. Appl. Phys.* **12**, 461 (1941).
40. A. S. Balchan and H. G. Drickamer, *Rev. Sci. Instrum.* **32**, 308 (1961).
41. F. P. Bundy, *J. Chem. Phys.* **38**, 631 (1963).
42. C. E. Weir *et al.*, *J. Res. Natl. Bur. Stand. Sect. A* **63**, 55 (1959).
43. K. J. Dunn, *J. Appl. Phys.* **48**, 1839 (1977).
44. _____ and F. P. Bundy, *ibid.* **49**, 5865 (1978).
45. A. L. Ruoff, in *High Pressure Science and Technology, 6th AIRAPT Conference, July 1977, Boulder, Colorado*, K. D. Timmerhaus and M. S. Barber, Eds. (Plenum, New York, 1979), vol. 2, p. 525.
46. H. K. Mao and P. M. Bell, *Science* **200**, 1145 (1978).
47. F. P. Bundy, *Rev. Sci. Instrum.* **46**, 1318 (1975).
48. _____ and K. J. Dunn, 7th AIRAPT International Conference on High Pressure, LeCreusot, France, August 1979, paper E IV 2.
49. T. R. Loree, C. M. Fowler, E. G. Zukas, F. S. Minshall, *J. Appl. Phys.* **37**, 1918 (1966).
50. K. J. Dunn and F. P. Bundy, *J. Chem. Phys.* **67**, 5048 (1977).
51. I. V. Berman, Zh. I. Binzarov, P. Kurkin, *Sov. Phys. Solid State* **14**, 2192 (1973).
52. F. P. Bundy and K. J. Dunn, in preparation.

promising development areas and the
prospects for achieving much higher
magnetic fields.

High-Field, High-Current Superconductors

J. K. Hulm and B. T. Matthias

Superconductors are electrical conductors in which electrical resistance vanishes and ohmic dissipation ceases below a critical temperature (T_c), which is typically a few degrees above absolute zero. This article is concerned with a special class of superconductors which have the unusual property of retaining their zero resistance state while simultaneously passing a high electrical current density ($\sim 10^6$ amperes per square centimeter) and being subjected to a high magnetic field (up to 50 teslas). In the jargon of this subject, these materials are known as high J_c, type II materials.

The coming availability of steady fields as high as 50 T has profound implications for both science and technology. Many new and exciting experiments will be possible in this frontier region (1). High-field superconducting magnet technology is already affecting other research fields—for example, biochemistry (nuclear magnetic resonance in high fields) and fundamental particle physics (new high-field accelerators with 100 percent duty cycles). The new technology is also being applied to large electrical generators, magnetic fusion reactors, and magnetohydrodynamic power systems.

Summary. This article deals with superconducting materials which have zero electrical resistance while carrying high electrical current densities (around 10^6 amperes per square centimeter) in high magnetic fields (up to 50 teslas). The technological importance of these materials is due to their use in the windings of loss-free electromagnets which generate high magnetic fields. Such magnets are the foundation for superconducting electrotechnology, a rapidly growing field whose applications include advanced electrical machines and fusion reactors. The article focuses primarily on the materials aspects of this new techology. A brief overview is given of the physical principles which underlie this special type of superconducting behavior, and some of the important basic parameters are examined. The technology required to adapt the materials to electromagnets is also discussed. A few concluding remarks concern future possibilities for materials that can be used in generating very high magnetic fields.

Why are high J_c, type II materials important? Quite simply, it is because they permit a major advance in electrotechnology. Present-day electric power technology is built around the copper-iron electromagnet, which dates back to the work of Volta, Oersted, and Faraday in the early years of the 19th century. Since the magnetic induction of iron saturates at about 20,000 gauss ($\equiv$ 2 T, where T is the tesla), this magnetic field level constitutes the practical upper limit for conventional power technology (Fig. 1). However, by using high J_c, type II superconductors to construct magnet windings, the 2-T barrier can be surpassed in a dramatic fashion. Superconducting magnets have already attained 18 T. It appears to be feasible to reach at least 50 T with known materials.

Although most of these devices are still in the early development stage, the era of superconducting electrotechnology seems to be dawning.

In view of the materials emphasis of this article, it is not possible to describe the wide range of superconducting electrical equipment which is now under development. Fortunately, the electromagnet constitutes a common element for most of these devices. We will therefore focus our discussion on superconducting magnet materials. We begin with an overview of the scientific principles underlying high J_c, type II superconducting behavior. This is followed by a description of the type of material necessary for magnet conductors and the various technologies required for conductor fabrication. Finally, we treat some of the more

General Principles

Superconductivity was discovered by Kamerlingh Onnes in 1911 (2), but high J_c, type II materials did not emerge clearly until 50 years later. A brief outline of the intervening events may help to clarify the differences between various types of superconducting material.

Onnes (2) observed the vanishing of resistance in pure metals at a critical temperature characteristic of each material—for example, mercury at 4.2 K, lead at 7.2 K, and so on. He later discovered that at a temperature well below T_c, the application of a magnetic field of a few hundred gauss would entirely quench superconductivity and restore the full normal state resistance. In pure metals, this quenching phenomenon occurred quite suddenly at a well-defined critical magnetic field (H_c). Such behavior is now classified as type I superconductivity.

In the 1920's it was discovered that metallic alloys did not exhibit the sharp magnetic quenching characteristic of pure metals. In particular, de Haas and Voogd (3) found that lead-bismuth alloys consisted of a mixture of superconducting and normal material at fields from a few hundred gauss up to 20,000 gauss. The effect was regarded as a metallurgical artifact caused by failure to achieve homogeneous samples; since these alloys also quenched into the normal state at low current densities ($\sim 10^2$ A/cm²), no technological interest developed.

Theoretical work (4, 5) in the Soviet Union after World War II led to an alternative explanation for the broad "mixed state" transition of alloys. It was proposed that for alloy materials, the surface energy between the superconducting and normal regions could be negative, which would encourage the formation of a mixture of superconducting and normal regions over a wide range of fields. This condition was characterized as superconductivity of the second kind, now more commonly labeled type II behavior.

During this same period, experimental work on transition metal compounds in the United States led to the discovery of

J. K. Hulm is manager of the Chemical Sciences Division, Westinghouse Electric Corporation Research and Development Center, Pittsburgh, Pennsylvania 15235. B. T. Matthias is director of the Institute for Pure and Applied Physical Sciences, University of California, San Diego, La Jolla 92093, and a member of the technical staff at Bell Laboratories, Murray Hill, New Jersey 07974.

the occurrence of high critical temperatures in the A15 crystal structure, in particular V_3Si (6) ($T_c = 17$ K) and Nb_3Sn (7) ($T_c = 18$ K). In 1961, Nb_3Sn (8) was found to be not only unquenchable at 88 kilogauss but also capable of simultaneously carrying a supercurrent density in excess of 10^5 A/cm².

The discovery of the remarkable properties of Nb_3Sn stimulated a great deal of work on type II superconducting behavior, from which the following picture has emerged. High magnetic field superconductivity occurs for those high T_c superconducting materials which are characterized by a relatively short coherence length in the superconducting state. The coherence length, ξ, is a measure of the maximum spatial rate of change of the superconducting order parameter. It is closely related to the size of the ground state wave functions in the superconductor. Coherence lengths for typical high-field materials are of the order of 100 angstroms or less, whereas considerably larger values are typical of pure metal superconductors such as In, Pb, or Sn.

Another important parameter is the penetration depth, λ, or the thickness of a surface layer carrying electric currents which prevent penetration of the magnetic field into the interior of the superconductor. For a high-field material, λ is

significantly greater than the coherence length, which gives rise to the negative interface energy between superconducting and normal regions mentioned earlier. A direct result is that at a lower critical field, H_{c1}, it becomes energetically favorable for the material to enter the mixed state, exhibiting the magnetization curve shown in Fig. 2.

Below H_{c1}, the material is a pure superconductor in the "Meissner state," in which, except for the penetration depth, the magnetic field is excluded from the interior of the material. Above H_{c1}, individual quantized flux bundles, each with a magnitude of 2×10^{-7} G-cm², penetrate the interior of the material. Each fluxoid is surrounded by a cylindrical supercurrent vortex, which is associated with the decay of magnetic field from the center of the fluxoid to its boundary. With increasing field, more and more fluxoids enter the material, forming a periodic array or lattice. At very high fields, the fluxoids become quite closely packed (B in Fig. 2) so that there is hardly any decay of field in the region between fluxoids. Thus, the material is almost completely penetrated by the field; the diamagnetic magnetization approaches zero, and at H_{c2} the supercurrent vortex structure collapses. The field H_{c2} is proportional to $1/\xi^2$ and can

become very large as the coherence length drops below 100 Å. Values of H_{c2} as high as 60 T have been reported (9).

Materials which satisfy the type II condition, $\lambda/\xi > \frac{1}{2}$, fall into two classes, intrinsic and impurity-dominated (10). A useful approximate criterion for intrinsic type II behavior is that T_c exceeds 8 K. In principle, any superconductor can be converted to impurity-dominated type II behavior if, by the addition of impurities or other "defects," the normal state resistivity can be raised to a sufficiently high value.

It should be realized that high-field superconductors are a subgroup of type II superconductors in which H_{c2} happens to be particularly high. The origin of a high H_{c2} value may be intrinsic, impurity-dominated, or include both effects acting in unison.

In addition to high T_c and H_{c2} values, other properties are essential for a good high-field magnet conductor. A carefully annealed homogeneous single crystal of high H_{c2} material is not useful as a magnet material, primarily because the fluxoid structure of such a material is only weakly tied to the crystal lattice. If a current is passed through such a material in the mixed state and a small threshold current density J_c (< 1 A/cm²) is exceeded, the Lorentz force $\mathbf{J} \times \mathbf{B}$ between the current and the fluxoids causes the latter to move through the lattice. Such flux flow produces observable resistance and power loss, which is believed to be connected with the motion of the fluxoid cores (11).

Fluxoid motion cannot be prevented above J_c, but J_c itself can be greatly increased through the deliberate introduction of imperfections into the crystal lattice. Dislocations, precipitates, and grain boundaries all play a role in enhancing J_c by "pinning" fluxoids to the atomic lattice (12). Figure 3 shows typical values of J_c at various fields for the two most widely used magnet materials.

That the levels of J_c indicated by Fig. 3 are reasonable for magnet performance can be seen as follows. The internal field of a long solenoid is approximately fJt, where J is the conductor current density, t is the winding thickness, and f is the packing ratio, which allows for insulation, cooling, mechanical support, and other non–current-carrying features of the winding. Experience dictates maximum f values around 0.1, so that for $J_c = 10^5$ A/cm², it is possible to reach 10 T with a solenoid of thickness 10 cm. Somewhat thicker windings can be endured at this field level, but costs and cryogenic requirements rapidly become

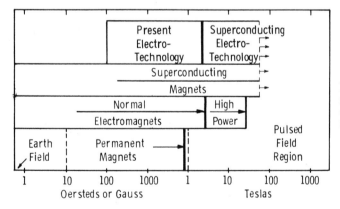

Fig. 1. Comparison of normal and superconducting electro-technology.

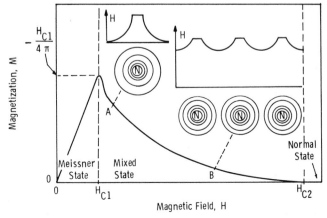

Fig. 2. Magnetization versus field for low J_c, type II superconductor.

objectionable if t is increased above 1 meter.

It might appear at first sight that much higher values of J_c would be beneficial in enabling thinner solenoids to generate the same field strength. Unfortunately, some difficulties arise in magnets as J_c is increased. To illustrate this, we consider a typical magnetization curve for a high J_c, type II material, as shown in Fig. 4.

Characteristic of the behavior of high J_c material is the occurrence of an additional hysteretic component of magnetization which is approximately equal to $\pm J_c r$, where r is the sample radius. The sign is plus for increasing field and minus for decreasing field. The extra magnetization is due to the fact that for a high J_c material, the magnetic field induces large-scale circulating currents in the bulk of the mixed state, which oppose the change of field. These induced currents rise to the critical value J_c; with further increase of magnetic field, dissipative flux flow occurs, the shielding currents collapse, and the magnetization moves toward the equilibrium type II value.

A type II material carrying a current of density J_c is said to be in the critical state (13). Unfortunately, this is a rather unstable state, primarily because any attempt to change H will cause dissipative flux motion and consequent heating. The rising temperature further depresses J_c locally, which causes more flux flow and more heating. This feedback mechanism can cause spontaneous collapse of the critical state magnetization (14), as illustrated by the oscillations in Fig. 4. The magnetization collapses suddenly and then gradually builds back up again as H changes. This phenomenon is known as flux jumping.

Most early high-field magnets experienced premature normalization difficulties (15). At an excitation current well below J_c, the self-field of the magnet induced flux jumps in the winding, releasing heat locally and causing a short length of conductor to enter the normal state. This normal zone then propagated through the entire winding, fed by the stored energy of the magnet.

Since the critical state magnetization term is proportional to the radius of the superconducting specimen, the energy released per unit volume by collapse of this type of magnetization within a magnet is also dependent on the radius of the superconducting wire. The flux jump problem can therefore be ameliorated by reducing the size of the superconductor (16). Magnet conductors are now constructed from fine filamentary super-conductors, separated from each other by normal metal, which serves to decelerate the transition to the normal state. The filament diameter is usually less than 50 micrometers.

Summarizing the discussion up to this point, there are three primary requirements which a high-field superconducting magnet conductor must satisfy. First, the material should exhibit relatively high T_c and H_{c2} values, at least exceeding 8 K and 10 T, respectively. Second, it must be susceptible to a method of preparation which results in a high concentration of defects, without appreciable reduction of T_c or H_{c2}. Third, but no less important, the preparation technique must parallel the superconductor with a pure metal stabilizer such as copper; fine subdivision of the superconductor is also desirable.

In addition to these primary requirements, there are other magnet design parameters which affect the conductor design. For example, the winding must withstand large mechanical stresses due to $\mathbf{J} \times \mathbf{B}$ forces. Thus, if possible, the conductor itself should be strong and its superconducting properties should not be appreciably degraded by stress. It may be necessary to use other high-strength materials as bracing or packing elements in the winding, at the expense of lower f. These may also be in conflict with the basic need to expose the conductor to maximum cooling by liquid helium. It is no exaggeration to say that good magnet design requires the best skills and cooperation of mechanical, electrical, and heat transfer engineers, combined with special demands on materials engineering for the conductor itself.

We will now turn our attention to materials which either satisfy the three primary conductor requirements or promise even more attractive properties in the future.

High T_c

All of the presently known high T_c superconductors were discovered exclusively by experimental investigation of the occurrence of superconductivity throughout the periodic system (Fig. 5). Niobium, at the center of the transition

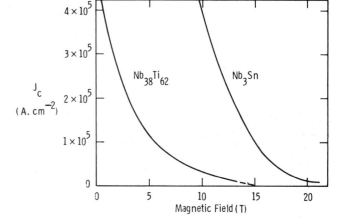

Fig. 3. Critical current density as a function of field, $Nb_{38}Ti_{62}$ and Nb_3Sn.

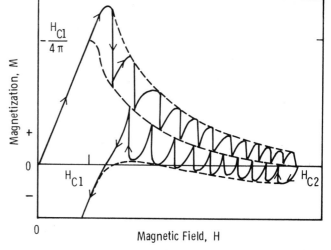

Fig. 4. Magnetization versus field for a high J_c type II superconductor.

77

metal group, exhibits the highest T_c value for an element, 9.25 K. It should be noted that many elements become superconducting at ultrahigh pressures, usually by the formation of new phases. Most of these are not stable at atmospheric pressure and thus at present have little technological value.

Studies of superconducting compounds were started before World War II by Meissner, Justi, and their co-workers (17, 18) who examined the carbides and nitrides of the transition metals. They managed to find several compounds with T_c above 10 K and in one case, NbN, just over 15 K.

In 1950 we began an effort to determine the critical conditions for the occurrence of superconductivity and, in particular, high T_c values. Our experiments led to the formulation of a rather clear-cut criterion. For high T_c among materials based primarily on the transition metals, the average electron concentration per atom, e/a, is the primary parameter. The definition of e/a is very simple. Counting all electrons outside a filled shell will give e/a for a single element, whereas for compounds the arithmetic average has to be taken. There are certain critical values of e/a which favor high T_c—for example, 4.7 and 6.4. The rule is valid for elements, their mutual solid solutions, and compounds formed between them (19). This e/a criterion is in general valid only for phases containing one or two elements.

Our experiments led to the discovery of superconductivity in the A15 structure (6), which has since yielded four distinct compounds with T_c above 18 K, as listed in Table 1. High T_c values are also exhibited by certain pseudobinary A15's (20).

The clear superiority of A15 compounds over all other contenders, at least as far as high T_c is concerned, has focused great attention on this structure. A test of the e/a rule for the known superconducting A15's is illustrated in Fig. 6, where the peaks mentioned above are apparent. The curves of Fig. 6 are essentially boundaries to the data. It seems that e/a values of 4.7 and 6.4 favor, but, because of various side effects, do not ensure high T_c values.

Most of the A15 superconductors with T_c values below about 19 K (Table 1) can be prepared by direct melting of the constituents. The materials with higher critical temperatures all require special techniques of synthesis. For example, T_c for the unique compound Nb_3Ge was brought up to 23 K by sputtering a thin film of the material onto a heated substrate (21). In the case of Nb_3Ga, quenching is necessary to achieve high

Table 1. Superconducting binary and pseudobinary compounds having the A15 crystal structure.

Compound	T_c (K)	$H_{c2}(0)$ (T)	$-(dH_{c2}/dT)$, $T = T_c$ (TK^{-1})
Nb_3Sn	18.0	29.6	2.4
Nb_3Al	18.7	32.7	2.5
Nb_3Ga	20.2	34.1	2.4
Nb_3Ge	23.2	37.1	2.4
$Nb_3Al_{0.7}Ge_{0.3}$	20.7	44.5	3.1
$Nb_3Al_{0.5}Ga_{0.5}$	19.0	31.6	2.4
V_3Si	17.0	34.0	2.9
V_3Ga	14.8	34.9	3.4

T_c (22). These various compounds are also susceptible to depression of T_c by disorder or by deviations from stoichiometry, which does not ease the problems of magnet conductor fabrication.

Future increases in T_c would appear to depend heavily on further advances within the A15 structure. The e/a correlation suggests that other intermetallic compositions might exhibit high T_c values if only they could by synthesized in the A15 structure. A good example is Nb_3Si, which does not form an A15 phase with conventional preparation (23). By explosive compression to megabar pressures, superconductivity has recently been reported in Nb_3Si between 18 and 19 K (24). Novel methods of preparation such as this may yield metastable forms of higher T_c material in binary or pseudobinary systems and are certainly to be encouraged.

High H_{c2}

Upper critical field data for several important high-field superconductors are plotted against temperature in Fig. 7. Both $Nb_{38}Ti_{62}$ and Nb_3Sn have been available in filamentary composite form for some time and are widely used. The other materials in Fig. 7 have not yet been developed as filamentary conductors, and consequently the highest field generated by a superconducting magnet to date is about 18 T. In the face of the unexploited materials potential shown in Fig. 7, to advocate further advances in H_{c2} may seem to be gilding the lily. Nevertheless, the prospect of a megagauss material is extremely exciting and is by no means ruled out by present knowledge.

We have already noted the subdivision of type II materials into two classes, the pure or intrinsic type and the impurity-dominated type. From Gorkov's work (10), it is found that $H_{c2}(0)$ (that is, H_{c2} at $T = 0$) for intrinsic materials is proportional to $(\gamma T_c)^2$, where γT is the linear

term in the normal state heat capacity at low temperatures. Similarly, $H_{c2}(0)$ for impurity-dominated material is proportional to $\rho_n \gamma T_c$, where ρ_n is the electrical resistivity in the normal state. With some approximations, the intrinsic and impurity regions can be separated by plotting $H_{c2}(0)$ against γT_c, as in Fig. 8.

The bottom curve of Fig. 8 denotes the intrinsic type II case, which seems to be well represented by pure Nb_3Sn (25). Each dashed curve in Fig. 8, labeled according to a specific level of normal resistivity in microhm-centimeters, represents the sum of the intrinsic and impurity components for that level of resistivity. The numbers in parentheses are the actual normal resistivities for the materials. A higher resistivity sample of Nb_3Sn is compared with the pure material. The increase in H_{c2} caused by impurities is small because of the downward shift of γT_c. It is well known that $Nb_{38}Ti_{62}$ is impurity-dominated, and $PbMo_6S_8$ is probably of the same type. Thus, the record high H_{c2} value of this Chevrel phase compound is primarily due to its unusually high electrical resistivity (26), which overcomes the handicap of a T_c value somewhat below that of the A15 compounds.

The model on which Fig. 8 is based ignores other possible difficulties in the search for higher critical fields. The most famous of these is the paramagnetic limit (27, 28), which suggests that type II behavior will not be energetically favored above a field, H_p, such that the magnetization energy associated with the spins of conduction electrons in the normal state exceeds the superconducting condensation energy. Experience has shown that there is, indeed, an appreciable paramagnetic lowering of $H_{c2}(0)$ in superconductors formed from elements in the first long period, such as Ti and V. However, the effect is less noticeable in materials formed predominantly from elements in the second and third long periods, such as Nb and Mo. This is usually attributed to the effect of spin-orbit scattering, which causes spin depairing in the superconducting ground state and thus tends to equalize the spin magnetization energy terms between the superconducting and normal states. Spin-orbit scattering is known to increase with increasing atomic mass, which is consistent with the experimental data mentioned above.

As Fig. 8 indicates, there is considerable scope for further advances in $H_{c2}(0)$, provided ways can be found to increase T_c, γ, or ρ_n. The normal resistivity offers perhaps the easiest line of attack on this problem.

High J_c Composites

Shortly after it was realized that conductors for magnets could be stabilized against flux jumping by means of fine filamentary construction, successful conductors of this type were produced from the Nb-Ti alloy system. This material is extremely ductile, which greatly facilitates the manufacturing process.

To produce this type of conductor, rods of Nb-Ti alloy are inserted into holes drilled in a large pure copper billet. The billet is then hot-extruded down to a size suitable for wiredrawing, which usually involves a reduction of about 50 to 1. Cold drawing is then carried out, with perhaps a few intermediate anneals, until the Nb-Ti filaments reach the desired size, usually less than 50 μm. The process ends with a long anneal at 375°C, which causes a second phase precipitation accompanied by a rearrangement of the dislocation cell structure. This results in a substantial increase of J_c, to levels around those shown in Fig. 3.

A monolithic, multifilamentary Nb-Ti conductor is shown in Fig. 9 (top conductor). This material contains about 1000 superconductor filaments 50 μm in diameter which have been exposed by etching away the copper matrix. The current-carrying capacity of this small conductor at 4.2 K and 5 T is about 10,000 A, which is comparable with the capacity of the larger bar of plain copper (bottom conductor in Fig. 9) at room temperature and 2 T in the field coil of a large generator.

Conductors based on Nb-Ti are generally excellent for applications up to about 8 T, but the T_c and H_{c2} performance of these alloys (Figs. 3 and 7) is quite limited. Regrettably, we know of no other ductile alloy systems which exhibit appreciably better critical superconducting parameters at present. The Nb-Ti lies close to the 4.7 e/a peak (29) and has a ρ_n of about 60 μohm-cm (30). There are ductile alloys near the 6.4 e/a peak, such as Mo-Re, but here the normal resistivity is much lower and gives rise to a disappointingly low H_{c2}. There seems to be a basic conflict between ductility and high T_c or H_{c2}.

To build magnets for fields appreciably above 8 T, intermetallic compounds must be used. Intermetallic superconductors cannot be drawn directly into wire, because of their extreme brit-

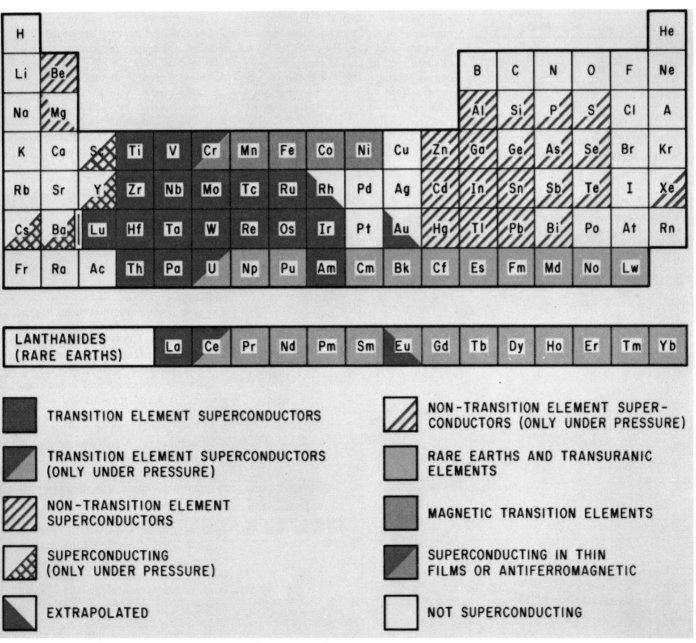

Fig. 5. The periodic system of superconductors.

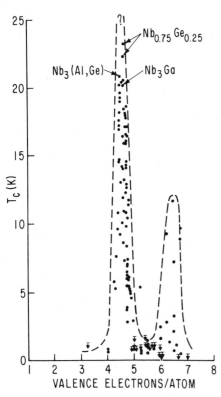

Fig. 6. Critical temperature as a function of electron-to-atom ratio for superconducting A15 compounds (36).

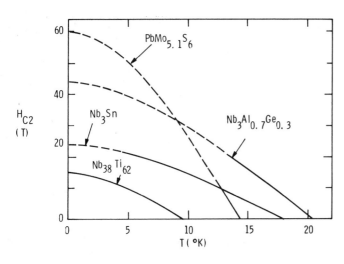

Fig. 7. Upper critical field versus temperature for high-field superconductors; solid curve, steady field data; dashed curve, pulsed field data.

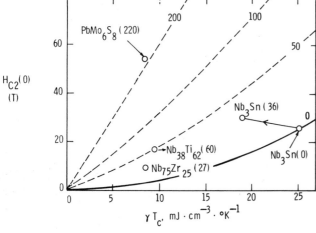

Fig. 8. Upper critical field as a function of γT_c for high-field superconductors.

tleness. Thus early efforts to produce conductors were focused on Nb_3Sn tapes produced either by directly reacting Nb with molten Sn or by depositing both components at high temperatures from metal halide vapors (chemical vapor deposition). Such tapes are useful for small working volume, high-field magnets; in larger systems the superconductor needs to be subdivided for stability.

Useful subdivided composites of V_3Ga and Nb_3Sn were first achieved by an ingenious diffusion process (31, 32). In this technique, fine filaments of Nb metal are embedded within a bronze (Cu-Sn) matrix, in a manner analogous to that already described for Nb-Ti filaments in copper. After drawing is completed, a high-temperature heat treatment causes tin to migrate out of the bronze into the niobium filament, forming an Nb_3Sn layer with an Nb core. One drawback for stabilization is that the residual copper matrix is not highly conducting because of its tin impurities. However, this can be partly corrected by including pure copper regions in the parent matrix, segregated from the bronze by means of a tantalum barrier. The J_c-H data for

Nb_3Sn in Fig. 3 were obtained for a diffusion-reacted composite of this type.

Multifilamentary composite conductors of both Nb_3Sn and V_3Ga are now commercially available. Unfortunately, the process requires special features in the phase diagram of the A15 compound which are not present in the case of the other high-field A15 binary and pseudo-binary systems of Table 1. Some other method must be found to achieve a suitable fabrication technique for these compounds.

One approach is to abandon the linear filament process and to attempt, instead, to produce a three-dimensional network of superconducting material with its interstices filled with stabilizing material (33). Several concepts have been tried to produce such a network—for example, straight powder metallurgy, liquid infiltration into a partially sintered niobium rod, and the so-called in situ process. The last method depends on the fact that niobium and copper are completely soluble in each other in the liquid phase, but have negligible mutual solubility in the solid phase, with no intermediate phases. If a suitable liquid mixture is cooled down, niobium is first precipitated out, followed by a nearly pure copper matrix. Tin can be added to the melt or infiltrated later, with a final high-temperature heat treatment to produce the A15 network. This fabrication technique has already yielded Nb_3Sn composites which have overall J_c values comparable to those obtained with the diffusion process. However, it is likely to have the same limitations as the diffusion process with respect to the types of A15 compound that can be handled. Powder metallurgy or liquid infiltration seems to provide the best hope for obtaining composites for the A15 compounds with higher H_{c2} values.

Whichever form of composite is considered, the superconductor J_c must approach 10^5 A/cm² for practical magnet construction. Nb-Ti alloys achieve this level by a combination of cold work and heat treatment; the fluxoids are pinned by dislocation clusters. For A15 compounds, the fluxoids are probably pinned by grain boundaries; it is fortunate that most of the processing techniques discussed above can be manipulated to yield small grain sizes and adequate J_c levels. It is not clear that this natural dispensation will extend to other materials of interest for use at very high fields; indeed, preliminary work on the Chevrel phases (34) has yielded poor J_c levels so far.

At present, the majority of super-

Fig. 9. Multifilamentary Nb-Ti composite conductor; beneath it is a copper bar.

conducting magnets in use or under development utilize some form of Nb-Ti multifilamentary composite. A15 tapes are utilized in a few high-field research magnets. A very large magnet, using an Nb_3Sn filamentary composite, is currently being developed by Westinghouse and Airco for Oak Ridge National Laboratory. This magnet is part of an experimental prototype system for a future tokamak fusion test reactor. It will offer a severe test of the viability of A15 composites.

Future Possibilities

Major efforts are now in progress to build large magnets in the range 5 to 12 T for application to fusion systems, high-energy particle accelerators, magnetohydrodynamic power ducts, electrical generators, and other large electrical power devices. By large, we mean a working volume of at least 1 cubic meter or a stored energy in excess of 1 megajoule, or both. Considerably less effort—indeed, hardly any at all—is being devoted to pushing into the unexplored zone above about 18 T. It must be admitted that much of the magnet technology which is being developed for the lower-field range will be extremely useful above 18 T. Unfortunately, this will not in itself be sufficient to open up the frontier.

Specific development work is needed on superconducting composite con-ductors suitable for use above 18 T. There are several promising candidates for this purpose, including the A15 compounds Nb_3Ge and $Nb_3Al_{0.7}Ge_{0.3}$ and Chevrel phase compounds such as $PbMo_{5.1}S_6$. The requirements include fluxoid pinning strength sufficient to yield J_c values above 10^5 A/cm², pure metal cladding or bonding to the conductor, and, if possible, fine subdivision of the superconductor. This program seems to present an exciting challenge to materials scientists and engineers.

Behind the conductor development program lies the more fundamental question of improving the basic superconducting parameters, particularly T_c and H_{c2}. We have indicated some of the possible directions of attack in this area. The possibility of forming new compounds in the transition metal region of the periodic system is by no means exhausted, particularly in view of the growing spectrum of new techniques of synthesis which have appeared in recent years. In particular, the field of true ternary compounds is in its infancy, not yet a decade old, but is already very promising.

In the case of T_c, past progress has depended entirely on experimental exploration for new materials. Theory has had very little to contribute, mainly offering explanations after the event rather than predictions beforehand. We see no reason for a change in this situation.

As regards H_{c2}, early experimental studies of low H_{c2} alloys (35) provided the basis for Abrikosov's remarkable theory of type II behavior. However, the discovery of high-field, high-current superconductors was made independently of theory and came about primarily from materials exploration. The Abrikosov-Gorkov theory has certainly provided excellent guidelines for further work in this area.

We have noted that high-field, high-current superconductors have opened up a new regime of electric power technology (Fig. 1). This regime will be further advanced if better materials can be found. In our view, there is a reasonably good prospect for a 30 K superconductor and a 1-megagauss superconductor in the future, either separately or perhaps together in the same material. It would seem to be of value to both science and technology to pursue these goals.

References and Notes

1. *High Magnetic Field, Research and Facilities* (National Academy of Sciences, Washington, D.C., 1979); see also *Phys. Today* 32, 21 (September 1979).
2. H. Kamerlingh Onnes, *Leiden Commun. No. 119B* (1911); *ibid.*, No. 120B; *ibid.*, No. 122B.
3. W. J. de Haas and J. Voogd, *ibid.*, No. 199C (1929); *ibid.*, No. 214B (1931).
4. V. L. Ginzburg and L. D. Landau, *Zh. Eksp. Teor. Fiz.* 20, 1064 (1950).
5. A. A. Abrikosov, *Sov. Phys. JETP* 5, 1174 (1957).
6. G. Hardy and J. K. Hulm, *Phys. Rev.* 87, 884 (1953).
7. B. T. Matthias, T. H. Geballe, S. Geller, E. Corenzwit, *ibid.* 95, 1435 (1954).
8. J. E. Kunzler, E. Beuhler, F. S. L. Hsu, J. H. Wernick, *Phys. Rev. Lett.* 6, 89 (1961).
9. S. Foner, E. J. McNiff, E. J. Alexander, *Phys. Lett. A* 49, 269 (1974).
10. L. P. Gorkov, *Sov. Phys. JETP* 9, 1364 (1960).
11. Y. B. Kim, C. F. Hempstead, A. R. Strnad, *Phys. Rev.* 129, 528 (1963).
12. P. W. Anderson, *Phys. Rev. Lett.* 9, 309 (1962).
13. C. P. Bean, *ibid.* 8, 250 (1962).
14. S. L. Wipf and M. S. Lubell, *Phys. Lett.* 16, 103 (1965).
15. J. K. Hulm, B. S. Chandrasekhar, H. Riemersma, *Adv. Cryog. Eng.* 8, 17 (1963).
16. C. Laverick, *Proc. Int. Symp. Magnet. Technol. Stanford* (1965), p. 560.
17. W. Meissner and H. Franz, *Z. Phys.* 65, 30 (1930).
18. G. Aschermann, E. Friederich, E. Justi, J. Kramer, *Phys. Z.* 42, 349 (1941).
19. B. T. Matthias, *Phys. Rev.* 97, 74 (1955).
20. ———, T. H. Geballe, L. D. Longinotti, E. Corenzwit, G. W. Hull, R. H. Willens, J. P. Maita, *Science* 156, 645 (1967).
21. J. R. Gavaler, *Appl. Phys. Lett.* 23, 480 (1973).
22. G. W. Webb, L. T. Vieland, R. E. Miller, A. Wicklund, *Solid State Commun.* 9, 1769 (1971).
23. D. Dew-Hughes and L. T. Luhman, in *Treatise on Materials Science and Technology* (Academic Press, New York, 1979), vol. 14, p. 437.
24. V. M. Pan *et al.*, *JETP Lett.* 21, 228 (1975).
25. T. P. Orlando, E. J. McNiff, S. Foner, M. R. Beasley, *Phys. Rev. B* 19, 4545 (1979).
26. Ø. Fischer, *Appl. Phys.* 16, 1 (1978).
27. A. M. Clogston, *Phys. Rev. Lett.* 9, 266 (1962).
28. B. S. Chandrasekhar, *Appl. Phys. Lett.* 1, 7 (1962).
29. J. K. Hulm and R. D. Blaugher, *Phys. Rev.* 123, 1569 (1961).
30. T. G. Berlincourt and R. R. Hake, *ibid.* 131, 140 (1963).
31. K. Tachikawa and T. Tanaka, *Jpn. J. Appl. Phys.* 6, 782 (1967).
32. A. R. Kaufmann and J. J. Pickett, *J. Appl. Phys.* 42, 58 (1971).
33. R. Roberge, *J. Magn. Magn. Mater.* 11, 182 (1979).
34. D. Dew-Hughes and L. T. Luhman, in *Treatise on Materials Science and Technology* (Academic Press, New York, 1979), vol. 14, pp. 435–436.
35. L. W. Schubnikow, W. I. Chotkewitsch, J. D. Schepelew, J. N. Rjabinin, *Phys. Z. Sowjetunion* 10, 165 (1936).
36. B. W. Roberts, *Natl. Bur. Stand. U.S. Tech. Note 825* (Suppl.) (1974), p. 7.
37. Work at La Jolla was supported by NSF grant DMR 77-08469.

New Magnetic Alloys

G. Y. Chin

Since the turn of the century, a large number of new magnetic materials have been developed to meet growing needs in telecommunications, electric power generation, and information processing. Today the value of the U.S. magnetics market is estimated to exceed $2 billion, three times more than a decade ago. We are constantly touched by some aspect

when operated at high induction (1.7 teslas); thus they promise greater economy in the transmission of electrical power.

Other new developments in magnetic materials during the past decade include magnetic bubbles and amorphous alloys, which are treated elsewhere in this issue by Giess (1) and Gilman (2), respective-

Summary. Three notable new developments in magnetic alloys are highlighted. These include rare earth–cobalt permanent magnets with maximum energy products up to 240 kilojoules per cubic meter; chromium-cobalt-iron permanent magnets that have magnetic properties similar to those of the Alnicos, but contain only about half as much cobalt and are sufficiently ductile to be cold-formable; and high-induction grain-oriented silicon steels that exhibit 20 percent less core loss as transformer core materials than conventional oriented grades.

of magnetics—making or receiving a telephone call, switching on a radio or TV set, turning on a washing machine or vacuum cleaner.

In this article three notable developments in magnetic alloys during the past decade are highlighted. These are the rare earth–cobalt and chromium-cobalt-iron permanent magnets and the high-induction, grain-oriented soft magnetic silicon steels. The rare earth–cobalt alloys have intrinsic coercivities more than 20 times and maximum energy products more than four times those of Alnico 5, the most widely used permanent magnet alloy to date. The chromium-cobalt-iron alloys essentially duplicate the magnetic properties of Alnico 5 at less than half the cobalt content of the latter. They also have the advantage of good ductility, which permits them to be cold-rolled, drawn, or stamped into finished shape, whereas most permanent magnets are brittle and must be formed by casting or powder technology. As transformer core materials, the high-induction, grain-oriented silicon steels exhibit a 20 percent decrease in core loss compared with conventional grain-oriented silicon steel

ly. Of course, important advances have also been made in the established magnetic materials; a summary of progress in several areas has recently been given by Jacobs (3).

Rare Earth–Cobalt Permanent Magnets

In the late 1950's and early 1960's, Nesbitt, Wernick, and their co-workers (4) and Wallace and his associates (5) prepared a series of rare earth (R)–transition metal compounds and studied their intrinsic magnetic properties. Among the compounds studied were RCo_5 and R_2Co_{17}, which now form the bases of the most prominent R-Co high-performance permanent magnet systems. Studies by others followed, but it was not until 1966 that R-Co magnet development really took off; in that year Hoffer and Strnat (6) reported on the extremely high magnetocrystalline anisotropy of YCo_5 and emphasized the potential of such compounds as permanent magnets. There are now more than 20 companies worldwide that offer rare earth–cobalt magnets commercially. These alloys are distinguished by their extremely high values of intrinsic coercivity, up to 3 million amperes per meter, and maximum energy

product, $(BH)_{max}$, up to 240 kilojoules per cubic meter. The latter represents the maximum energy storage per unit volume and is the figure of merit most often used for permanent magnet materials. Figure 1 shows the spectacular progress in quality of the rare earth alloys in comparison with other permanent magnet materials over the years.

The RCo_5 compounds have the $CaCu_5$ type of hexagonal crystal structure; the R_2Co_{17} compounds also have a hexagonal crystal structure, with either the Th_2Ni_{17} type or the Th_2Zn_{17} type of modification. Table 1 lists the Curie temperature (T_c) and room-temperature values of the saturation magnetization ($4\pi M_s$) and magnetocrystalline anisotropy constant (K_1) for most of these compounds. The RCo_5 compounds have moderate values of $4\pi M_s$ (~1 T) but extremely large values of K_1 (>1 MJ/m³). The latter property is primarily responsible for the exceptionally large coercivity. In this regard, $SmCo_5$ is the most outstanding of all RCo_5 magnets by virtue of having the largest value of K_1 (17 MJ/m³). The R_2Co_{17} compounds have higher $4\pi M_s$ values than the corresponding RCo_5 series and hence might have greater energy products. However, as Table 1 shows, the magnitude of K_1 for the "2-17" compounds is considerably smaller than that for the corresponding "1-5" series. Furthermore, with the exception of Sm, Er, and Tm, all 2-17 binary compounds have $K_1 < 0$—that is, an easy (0001) plane. For these compounds, magnetization reversal becomes relatively easy and low coercivity is expected. There has been some success at modifying the sign and magnitude of K_1 by partial substitution for Co by other transition metals such as Fe, Cr, and Mn.

The magnetism of the R-Co compounds is due to the interatomic exchange between the spins of the two sublattices plus the spin-orbit coupling within the rare earth atoms. In the lighter rare earth series—Ce, Pr, Nd, and Sm—the spins of the R and Co atoms are aligned parallel. The values of $4\pi M_s$ are thus high. In the others they are aligned antiparallel, and the values of $4\pi M_s$ tend to be low. Yttrium is nonmagnetic and hence the magnetic induction comes from Co alone. The magnetocrystalline anisotropy also comes from two sources, one originating in the itinerant electrons of the Co sublattice and one due to the crystalline electric field of the rare earths. A broad summary ranging from basic magnetism to the technology of the rare earth magnets is given by Menth et al. (7).

The author is head of the Physical Metallurgy and Ceramics Research and Development Department, Bell Laboratories, Murray Hill, New Jersey 07974.

There are at present two dominant groups of R-Co alloys in commercial production. One group, the earlier and more established, is based on single-phase SmCo$_5$; the other, more recent group is based on precipitation-hardened alloys of the Sm$_2$(Co,Cu)$_{17}$ type. The best magnetic properties are obtained in Sm alloys, and these are produced in the greatest quantity.

SmCo$_5$-type single-phase magnets. Table 2 shows some representative magnetic properties of commercial SmCo$_5$ magnets compared with those of other materials. Typical demagnetization curves are shown in Fig. 2. Although values of $(BH)_{max}$ for laboratory samples as high as 200 kJ/m^3 have been reported, typical commercial samples are in the range 130 to 160 kJ/m^3.

Detailed studies have shown that the mechanism of coercivity in single-phase SmCo$_5$ magnets is one of domain nucleation or wall pinning at grain boundaries. Wall motion within the grains is relatively easy. To minimize the existence of domain walls within the grain by spontaneous nucleation, and thus to achieve high coercivity, the alloys must be ground into fine particles (1 to 10 micrometers). The particles are then aligned and compacted in a strong magnetic field and are sintered by powder metallurgical techniques. Plastic-bonded magnets are also in production, either as rigid bodies with thermosetting resins or as flexible parts with thermoplastic resins or rubbers. Energy products are in the range 3 to 10 kJ/m^3.

In addition to those containing Sm, other RCo$_5$ single-phase permanent magnets have been prepared, both in the laboratory and commercially. These include Sm in combination with Ce, Ce-rich misch metal (~55 percent Ce, 25 La, 13 Nd, and 5 Pr), and Gd. The Ce and misch metal additions lower the magnetic properties, but the raw material price is substantially lower than that of Sm. The Gd addition is used to decrease the temperature coefficient of remanence, a useful feature in temperature-stable devices.

Precipitation-hardened Sm$_2$(Co,Cu)$_{17}$ type alloys. Not long after the initial development of single-phase SmCo$_5$-type permanent magnets, Nesbitt *et al.* (8) and Tawara and Senno (9) discovered that partial substitution for Co by Cu in SmCo$_5$ and CeCo$_5$ can lead to high coercivity (to 2 MA/m) on suitable heat treatment near 400° to 500°C. Furthermore, the Cu-substituted magnets do not have to be ground into fine particles to develop magnetic hardness. For this reason, these magnets can be prepared as

Table 1. Magnetic properties of RCo$_5$ and R$_2$Co$_{17}$ compounds; $4\pi M_s$ and K_1 values at 25°C are given (31).

Rare earth	RCo$_5$			R$_2$Co$_{17}$		
	$4\pi M_s$ (T)	T_c (°C)	K_1 (MJ/m^3)	$4\pi M_s$ (T)	T_c (°C)	K_1 (MJ/m^3)
Ce	0.85	374	5.3	1.15	800	−0.6
Pr	1.12	612	8.1	1.38	890	−0.6
Nd	1.20	630	0.7	1.39	900	−1.1
Sm	0.97	724	17.2	1.20	920	3.3
Gd	0.19	735	4.6	0.73	930	−0.5
Tb	0.24	707		0.68	920	−3.3
Dy	0.30	693		0.70	910	−2.6
Ho	0.53	727	3.6	0.83	920	−1.0
Er	0.63	713	3.8	0.90	930	0.41
Tm	0.67	747		1.13	920	0.50
Yb						−0.38
Lu				1.27	940	−0.20
La	0.91	567	5.9			
Y	1.06	648	5.2	1.25	940	−0.34
Th			2.6			−0.53

Table 2. Properties of some permanent magnets.

Material	Remanent induction, B_r (T)	Coercive force, H_c (kA/m)	Maximum energy product, $(BH)_{max}$ (kJ/m^3)	Reference
Alnico 5 (8Al-14Ni-24Co-3Cu-Fe)*	1.28	50	44	(32)
Alnico 9 (7Al-15Ni-35Co-4Cu-5Ti-Fe)	0.82	130	72	(32)
Ceramic 5 [(Ba,Sr)Fe$_{12}$O$_{19}$]	0.38	190	27	(32)
SmCo$_5$	0.87	640/1600†	144	(32)
Sm(Co$_{0.76}$Fe$_{0.10}$Cu$_{0.14}$)$_{6.8}$	1.04	480/500†	210	(10)
Sm(Co$_{0.68}$Cu$_{0.10}$Fe$_{0.21}$Zr$_{0.01}$)$_{7.4}$	1.10	510/520†	240	(11)
23Co-28Cr-1Si-Fe	1.30	46	42	(14)
15Co-23Cr-3V-2Ti-Fe	1.35	44	44	(17)
11.5Co-33Cr-Fe	1.20	60	42	(16)
5Co-30Cr-Fe	1.33	42	42	(28)
70Mn-29.5Al-0.5C	0.56	180	44	(29)

*In the composition, numbers are percentages and the balance is Fe.　　†Intrinsic coercive force, H_{cJ}.

Fig. 1. Progress in permanent magnet quality since 1880 as indicated by the value of the maximum energy product achieved for various material systems. [Adapted from (3)]

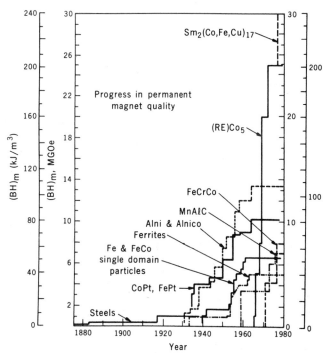

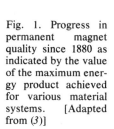

83

cast. However, current commercial practice is still based on powder metallurgy, as magnets so prepared are mechanically stronger and have better magnetic alignment.

Electron microscopy studies have revealed that low-temperature heat treatment of the Cu-substituted magnets leads to homogeneous fine-scale precipitation (~15 nanometers) of a second phase coherent with the RCo_5 structure. Magnetic hardening results from pinning of domain walls at these particles. Hence the coercivity is independent of the sample dimensions.

The $Sm(Co,Cu)_5$ magnets have low values of saturation magnetization because of the copper replacing cobalt. Nesbitt *et al.* (*8*) were able to increase the saturation magnetization by the addition of Fe. In recent years, further increases in saturation magnetization have been made possible by increasing the ratio of (Co,Cu,Fe) to Sm toward the value 17:2 at the sacrifice of coercivity. A $Sm(Co_{0.76}Fe_{0.10}Cu_{0.14})_{6.8}$ alloy, with values of $B_r = 1.04$ T, $H_{cJ} = 500$ kA/m, and $(BH)_{max} = 210$ kJ/m³ (Table 2), is representative of this class (*10*). Very recently, the addition of small amounts of Zr (*11*) or Hf (*12*) has enabled a further increase in saturation magnetization

with an additional increase in the (Co,Cu,Fe):Sm ratio toward the value 17:2 and a further increase in the Fe content. The addition of Zr or Hf stabilizes the coercivity that would otherwise be degraded. As a result, additional increases in energy product were achieved. As shown in Table 2, this class includes a $Sm(Co_{0.68}Cu_{0.10}Fe_{0.21}Zr_{0.01})_{7.4}$ alloy with $B_r = 1.10$ T, $H_{cJ} = 520$ kA/m, and $(BH)_{max} = 240$ kJ/m³ (*11*). The alloy containing Zr has the highest energy product of all permanent magnet alloys offered commercially to date.

Applications. The exceptionally large maximum energy products and coercivities of the rare earth magnets permit their use in devices where small size and superior performance are desired. Magnets for electronic wrist watches and for traveling wave tubes—once the domain of expensive platinum-cobalt magnets—are now largely made of rare earth alloys. Rare earth magnets have been used for a number of medical devices, ranging from thin motors in implantable pumps and valves to holding magnets for artificial teeth and for aiding eyelid motion. Direct-current and synchronous motors and generators have been designed with rare earth magnets with an overall reduction in size. A very succinct summary

has been given by Strnat (*13*). The use of rare earth magnets in automotive accessory motors was recently considered as a means of reducing the size of the motors and the mass of cars and achieving greater fuel economy. If this were done there would be huge demands for such magnets. However, this effort was dealt a serious blow when, in addition to the high cost of samarium, the price of cobalt jumped drastically in 1978 as a result of political instability in Zaire, which supplies 70 percent of the free world's cobalt. The general prognosis is that the price of the rare earth magnets will always be relatively high; however, because of their exceptional magnetic properties, they will be used in a large number of low-volume specialized applications.

Chromium-Cobalt-Iron Permanent Magnets

Overshadowed by the impressive progress made with the rare earth alloys has been the development of another new family of permanent magnet alloys. In 1971 Kaneko and his colleagues at Tohoku University announced the discovery of ductile permanent magnet alloys in the Cr-Co-Fe system (*14*). Since then, continued progress has been made in the development of these alloys, by the Tohoku group and more recently at Bell Laboratories. The magnetic properties of the Cr-Co-Fe alloys are remarkably similar to those of the Alnicos, with values in the ranges $B_r = 1.0$ to 1.3 T, $H_c = 15$ to 60 kA/m, and $(BH)_{max} = 10$ to 45 kJ/m³. Unlike the brittle Alnicos, which must be cast and ground to finished shape or shaped by powder metallurgy techniques, these new alloys are cold-formable at room temperature. Hence normal metal-forming operations such as rolling, wire drawing, and stamping can be done relatively easily. In addition, equivalent magnetic properties can be attained with much less of the expensive constituent cobalt.

Metallurgy and magnetic behavior. The Cr-Co-Fe alloys of interest as permanent magnets are in the range 25 to 35 percent Cr and up to about 25 percent Co. At temperatures exceeding about 1200°C, depending on composition, the alloys exist in the ferromagnetic body-centered cubic (α) phase. In the intermediate temperature range of about 700° to 1200°C, again depending on composition, the α phase tends to coexist with a nonmagnetic face-centered cubic phase, γ. At high-Cr end a brittle σ phase may

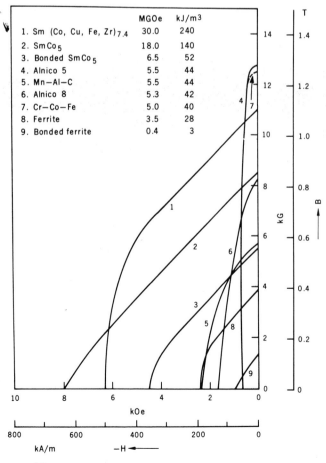

	MGOe	kJ/m³
1. Sm (Co, Cu, Fe, Zr)$_{7.4}$	30.0	240
2. SmCo$_5$	18.0	140
3. Bonded SmCo$_5$	6.5	52
4. Alnico 5	5.5	44
5. Mn−Al−C	5.5	44
6. Alnico 8	5.3	42
7. Cr−Co−Fe	5.0	40
8. Ferrite	3.5	28
9. Bonded ferrite	0.4	3

Fig. 2. Demagnetization curves of selected permanent magnet materials. Values of the maximum energy product are tabulated above the curves. [From (*31*)]

also appear in this temperature range. The α phase is retained when the alloy is rapidly cooled from a high temperature. It then undergoes spinodal decomposition at a lower temperature to a highly ferromagnetic Fe-rich α_1 phase and a less magnetic Cr-rich α_2 phase, both of which maintain the body-centered cubic structure. This decomposition temperature is about 550°C for the Fe-30Cr binary (Fe with 30 percent Cr) and increases to about 650°C for the Fe-30Cr-20Co ternary alloy. Thus a very important function of the addition of cobalt is to raise the decomposition temperature so that decomposition can proceed within practical heat-treating time intervals, since it is the decomposition product that yields the optimum permanent magnet properties.

As revealed by transmission electron microscopy, the mechanism for coercivity in the Cr-Co-Fe alloys is pinning of domain walls by the spinodally decomposed particles. Two very important structural parameters in this case are differences in saturation magnetization as a result of composition differences between the two decomposed phases, and particle size. Both are affected by heat-treatment temperature and time. Jin *et al.* (*15*) recently developed a two-step aging technique that optimizes both these parameters. The technique consists of initial rapid cooling from above the decomposition temperature to a suitable lower temperature to establish an optimum particle size for the decomposed phase of ~50 nm, followed by slower cooling to maximize the composition difference between the two phases.

Elongated and aligned particles are extremely important in increasing the energy product by increasing values of B_r and H_c, and by "squaring up" the shape of the demagnetization curve. The elongation and alignment can be affected by heat treatment in a magnetic field (*14*)—a well-known technique in the production of anisotropic Alnicos. On the other hand, Jin (*16*) took advantage of the ductility of the Cr-Co-Fe alloys to uniaxially deform an alloy that had been deliberately overaged. In this way, he was able to mechanically elongate and align the particles without magnetic field treatment (Fig. 3). As a result, values of $B_r = 1.2$ T, $H_c = 60$ kA/m, and $(BH)_{max} = 42$ kJ/m³—comparable to those of Alnico 5 and several times better than those without the deformation—have been achieved in an Fe-33Cr-11.5Co alloy. Figure 4 shows the improvement in magnetic properties of this alloy through "deformation-aging" and

compares them with those of Alnico 5.

Some representative properties of commercially available materials and laboratory specimens are listed in Table 2. One group is concentrated in the high-Co regime with a typical composition of 23Co-28Cr-1Si-Fe, along the lines of Kaneko's early studies. A small amount of Si (~1 percent) is said to improve the ductility, since formability decreases with Co content. Later, attention shifted to a medium Co content of ~15 percent, often with small additions such as Nb, Al, V, Ti, and Zr—so-called α-formers—to suppress the formation of the undesirable nonmagnetic γ phase. A typical alloy in this range is 23Cr-15Co-3V-2Ti-Fe (*17*). More recently, very favorable permanent magnet properties have been obtained near 10 percent Co (*16*). Unlike some 10 percent Co commercial alloys that were developed earlier and contained additions of a fourth element such as Si and Ti (*18*), the more recent alloys are pure ternaries.

Applications. One major application of the Cr-Co-Fe alloys is in ring armature-type telephone receivers. The current design for this type of receiver, which is produced at the rate of about 10 million annually for all general-purpose handsets, makes use of a cup-shaped permanent magnet to provide the required d-c bias magnetic field. Up to now the permanent magnet has been 20 Remalloy (20Mo-12Co-Fe), which has nominal values of $B_r = 0.95$ T, $H_c = 28$ kA/m, and $(BH)_{max} = 10$ kJ/m³. Remalloy is semi-

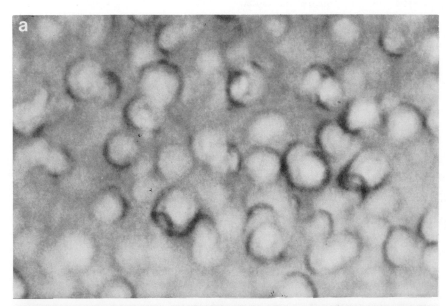
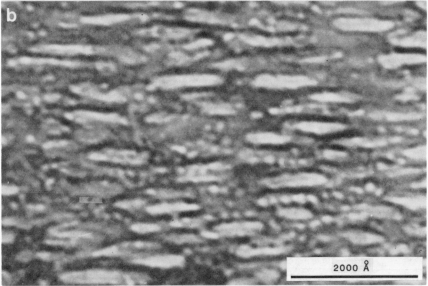

Fig. 3. Transmission electron micrographs of a Fe-33Cr-11.5Co alloy. (a) Spherical particles obtained by continuous cooling from solution-treated state. (b) Elongated structure obtained after uniaxial deformation by wire drawing of (a); such a structure is conducive to superior permanent magnet properties (see Fig. 4). [Micrographs by S. Mahajan from (*16*)]

brittle and fabrication in the cup shape requires blanking and drawing at 1250°C, a slow and energy-intensive process. The search for a low-cost cold-formable ductile magnet led Chin et al. (19) to develop a 15Co-28Cr-1Al-1/4Zr alloy called Chromindur I. Subsequently, Jin et al. (15) developed a low-cobalt ternary (Chromindur II; 10.5Co-28Cr-Fe) with simplified melting and fabrication practices. A team from Bell Laboratories and Western Electric has now successfully introduced Chromindur II in the commercial production of receiver magnets, at a projected annual savings of 10,000 kilograms of cobalt. Other uses of Cr-Co-Fe alloys in telephone apparatus as potential substitutes for Alnicos are outlined in (20).

Because of their similarity in magnetic properties to the Alnicos, particularly Alnico 5, and their added advantages of ductility and lower Co contents, the Cr-Co-Fe alloys are possible substitutes for Alnicos in a number of situations. In addition, because their properties are superior to those of other available ductile alloys, such as Cunife (60Cu-20Ni-Fe) and Vicalloy (10V-52Co-Fe), the Cr-Co-Fe alloys may replace some of these as well.

High-Induction Grain-Oriented Silicon Steel

Unlike the R-Co and Cr-Co-Fe alloys, silicon steel is not a new product; it has been around since the turn of this century. However, since on a tonnage basis silicon steel is used far more than other magnetic materials, any breakthrough in improvement must be applauded.

Silicon steel containing up to about 4 percent Si is divided into two grades, nonoriented and oriented, according to whether the steel is processed to exhibit a substantial preferred crystallographic orientation, or texture. Since the steel is used as the core in motors, generators, and transformers operating at standard low frequencies such as 50 or 60 hertz, the so-called core loss has been the single material parameter of utmost concern. This loss is related to the generation of eddy currents induced by the moving magnetic domain walls in response to the a-c excitation and represents precious energy that is wasted as heat. For silicon steel the optimum domain structure for minimum loss is in the $<100>$ direction, and one major breakthrough occurred in 1935, when Goss (21) developed a technique for processing low-loss silicon steel sheet with a $\{110\}<001>$ oriented texture, with $\{110\}$ planes predominantly parallel to the sheet surface and an $<001>$ direction in the rolling direction. This is the so-called cube-on-edge texture. Today, U.S. consumption of silicon steel is split between oriented and nonoriented grades, amounting to about 310 million kilograms each. The oriented grade, containing about 3 percent Si, is used almost exclusively in large power and distribution transformers, where low loss is particularly important. Until recently, the typical oriented grade had a core loss of about 1.15 watts per kilogram. Even at a low estimate of $\sim\$1000$ per kilowatt in capitalized value, which includes the capital outlay to generate and deliver extra power and the cost of power plant fuel and operation, the core loss amounts to nearly $\$400$ million annually. Therefore there is a large incentive to cut down losses, even by a small percentage.

Historically, with each improvement in core loss the steel tends to be used at a higher level of magnetic flux density, since this means a reduced amount of steel and hence transformer size for the same transformer core loss. Core loss level increases with flux density. Thus, for a number of years, oriented silicon steel has been operated at 1.5 T, in comparison with 1.0 T for the lossier nonoriented grade. In 1968 another breakthrough came with the commercial introduction by Nippon Steel (22) of new grades of oriented steel that had an exceptionally sharp cube-on-edge texture and low loss at high flux densities. As a result, operating core levels are now quoted at 1.7 T for these steels. Table 3 lists values of core loss of the new grades, often referred to as high-induc-

Table 3. Core loss of grain-oriented silicon steels at 50 Hz and 1.7 T, in watts per kilogram.

Type	Grade	Thickness (mm)		
		0.27	0.30	0.35
High-induction	M0H	0.99	1.05	
	M1H	1.04	1.11	1.16
	M2H	1.11	1.17	1.22
Conventional	M4	1.27		
	M5		1.39	
	M6			1.57

Table 4. Comparison of manufacturing processes for grain-oriented silicon steel.

Conventional	*Nippon Steel*
Steelmaking (MnS)	Steelmaking (AlN + MnS)
Hot rolling (1370°C)	Hot rolling
Annealing (800° to 1000°C)	Annealing (950° to 1200°C)
Cold rolling (70 percent)	Cold rolling (85 percent)
Annealing	
Cold rolling (50 percent)	
Decarburizing (800°C, wet H_2 + N_2)	Decarburizing
Box annealing (1200°C, dry H_2)	Box annealing
Kawasaki Steel	*General Electric–Allegheny Ludlum*
Steelmaking (Sb + MnSe or MnS)	Steelmaking (B + N + S or Se)
Hot rolling	Hot rolling (1250°C)
Annealing	Annealing
Cold rolling	Cold rolling (> 80 percent)
Annealing	
Cold rolling (65 percent)	
Decarburizing	Decarburizing
Box annealing (820° to 900°C, then 1200°C, dry H_2)	Box annealing

Table 5. Energy products of some Co-bearing permanent magnets (28).

Alloy	Co (%)	$(BH)_{max}$ (kJ/m³)	$(BH)_{max}$ per unit Co (based on Alnico 5)
Alnico 5	24	44	1.0
Alnico 9	35	72	1.1
SmCo$_5$	63	144	1.3
Sm$_2$(Co,Cu,Fe)$_{17}$	50	240	2.7
5Co-30Cr-Fe	5	42	4.7
3Co-32Cr-Fe	3	33	6.0

86

tion grain-oriented silicon steel, along with conventional oriented grades. The decrease in core loss is about 20 percent, a truly remarkable development. New grades of high-induction steel were also developed by Kawasaki Steel in 1974 (23) and by Allegheny Ludlum in 1977 (24). The latter effort was based on laboratory processes developed at General Electric (25).

In the conventional commercial process for making oriented steel, outlined in Table 4, the cast ingot is hot-rolled near 1370°C to a thickness of about 2 millimeters, annealed at 800° to 1000°C, and then cold-rolled to the finished thickness of 0.27 to 0.35 mm in two steps of about 70 and 50 percent, respectively. A recrystallization anneal is sandwiched between the two cold-rolling steps. The cold-rolled strip is first decarburized near 800°C as carbon is detrimental to the magnetic properties. This step also results in a primary recrystallized structure containing some grains of the desired cube-on-edge orientation. A final step involves box annealing at > 1200°C to form an essentially all cube-on-edge texture by secondary recrystallization as these grains cannibalize their neighbors.

A very important concept in the secondary recrystallization process involves so-called grain-growth inhibitors. In the conventional process, the manganese and sulfur that are normally present in steelmaking form MnS inclusions, which restrict grain growth during primary recrystallization. Then during secondary recrystallization the inclusions are dissolved, permitting the preferential growth of the cube-on-edge grains. In the Nippon process, AlN as well as MnS is used as a grain-growth inhibitor. A more potent inhibitor, the AlN permits the adoption of a one-stage cold reduction of large deformation (85 percent), resulting in a sharper grain orientation in the final steel. Without the AlN, such a large deformation would also have enhanced the undesirable growth of the primary recrystallized grains.

In the Kawasaki process, antimony, added along with MnSe or MnS, acts as an extra grain-growth inhibitor by segregating to the grain boundaries. Additional grain orientation sharpening also comes from the use of a two-step box anneal—a low-temperature, long-time anneal (820° to 900°C for 5 to 50 hours) followed by the usual high-temperature (1200°C) treatment. Similarly, boron and nitrogen together with sulfur or selenium are adopted in the General Electric–Allegheny Ludlum process for extra grain-growth inhibition. It is thought that the

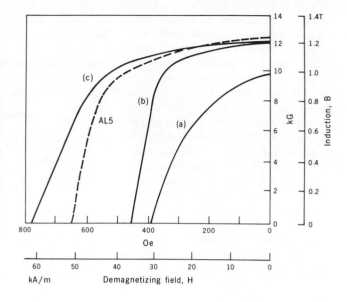

Fig. 4. Demagnetization curves of a Fe-33Cr-11.5Co alloy showing the improvement in permanent magnet properties obtained with the deformation-aging technique (16): (a) aged without intervening deformation step, (b) aged with intervening rolling step, and (c) aged with intervening wire-drawing step. AL5 refers to commercial Alnico 5.

B, N, and S (or Se) segregate to the grain boundaries, as does the Sb in the Kawasaki process. One advantage claimed for the Allegheny steel process is that since the Mn and S levels are lower than the others, the MnS solubility temperature is lower and hence the initial hot-working temperature can be decreased to below 1300°C. This is advantageous in terms of lower fuel costs and added mill life.

It is somewhat disappointing that the anticipated widespread use of these new high-induction oriented Si steels has not yet occurred, even though the price premium is slight, generally less than 10 percent. Because substantial cost penalties are associated with lossy transformers, the recent trend of designers has been to reduce loss by returning to a design with lower flux densities at the expense of a larger unit. At flux densities of 1.5 T and lower, there is hardly any difference in core loss between the new high-induction steel and the conventional oriented steel. For this reason, a great deal of current interest has focused on the amorphous soft magnetic alloys. Although their highest value of saturation induction is only about 1.7 T, considerably below the 2.0 T for 3 percent Si steel, these amorphous alloys, prepared in continuous ribbon form by rapid quenching from the melt, exhibit less than half the core loss of oriented Si steel. These new alloys are discussed in detail by Gilman (2).

Future Prospects

Since the developments described in this article are relatively new, further improvements are expected. In the area of rare earth–cobalt alloys, attention is

clearly focusing on the Cu-substituted R_2Co_{17} precipitation-hardened alloys, as these have exhibited the highest values of $(BH)_{max}$ thus far achieved on a commercial scale. Although equally high values of $(BH)_{max}$ have been achieved in single-phase Cu-free R_2Co_{17} alloys in the laboratory (26), reduction to commercial practice has proved difficult, apparently because of the need for critical control of processing variables. As noted earlier, because of the high cost of both cobalt and rare earths, the rare earth alloys will mainly be used in special applications.

The Cr-Co-Fe alloys are attractive both because of their ductility and because of their similarity to the popular Alnicos in magnetics. Values of $(BH)_{max}$ ~ 80 kJ/m³ have been achieved in a 23 percent Co alloy (27). This is the highest value so far reported in Cr-Co-Fe–based alloys and already surpasses that of Alnico 9 (~ 72 kJ/m³), the best of the commercial Alnicos. In addition, the steep rise in Co price has spurred intense research in very low Co or Co-free alloys. This search recently culminated in the attainment of $(BH)_{max}$ ~ 42 kJ/m³ in several 5 percent Co ternary and quarternary Cr-Co-Fe alloys (28). As shown in Table 5, these alloys represent the highest values of $(BH)_{max}$ per unit of cobalt to date in Co-bearing alloys, far exceeding even those of the precipitation-hardened R_2Co_{17} types. It should be recognized, of course, that $(BH)_{max}$ is only one of several design criteria for a magnetic component.

In the past 2 years there has been a substantial switch from Alnicos to ceramics in motor and loudspeaker designs because of the instability of the Co price and supply (20). Since the Cr-Co-Fe alloys are similar to the Alnicos in many respects, particularly the attractive features of high B_r and good temperature stability, they may offer a third alternative in material selection.

Similarly, very attractive Co-free Mn-Al-C magnets with $(BH)_{max} \sim 45$ kJ/m³ have been developed (29). These magnets have the coercivities of the ceramic magnets but have substantially higher values of B_r (Table 2). It is thus conceivable that the Mn-Al-C magnets could replace some ceramics as well as Alnicos. However, although cheap in raw material price, these magnets are expensive to process at present, requiring the use of hot extrusion. In addition, they have low Curie temperatures ($\sim 320°C$); therefore their temperature stability, while better than that of the ceramics, is substantially below those of the Alnicos and Cr-Co-Fe alloys.

As for the high-induction oriented silicon steel, further progress may depend on the willingness of designers to reemphasize designs with small volume and high flux density, and perhaps on the progress in the competing development of low-loss amorphous alloys. On the basis of the estimate given by Taguchi et al. (30), it appears that a further 50 percent decrease in core loss from the present low values might be achievable in the high-induction silicon steels.

References

1. E. A. Geiss, *Science* **208**, 938 (1980).
2. J. J. Gilman, *ibid.*, p. 856.
3. I. S. Jacobs, *J. Appl. Phys.* **50**, 7294 (1979).
4. E. A. Nesbitt and J. H. Wernick, *Rare Earth Permanent Magnets* (Academic Press, New York, 1973), chap. 3.
5. W. E. Wallace, *Rare Earth Intermetallics* (Academic Press, New York, 1973), chap. 2.
6. G. Hoffer and K. H. Strnat, *IEEE Trans. Magn.* **MAG-2**, 487 (1966).
7. A. Menth, H. Nagel, R. S. Perkins, *Annu. Rev. Mater. Sci.* **8**, 21 (1978).
8. E. A. Nesbitt, R. H. Willens, R. C. Sherwood, E. Buehler, J. H. Wernick, *Appl. Phys. Lett.* **12**, 361 (1968).
9. Y. Tawara and H. Senno, *Jpn. J. Appl. Phys.* **7**, 966 (1968).
10. H. Senno and Y. Tawara, *ibid.* **14**, 1619 (1975).
11. T. Ojima, S. Tomizawa, T. Yoneyama, T. Hori, *IEEE Trans. Magn.* **MAG-13**, 1317 (1977).
12. T. Nezu, M. Tokunaga, Z. Igarashi, in *Proceedings of the 4th International Workshop on Rare Earth–Cobalt Permanent Magnets and Their Applications* (Society for Non-Traditional Technology, Tokyo, 1979), p. 437.
13. K. J. Strnat, in *ibid.*, p. 8.
14. H. Kaneko, M. Homma, K. Nakamura, *AIP Conf. Proc.* **5**, 1088 (1971); H. Kaneko, M. Homma, K. Nakamura, M. Miura, *IEEE Trans. Magn.* **MAG-8**, 347 (1972).
15. S. Jin, G. Y. Chin, B. C. Wonsiewicz, *ibid.* **MAG-16**, 139 (1980).
16. S. Jin, *ibid.* **MAG-15**, 1748 (1979).
17. H. Kaneko, M. Homma, T. Minowa, *ibid.* **MAG-12**, 977 (1976).
18. M. Iwata, Y. Ishijima, O. Fujita, U.S. Patent 3,982,972 (28 Sept. 1976); M. Iwata and Y. Ishijima, U.S. Patent 3,989,556 (2 November 1976).
19. G. Y. Chin, J. T. Plewes, B. C. Wonsiewicz, *J. Appl. Phys.* **49**, 2046 (1978).
20. G. Y. Chin, S. Sibley, J. C. Betts, T. D. Soblabach, F. E. Werner, D. L. Martin, *IEEE Trans. Magn.* **MAG-15**, 1685 (1979).
21. N. P. Goss, *Trans. Am. Soc. Met.* **23**, 515 (1935).
22. S. Taguchi, T. Yamamoto, A. Sakakura, *IEEE Trans. Magn.* **MAG-10**, 123 (1974).
23. I. Goto, I. Matoba, T. Imanaka, T. Gotoh, T. Kan, *Proceedings of the EPS Conference on Soft Magnetic Materials* (University College, Cardiff, Wales, 1975), vol. 2, p. 262.
24. F. A. Malagari, *IEEE Trans. Magn.* **MAG-13**, 1437 (1977).
25. H. C. Fiedler, *ibid.*, p. 1433.
26. H. Nagel, *AIP Conf. Proc.* **29**, 603 (1976).
27. S. Jin, N. V. Gayle, J. E. Bernardini, paper presented at the Intermag Conference, Boston, 21 to 24 April 1980.
28. M. L. Green, R. C. Sherwood, G. Y. Chin, J. H. Wernick, paper presented at the Intermag Conference, Boston, 21 to 24 April 1980.
29. T. Ohtani et al., *IEEE Trans. Magn.* **MAG-13**, 1328 (1977).
30. S. Taguchi, A. Sakakura, F. Matsumoto, K. Takashima, K. Kuroki, *J. Magn. Magn. Mater.* **2**, 121 (1976).
31. G. Y. Chin and J. H. Wernick, in *Encyclopedia of Chemical Technology* (Wiley-Interscience, New York, ed. 3, in press).
32. *Standard Specification for Permanent Magnet Materials* (Magnetic Materials Producers Association, Evanston, Ill., 1978).

Heterogeneous Catalysts

Arthur W. Sleight

A catalyst is defined as a substance that accelerates a chemical reaction without itself being consumed. The degree of acceleration can be enormous, sometimes more than a factor of 10^{10}. However, the ability of certain catalysts

foundation. Our basic understanding of how catalysts function is very limited. Materials scientists have largely ignored the field of heterogeneous catalysts, partly because the task of bringing order to the field appears overwhelming. The

Summary. Proven catalysts exist in many different forms and with many different kinds of composition. An understanding of how catalysts function is beginning to emerge in a few areas, and some superior catalysts have been developed as a result of this knowledge. Applications of new techniques and disciplines should lead to impressive advances in the years ahead.

to direct a chemical reaction is even more impressive and useful. For example, passing propylene and oxygen over bismuth oxide produces hexadiene and benzene, whereas passing the same mixture under the same conditions over a molybdate catalyst produces acrolein.

Catalysts used in commercial processes do not, in fact, remain unaltered. They have lifetimes that vary from seconds to years. Catalysts with very short lifetimes can be useful if appropriate regeneration schemes are included. For example, zeolite catalysts used for the cracking of petroleum have effective lifetimes of only a few seconds but are regenerated on a continuous basis. Catalysts that last weeks to years are also sometimes regenerated; at other times they are discarded. Many catalysts in effect are regenerable reagents. For these catalysts dynamic processes in the bulk as well as at the surface are important.

Catalysts play critical roles in energy production, transformation, storage, and utilization. They are also essential for pollution abatement, production of numerous chemical products, and in many other ways. The field of heterogeneous catalysis has become a technologically advanced field without a sound scientific

situation is currently changing because of the advent of many new analytical techniques and the realization of the social and economic importance of the field. Society's ability to cope with pending energy shortages during the next decades may well depend on our ability to design new and better catalysts. We can hope to provide such catalysts only if we gain an improved understanding of current catalysts.

Catalysts exist in many different forms. They can be metals, alloys or multimetallics, sulfides, oxides, nitrides, carbides, or mixtures of these. Some are single phase; others are composed of several phases. Catalysts can be used in bulk form such as gauze. Many are sufficiently crystalline to produce good diffraction patterns. Other catalysts are amorphous or are composed of particles too small to diffract. Catalysts can be electrically insulating, semiconducting, or metallic. Surface properties of catalysts are always important; frequently, bulk properties are also important.

There are hundreds of catalysts and hundreds of different reactions for which catalysts are used. Rather than attempting to survey the whole field in this article, I will present the current understanding of catalysts in several different and important areas. Recent advances in these areas have been impressive, and there is reason to be optimistic that advances in the next few years will be even more profitable.

Metal Catalysts

Many different transition metals are used as catalysts. Since most of these are easily oxidized, their use tends to be restricted to reactions of a reducing nature, for example, hydrogenation. Metals used as oxidation catalysts generally take up considerable amounts of oxygen. For example, when platinum is used as an oxidation catalyst, its surface may be largely platinum oxide (*1*).

Metal catalysts are sometimes used in bulk form as pure metal or as alloys. More commonly, metals are dispersed on supports such as silica, alumina, or carbon. If the metal is very expensive, and this is frequently the case, then such dispersion is of economic importance. The dispersion can be very great with essentially every metal atom available for catalysis.

Although a support may be used for some practical reasons such as economics or physical strength, the support may also play a critical role in the catalytic reaction.

1) Some cations tend not to be reduced to the metal on certain supports. This behavior is often encountered when iron is supported on oxides such as alumina.

2) The pore structure of the support can have considerable influence during catalysis. This occurs when a metal is supported within the pores of a zeolite structure, for example, but even the pore structure of an amorphous support can influence catalytic properties (*2*).

3) The morphology of a metal particle may be different on different supports. For example, a metal may wet some supports but not others.

4) The pattern of metal particles on the support may be governed by the surface structure of the support.

5) The supported metal may interact chemically with the support. For example, there is evidence that platinum will take aluminum into solid solution when a platinum on alumina catalyst is heated to high temperatures under hydrogen (*3*).

6) A spillover mechanism may be in operation. There are several different variations of this; the best known example is hydrogen spillover (*4*). Hydrogen is adsorbed and dissociated on the metal particles. This activated hydrogen then spills over on the oxide support such as silica or alumina (*5*). Organic molecules adsorbed on the oxide surface may then react with the activated hydrogen.

7) A metal-support interaction cur-

The author is a research supervisor in the Central Research and Development Department, E. I. du Pont de Nemours & Company, Wilmington, Delaware 19898, and adjunct professor in the Department of Chemical Engineering at the University of Delaware, Newark 19711.

rently receiving wide attention involves a charge-transfer mechanism. When two materials are placed in contact with each other, there is a flow of electrons from one material to the other because their Fermi levels have different energies. Thus one material assumes a positive charge, and the other a negative charge. In the case of metals in contact with insulating oxides, the electron flow will generally be from the metal to the oxide. The amount of charge transfer is difficult to measure or to predict theoretically, but it appears to depend on metal particle size as well as surface states, such as Lewis acid centers, on the support. Results from both XPS (x-ray photoelectron spectroscopy) and EXAFS (extended x-ray absorption fine structure) have indicated that metals such as platinum take on a positive charge when they are well dispersed on insulating supports such as silica, alumina, and zeolites (6). The effect is such that one may regard a metal as having moved to the left within a row of the periodic table. The amount of charge transfer is limited by the insulator because it does not have an abundance of empty states of appropriate energy to accept the electrons. Thus this type of charge transfer, although real, may be regarded as weak relative to some recently discovered effects.

A strong metal-support interaction is observed when a platinum group metal is supported on reduced TiO_2 (7). Such materials are prepared by intially supporting the metal on fully oxidized TiO_2. Subsequent hydrogen reduction at low temperature ($\sim 200°C$) leads to a material showing normal hydrogen chemisorption. However, hydrogen reduction at higher temperatures ($\sim 400°C$) leads to a great decrease in the ability of the catalyst to chemisorb hydrogen. Detailed studies have indicated that this is not due to a loss of metal surface area (7). Nevertheless, the TiO_2 becomes reduced, and the metal particles take on a negative charge (8). The apparent reason for this direction of the charge flow is indicated in Fig. 1. When TiO_2 is reduced, its Fermi level energy is increased until it is actually above that of the metal. Thus there is a flow of electrons from the TiO_{2-x} to the metal. This effect is stronger than the more traditional and opposite charge flow because in this case there are many available electron acceptor states on the metal. One is now in a sense moving the metal to the right within a row of the periodic table. The possibilities for this charge-transfer type of support interaction are very exciting. We can control the charge of metal particles over a considerable range, and the systematic

study of resultant catalytic effects has just begun.

The catalytic behavior of a metal can differ significantly on different crystallographic faces (9). There is also evidence that some reactions on metals occur primarily at the edges of crystallites (9). Thus one might expect that catalytic properties will depend on crystallite size even if the reaction rate is normalized to metal surface area. Just such an effect is generally found for reactions that require carbon-carbon bond cleavage. The specific activity for hydrogenolysis reactions can decrease by orders of magnitude with increasing particle size. In contrast, for reactions involving carbon-hydrogen bonds, such as hydrogenation and dehydrogenation, the specific activity usually remains essentially constant with varying metal particle size (10). Presumably in this case the reaction occurs as rapidly on crystallite faces as on the edges.

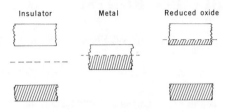

Fig. 1. Schematic energy level diagram for an insulator (for example, Al_2O_3), a metal (for example, Pt), and a reduced oxide (for example, TiO_{2-x}). Filled states are shaded, and the Fermi levels are indicated by broken lines.

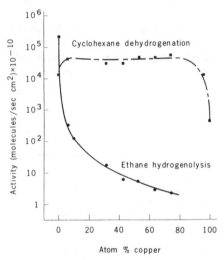

Fig. 2. Activity of Ni-Cu alloy catalysts for the hydrogenolysis of ethane to methane. [From Sinfelt (12)]

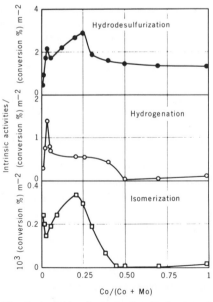

Fig. 3. Activities for hydrodesulfurization of thiophene, hydrogenation of cyclohexene, and isomerization of cyclohexane to methylcyclopentane. [After Delvaux et al. (32)]

Another variation on metal catalysts is the use of two or more metals on the same support (11, 12). These are frequently referred to as bimetallic or multi-metallic catalysts since the term alloy is inappropriate for isolated clusters made up of only a few metal atoms. Furthermore, the different metals may not even be in the same cluster. Originally, alloy-type catalysts were envisioned as a way of obtaining a continuous variation of properties between two metals. In fact, the catalytic properties of alloys do not generally vary linearly with changing composition (Fig. 2). The reason in the case of ethane hydrogenolysis is fairly well understood (12). This reaction is many orders of magnitude faster on nickel than on copper, and it is known that the surface of nickel-copper alloys is enriched with copper relative to the bulk. The surface composition of alloys is generally different from that of the bulk in a manner that is largely predictable (13). An interesting aspect of the surface composition is that it can be changed during catalysis because of strong chemisorption bonds of certain molecules on a specific metal of the alloy (13).

A reforming reaction is carried out in oil refineries to improve the octane rating of gasoline-range materials and to produce aromatics. Catalysts for this reaction were originally chromium oxide and molybdenum oxide supported on alumina. About 30 years ago platinum on alumina was found to be a much more active catalyst and to possess improved selectivity, and until about 10 years ago this was the catalyst of choice for reforming. In the late 1960's the Chevron Research Company developed a catalyst composed of platinum-rhenium on alumina, and this has properties superior to those of platinum on alumina (11). The initial activity of the platinum-rhenium catalyst is, in fact, no greater than that of the pure platinum catalyst. However, the high activity is maintained for a much longer time with the bimetallic catalyst. This development has had major technological consequences, and more recently a platinum-iridium bimetallic catalyst for reforming has been developed at Exxon by Sinfelt (14, 15). There remain many reactions for which multimetallic catalysts have not been evaluated. The number of possible metal combinations taken together with support variations leads to enormous flexibility. Undoubtedly other practical applications for such catalysts will be found. One promising area for the use of these catalysts is in the hydrogenation of carbon monoxide.

Sulfide Catalysts for

Hydrodesulfurization

The hydrodesulfurization reaction is of great importance because natural gas, petroleum, and coal all contain sulfur which must be removed. This reaction can be schematically represented as

$$-CH-S- + H_2 \rightarrow -CH_3 + H_2S$$

Other reactions, such as hydrogenation or hydrodenitrogenation, usually occur simultaneously.

The most common catalyst for this reaction is frequently referred to as a cobalt molybdate catalyst. (Sometimes nickel is substituted for cobalt, and sometimes tungsten is substituted for molybdenum.) However, labeling this catalyst as cobalt molybdate is highly misleading. There are two forms of $CoMoO_4$, but neither of them is present in the working catalyst. Furthermore, despite the fact that normal preparation methods involve oxide precursors, the synthesis procedures giving best results do not involve either of the cobalt molybdates as precursors (16). The actual working catalyst is essentially a mixture of Co_9S_8 and MoS_2 supported on γ-Al_2O_3.

The most intriguing aspect of the cobalt-molybdenum-sulfide catalyst for hydrodesulfurization is the existence of a synergistic effect when both Co_9S_8 and MoS_2 are present. The intrinsic catalytic activity of the mixture is clearly higher than that of either end member (Fig. 3). This is true not only for the hydrodesulfurization reaction but also for the hydrogenation and isomerization reactions. The data in Fig. 3 are for unsupported sulfides, but the same type of synergistic effect exists for catalysts supported on γ-Al_2O_3.

There have been many attempts to ra-

tionalize the promoter effect of cobalt on MoS_2. Five models are discussed in a recent review by Delmon (16). One view is that the electrical properties of MoS_2 are altered by cobalt doping and that this in turn affects catalyst properties (17). Undoubtedly some doping of MoS_2 by cobalt occurs with a resulting effect on electrical properties. However, there has been no clear correlation with catalytic properties, and the maximum doping level of cobalt in MoS_2 is far below the cobalt content where the synergistic effect is most prominent.

Molybdenum disulfide has a layer-type structure, and such structures are frequently amenable to intercalation. However, it is known that cations such as cobalt and nickel do not move between the layers of MoS_2 to any significant extent. It has been proposed that there is some intercalation of these layers by cobalt or nickel at the surfaces of MoS_2 crystallites (18). This may well occur, and there may be implications for catalysis. However, as with the doping effect, the amount of cobalt intercalated by this mechanism is far short of that required for a completely satisfactory explanation of the synergistic effect.

Many might conclude that some ternary sulfide of cobalt and molybdenum is formed. Indeed, ternary compounds such as $CoMo_2S_4$ are known, but extensive and sophisticated studies have failed to find such compounds in catalysts used under actual reactor conditions.

One of the interesting aspects of the cobalt-molybdenum-sulfide catalysts is that the maximum synergistic enhancement occurs at different cobalt contents for different reactions (Fig. 3). This implies that different sites are involved for the different reactions, yet catalysis at these different sites is still strongly related to the synergistic effect.

Although the cobalt-molybdenum-sulfide catalyst has been extensively studied for many years in many different laboratories, the nature of the synergistic effect between Co_9S_8 and MoS_2 is not understood. One or both phases may be affected by the presence of the other. Alternatively, some sort of spillover mechanism may be involved. As with any synergistic effect of this type, catalytic properties depend not only on composition but also on the way in which these elements are put together. The microstructure of the resulting catalyst is of utmost importance.

Zeolites

Zeolites are crystalline aluminosilicates with open framework structures. These structures are so open that various organic molecules can actually diffuse in and out of the tunnels of the structures. Thus, for catalysis, a shape selectivity results since certain molecules or transition states are too large for the pores of the zeolite structures. Another important

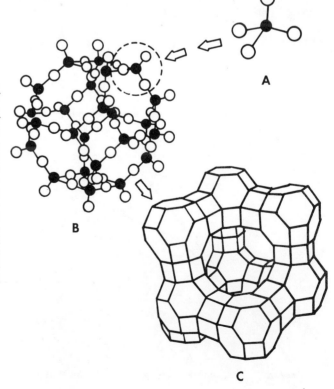

Fig. 4. The framework structure of faujasite. (A) Tetrahedral arrangement of silicon (or aluminum) atoms sharing oxygen atoms. (B) Sodalite unit consisting of 24 SiO_4^- and AlO_4^- tetrahedra. (C) Zeolite superstructure consisting of tetrahedral arrangement of sodalite units connected by oxygen bridges forming hexagonal prisms. [From Haynes (33)]

property of some zeolite catalysts is their exceptional acidity.

Zeolite structures are based on a framework containing silicon-oxygen tetrahedra that share corners only and are usually mixed with aluminum-oxygen tetrahedra (Fig. 4). The coordination number of all the oxygen atoms is two, and the composition of the framework is $(Si^{4+}, Al^{3+})O_2$. When no Al^{3+} is present the network is electrically neutral. However, Al^{3+} is generally present, and the network has an overall negative charge which is balanced by interstitial cations or protons.

There are about 35 naturally occurring zeolites, of which about 25 have been synthesized; another 100 or so new zeolites have been synthesized during the last 30 years (19). The very successful gel synthesis route was developed in the late 1940's by Barrer (20). The gel is produced at a high pH from an aqueous mixture of Si^{4+}, Al^{3+}, and various alkali cations. Crystallization is normally carried out at 200° to 400°C in autoclaves.

Nucleation of crystals can be rapid, irreversible, and difficult to control.

Impressive developments in zeolite synthesis have resulted recently from the use of quaternary ammonium hydroxides as a source of cations. This approach led to the first laboratory synthesis of certain naturally occurring zeolites. It has also led to the synthesis of some new and technically important zeolites such as Mobil's ZSM-5 (21). Another approach has been to use quaternary ammonium polymers as organic cations (22). Stacking faults in some zeolites cause much of the pore space to be unavailable for catalysis, but by using long-chain polymer cations these faults can be prevented so that all the pore space becomes available.

One of the most useful zeolite structures is that of the naturally occurring zeolite faujasite. The way in which this structure is built up from the tetrahedral unit is indicated in Fig. 4. The silica-to-alumina ratio is variable from about 2 to 6. Zeolites synthesized with this structure are generally referred to as either zeolite X(low silica) or zeolite Y(high silica). The free pore openings of the faujasite structure are the largest among zeolites, about 7.4 angstroms. Such pores are large enough for molecules such as the xylenes to diffuse in and out. Other zeolites have smaller pores ranging down to about 4 Å.

Acid zeolites are prepared by replacing the alkaline cations with protons. This can sometimes be accomplished by ion exchange with mineral acids. However, some zeolites, especially those with high aluminum content, are attacked by acids. Thus a better method of preparing the acid form of zeolites is by ion exchange with NH_4^+ followed by heating to drive off NH_3. Different acid sites exist in such zeolites, and some possess the acid strength comparable to that of sulfuric acid. The acid form of the faujasite zeolite was introduced into the cracking catalyst in the 1960's. Its catalytic activity is several orders of magnitude higher than that of the amorphous aluminosilicates formerly used. The use of this zeolite is currently saving the United States several hundred million barrels of crude oil per year.

Another impressive aspect of zeolite catalysts is their shape selectivity, which can be reflected in different ways.

1) If a mixture of different gases is passed over a zeolite catalyst only the smaller molecules will enter the pores and react. Thus the zeolite selectively catalyzes reactions of certain gases in the mixture.

2) On passing a gas over a zeolite

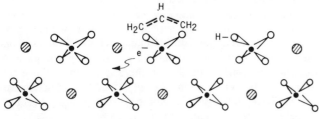

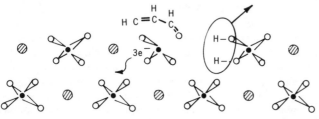

Dissociative adsorption

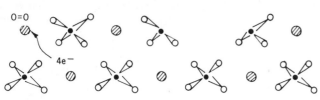

Redox and desorption

Oxidation of catalyst

⊘ = Bi   ● = Mo   ○ = Oxygen

Fig. 5. Schematic of propylene oxidation showing dissociative adsorption, redox and desorption, and reoxidation of the catalyst. [After Sleight (28)]

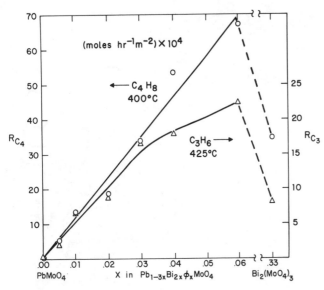

Fig. 6. Activity (R) for propylene oxidation to acrolein and butene oxidation to butadiene. [After Sleight and Linn (27)]

catalyst, some products of the reaction may be too large to leave the zeolite pores. Only the smaller molecules leave; the larger molecules stay behind and may continue to react.

3) There are also cases in which both the reactant and product gases can easily diffuse in and out of the zeolite pores but where the transition state to produce a certain product involves an intermediate which is too large for any cavity within the pore structure of the zeolite. Thus a particular reaction is effectively blocked, and shape selectivity results.

Recently, there has been a striking success based on shape-selective catalysis by a zeolite. The new synthetic zeolite ZSM-5 is an excellent catalyst for the conversion of methanol to gasoline (21). Pilot plants for producing gasoline by this method are now being built.

Many different cations can be exchanged into zeolites. This is frequently done to adjust the acidity in zeolites that are used for cracking. Transition metal cations exchanged into zeolites have other functions (23). These cations are sometimes reduced to the metal for catalyst studies or applications. Different types of activity (for example, hydrogenation activity) are introduced in this manner. Unusual and useful properties result from having such catalytic functions within the pores of a zeolite structure.

The future of catalysis by zeolites is very promising. We have only barely begun the task of evaluating zeolites for the multitude of reactions of interest. There are many different zeolites, and there are various ways in which the properties of a given zeolite can be altered. Fine tuning of the selectivity and activity is possible by many different methods, and new zeolite structures continue to be discovered.

Oxides for Selective Oxidation

Oxide catalysts are used in a variety of catalytic reactions. For the oxidation of hydrocarbons, oxides may be nonselective, complete combustion catalysts such as Co_3O_4. Alternatively, they may very selectively oxidize methanol to formaldehyde, propylene to acrolein, or butene to butadiene, for example. There are always other reactions, such as complete combustion, that might occur under the same conditions, but a good selective catalyst effectively blocks these other pathways. The most studied of the selective oxidation catalysts are the molybdates, and only these catalysts will be considered here, with olefin oxidation

reactions being used as examples. Bismuth molybdates were the first catalysts discovered for the selective oxidation of propylene to acrolein (24).

$$CH_3CH=CH_2 + O_2 \rightarrow$$
$$CH_2=CHCHO + H_2O$$

Experiments with the use of oxygen-18 have established that the oxygen in the products originates primarily from the lattice of the catalyst (25). This lattice oxygen is readily available; thus, the rate of propylene oxidation is independent of the partial pressure of oxygen. This is an example of a catalyst that is, in fact, a reactant that is continually being regenerated in situ by gas-phase oxygen. Studies of the effect of deuteration of propylene on the rate of the reaction indicate that the rate-limiting step for propylene oxidation over bismuth molybdate catalysts is the dissociative chemisorption of propylene (26).

$$CH_3CH=CH_2 + O_s \rightarrow$$
$$CH_2-CH-CH_2 + O_sH$$

This produces a surface allylic species, and a surface oxygen (O_s) is converted to a hydroxyl group. There is a second hydrogen abstraction, and acrolein desorbs taking a lattice oxygen (Fig. 5). Water also desorbs, and the catalyst is reduced by four electrons per propylene molecule converted to acrolein. One oxygen molecule then restores the lost electrons and the lattice oxygen. After many studies, the roles of bismuth and molybdenum in

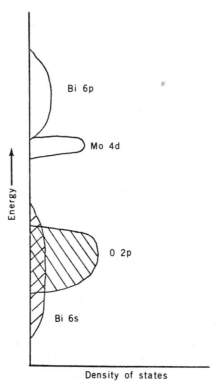

Fig. 7. Schematic energy level diagram for a bismuth molybdate.

this process have become reasonably well defined.

Model studies of propylene and butene oxidation over oxides with a scheelite structure were carried out at DuPont (27, 28). The scheelite structure is very simple compared to the bismuth molybdate structures, and cation vacancy defects are readily introduced. Furthermore, the bismuth-to-molybdenum ratio can be varied while retaining the same basic structure. Molybdates with a scheelite structure may be inactive, or they may be as active and selective as the bismuth molybdates (Fig. 6). In many studies the bismuth and defect concentrations were varied independently. Both the defects and bismuth enhance the reaction, but in different ways. For those catalysts with little or no bismuth, the ability of the catalyst to reoxidize itself is very poor and the rate of reaction depends on the partial pressure of oxygen. This and other data indicate that bismuth cations are the sites for catalyst reoxidation.

The site for propylene dissociative chemisorption is probably a molybdate group. This is an electron donor-acceptor interaction with analogies in homogeneous oxidation reactions. Bismuth may play some role in propylene chemisorption, but this role is secondary to its role in reoxidation of the catalyst. It has been suggested by Grasselli (29) that bismuth may be the site at which the abstracted hydrogen is converted to a hydroxyl group. Very recent experiments at SOHIO indicate that the bismuth role is more important for the second hydrogen abstraction (30). Bismuth may also play a role in propylene chemisorption because of an inductive or electronic effect. In any case this role of bismuth can be fulfilled by other cations or by defects in the case of the scheelite structure.

Since we have a model in which reduction occurs at a molybdate group and reoxidation of the catalyst occurs at a bismuth site, a special electronic structure must exist. A schematic of the energy diagram of a bismuth molybdate is shown in Fig. 7. There are two degeneracies of energy levels. The filled bismuth 6s band overlaps the filled oxygen 2p band, and this may or may not have implications for catalysis. The overlap or hybridization of the molybdenum 4d and bismuth 6p bands has profound implications for catalysis. The working catalyst is always in a slightly reduced state. During reduction, electrons are introduced into the molybdenum 4d levels. However, in view of the mixing of these levels, electrons in the bismuth 6p band are always available at any bismuth surface

93

site where the catalyst is reoxidized. Such valence degeneracy is a necessary requirement for catalysts where reduction and oxidation occur at different sites with different cations.

Another aspect of these catalysts is that the lattice oxygen must migrate from the site of reoxidation to the site where oxygen was removed from the lattice. We know that for the most selective oxidation catalysts this process is sufficiently fast so that it does not affect the rate of olefin oxidation. It is generally believed that oxygen mobility in bismuth molybdates is very high, but there are very few reliable data. It may be of significance, however, that the highest oxygen mobility yet observed is for bismuth oxide doped with tungsten or molybdenum (31).

There are many different requirements for a good catalyst for selective oxidation of olefins. There must be appropriate surface sites for allyl and hydroxyl formation and for catalyst reoxidation. There must be a special electronic structure involving valence degeneracy, and there must be high oxygen mobility. Optimization of all of these factors is a fascinating challenge for a materials scientist. Here, as with most catalysts, the developments to date have been largely empirical with subsequent rationalizaton, but recent understanding is improving this situation.

Conclusions

Heterogeneous catalysis has traditionally been a largely empirical science. Many obviously critical experiments have never been performed. This has encouraged speculation on mechanisms to run rampant. Although reliable data are now emerging at an encouraging rate, much current speculation is at odds with these observations. Many simplistic explanations of catalyst behavior have not stood the test of time. The true explanation in catalysis is generally very complex, and it may encompass several of the simplistic interpretations. A good catalyst combines many important factors, as has been illustrated for the selective oxidation catalysts. All too often we do not even have a good understanding of what these various factors are. Theoretical calculations can occasionally rationalize what we know from experiment. There is hope that such calculations may soon have some predictive value.

In view of the complexity of heterogeneous catalysts, many workers shrink from the challenge of bringing understanding to this field. However, there are now many techniques to study surface composition and structure. Advances in analytical electron microscopy make it possible to see features of catalysts that we could only ponder on a few years ago. These techniques are of immense importance, but breakthroughs will require significant contributions from scientists of different disciplines.

References and Notes

1. J. J. Ostermaier, J. F. Katzer, W. H. Manogue, *J. Catal.* **41**, 277 (1976).
2. T. Inui, T. Sezume, K. Miyaji, Y. Takegami, *J. Chem. Soc. Chem. Commun. No. 873* (1979).
3. G. J. Den Otter and F. M. Dautzenbert, *J. Catal.* **53**, 116 (1978).
4. P. A. Sermon and G. C. Bond, *Catal. Rev.* **8**, 211 (1974).
5. D. Bianchi *et al.*, *J. Catal.* **59**, 467 (1979).
6. J. R. Katzer, J. Lorntson, T. Fukushima, J. M. Schults, private communication; F. Lytle, private communication; K. F. Foger and J. R. Anderson, *J. Catal.* **54**, 318 (1978).
7. S. J. Tauster, S. C. Fung, R. J. Garten, *J. Am. Chem. Soc.* **100**, 170 (1978); R. T. K. Baker, E. B. Prestridge, R. L. Garten, *J. Catal.* **59**, 293 (1979).
8. S. J. Tauster, S. C. Fung, R. L. Garten, R. T. K. Baker, J. A. Horsley, 1978th American Chemical Society National Meeting, Washington, D.C., September 1979, abstr. INOR 167. Recent Auger and photoemission investigations of Pt on reduced $SrTiO_3$ [M. K. Bahl, S. C. Tasi, Y. W. Chung, *Phys. Rev. Sect. B* **21**, 1344 (1980)] also indicate that the charge transfer is from $SrTiO_{3-x}$ to Pt. The electrons from $SrTiO_{3-x}$ may come largely from surface states (Ti^{3+}) rather than from the bulk.
9. G. A. Somorjai, *Science* **201**, 489 (1978).
10. M. Boudart, *Adv. Catal.* **20**, 153 (1969).
11. R. L. Jacobson *et al.*, *Proc. Am. Pet. Inst. Sec. 3* **49**, 504 (1969).
12. J. H. Sinfelt, *Science* **195**, 641 (1977).
13. F. L. Williams and D. Nason, *Surf. Sci.* **45**, 377 (1974).
14. J. H. Sinfelt, U.S. Patent 3,953,368 (1976).
15. _____ and G. H. Via, *J. Catal.* **56**, 1 (1979).
16. B. Delmon, *J. Less Common Met.*, in press.
17. R. R. Wentrak and H. Wise, *J. Catal.* **51**, 80 (1978).
18. R. J. H. Voorhoeve and J. C. M. Struver, *ibid.* **23**, 243 (1971).
19. H. Robson, *Chem. Technol.* (1978), p. 176.
20. R. M. Barrer, *Molecular Sieves* (Society of the Chemical Industry, London, 1968), p. 39.
21. S. L. Meisel, J. P. McCullough, C. H. Lechthaler, P. B. Weisz, *Chem. Technol.* **6**, 86 (1976).
22. L. D. Rollman, *Adv. Chem.* **173**, 387 (1979).
23. P. Gallezot, *Catal. Rev.* **20**, 121 (1979).
24. J. D. Idol, U.S. Patent 2,904,580 (1959); F. Veatch, J. L. Callahan, E. C. Milberger, R. W. Foreman, *Proc. Int. Congr. Catal. Paris* **2**, 2647 (1960).
25. G. W. Keulks, *J. Catal.* **19**, 232 (1970).
26. C. R. Adams and T. Jennings, *ibid.* **2**, 63 (1963); W. M. H. Sachtler, *Recl. Trav. Chim. Pays-Bas* **82**, 243 (1963).
27. A. W. Sleight and W. J. Linn, *Ann. N.Y. Acad. Sci.* **272**, 22 (1976).
28. A. W. Sleight, in *Advanced Materials in Catalysis*, J. J. Burton and R. L. Garten, Eds. (Academic Press, New York, 1977).
29. R. K. Grasselli, paper presented at the Fourth International Symposium on Catalysis, Roermond, Netherlands, 1978; J. D. Burrington and R. K. Grasselli, *J. Catal.* **59**, 79 (1979).
30. J. D. Burrington, C. T. Kartisek, R. K. Grasselli, *Am. Chem. Soc. Div. Pet. Chem. Prepr.* (September 1979), p. 1034.
31. T. Takahoshi and H. Iwahara, *J. Appl. Electrochem.* **3**, 65 (1973).
32. G. Delvaux, P. Grange, B. Delmon, *J. Catal.* **56**, 99 (1979).
33. H. W. Haynes, Jr., *Catal. Rev.* **17**, 276 (1978).

A new growth technique, "edge-supported pulling," has been developed in the field of photovoltaics. A sheet of silicon ⟶ *is solidified between graphite filaments. Solar cell efficiencies above 13 percent air mass 1 have been attained on cells made from this material. The process offers the advantages of technological simplicity, minimal skill requirement on the part of the operator, and lower cost than other sheet methods. [Solar Energy Research Institute, Golden, Colorado]*

Photovoltaic Materials

Evelio A. Perez-Albuerne and Yuan-Sheng Tyan

During the last quarter of a century photovoltaic generation of electricity has been proved technically feasible and reliable, and photovoltaic cells became the standard source of power for space vehicles and satellites. Adoption of this technology for widespread terrestrial use has not been feasible, however, since the photovoltaic cells used in space during the 1960's were at least 1000 times too

and to use continuous, automated, low-cost fabrication processes akin to those encountered in the printing and photographic industries.

The materials and device properties required to successfully meet these goals are different for each approach and are intimately related to the fabrication process chosen. In this article, we review the materials requirements of typical sol-

Summary. Solid-state photovoltaic cells are feasible devices for converting solar energy directly to electricity. Recent cost reductions have spurred an incipient industry, but further advances in materials science and technology are needed before photovoltaic cells can compete with other sources for the supply of large amounts of energy. In this article energy loss mechanisms in solid-state photovoltaic cells are examined and related to materials properties. Various systems under development are reviewed which illustrate some key concepts, opportunities, and problems of this most promising emerging technology. Areas where contributions from innovative materials research would have a significant effect are also indicated.

expensive to compete with other methods of electricity generation. The challenges of producing competitive and reliable electrical power in a terrestrial environment have dominated photovoltaic research and development during the past decade and are expected to continue to do so.

Cost has been significantly reduced in the last few years by reducing the manufacturing costs of solar cells substantially identical to those used in space. These cells are produced from sliced wafers of single-crystal silicon, processed by standard methods of semiconductor fabrication and assembled into flat-plate collectors. Since it is generally accepted that this technology will not reach the cost goals required for widespread use, systems based on solar concentrators and thin-film flat plates are also being investigated. In the former, costs are reduced by use of relatively inexpensive materials to concentrate sunlight on a small area of costly but highly efficient cells. The second approach seeks to reduce materials costs to an insignificant level

id-state photovoltaic cells and how they are related to the constraints and demands imposed by the various configurations under study. A view in greater depth can be obtained from comprehensive review articles and books (*1*).

Operation of Photovoltaic Cells

The key component of a photovoltaic cell is a semiconductor that absorbs light energy by exciting an electron from the valence band to the conduction band and leaving a positive hole behind. The electron and hole so generated eventually recombine, giving up the acquired energy to the lattice as heat or emitting light. In a photovoltaic cell, however, a region of high electric field is provided within the semiconductor, so that most of the photogenerated electrons and holes are separated on reaching this region and thus prevented from recombining. The flow of these charges through an external load produces useful work and in this way completes the process of direct conversion of light to electricity.

Intimate contact of two materials, at least one of which is a semiconductor, produces the high-field region if the

chemical potentials of electrons in the two materials are different. Such a structure—a semiconductor junction—can be obtained in various ways in solid-state devices. A *p-n* junction occurs between oppositely doped semiconductors, so that there are excess holes (*p* type) on one side of the junction and excess electrons (*n* type) on the other. The major components of the semiconductors on either side of the junction can be the same (homojunction) or different (heterojunction). Certain metal-semiconductor contacts (Schottky barriers) likewise produce high-field regions in the semiconductor side of the junction. Variations of these structures are achieved when a thin insulating layer is interposed between the two active regions, leading to semiconductor-insulator-semiconductor or metal-insulator-semiconductor configurations. All of these junctions have been used in promising photovoltaic devices, but the materials requirements of each depend on the predominant optical and electronic processes characteristic of that configuration.

Next we focus attention on the mechanisms that limit the efficiency of the charge generation and separation processes described above, and their relation to materials properties. Figure 1 is a schematic view of a solid-state photovoltaic cell.

The current output of solar cells is limited by the number of carriers (electrons and holes) generated by the incident light. Losses inevitably occur when light is prevented from reaching the active semiconductor, and these must be minimized by optimization of front electrode design and reduction of reflection losses. Absorption of light by the active semiconductor requires that the light energy be at least equal to the magnitude of the energy gap separating the valence band from the conduction band of the semiconductor (the band gap). The low-energy fraction of the available solar spectrum is therefore lost without being absorbed, since it falls below the band-gap energy of most semiconductors. A fraction of the light having the proper energy can also be lost if the semiconductor is not thick enough to allow complete light absorption—that is, thinner than the optical absorption length of the semiconductor.

The losses most sensitive to materials quality and device structure are those incurred when photogenerated electrons and holes recombine before they are separated in the high-field region. Figure 2 shows energy band diagrams of a *p-n* homojunction solar cell under three conditions. These diagrams present the elec-

Evelio A. Perez-Albuerne is a senior laboratory head and Yuan-Sheng Tyan is a research associate at the Research Laboratories, Eastman Kodak Company, Rochester, New York 14650.

tron energy as a function of depth into the cell. Solid lines indicate the energy at the edges of the conduction and valence bands, and the dashed line indicates the Fermi level—that is, the chemical potential of the electrons in the semiconductor. Figure 2A illustrates the condition when the terminals of an illuminated cell are directly connected to each other (short circuit) and shows the most important recombination paths. Bulk recombination of carriers (process 1) takes place away from surfaces and interfaces and generally occurs at impurity sites and structural defects in the semiconductor. Minimizing bulk recombination is important for efficient cell operation, since many carriers are generated outside the narrow high-field region and must diffuse to it in order to be separated. The diffusion length is the distance photogenerated carriers diffuse without recombination and ideally is long enough to allow all such carriers to reach the high-field region. Weakly absorbing materials therefore require longer diffusion lengths for efficient carrier collection.

Unique electronic states arise in the vicinity of compositional and structural discontinuities in solids. These states often act as effective recombination centers; consequently, surfaces and interfaces are critical regions in photovoltaic devices. Recombination at the front (illuminated) surface (process 2) is a loss mechanism particularly important in *p-n* homojunction cells, where many carriers are generated in the front active element (Fig. 1). In these cells the effect of front-surface recombination can be reduced by decreasing the thickness of the front active element or by "passivating" the surface by chemical treatments or addition of a compatible, transparent semiconductor layer in front of the cell (heterostructure). In heterojunction cells, the front active element can be made of a semiconductor with a wide band gap, which will absorb only the small fraction of the incident radiation of energy larger than the band gap. In this way most carriers are generated in the back active element and thus are not susceptible to front-surface recombination.

Recombination at the back surface (process 3) becomes important when the thickness of the back active element is comparable to both the optical absorption length of the semiconductor and the carrier diffusion length. In this case a significant number of carriers are generated near the back surface, where the recombination rate is high. The effect of back-surface recombination can be reduced by using a back-surface-field structure, in which the region of the back active ele-

ment adjacent to the electrode is heavily doped. This produces a field that effectively confines the photocarriers to the bulk of the back active element and keeps them away from the electrode.

Defects arising at the interface between two dissimilar materials provide another important recombination path. Interface recombination (process 4) is an important loss mechanism in heterojunctions and Schottky barriers, and is reduced by matching certain key properties of the materials involved, such as lattice parameter, thermal expansion, energy of the conduction and valence band edges, and so on. A critical interface is also found at the grain boundaries

of polycrystalline materials. Recombination at these centers is a crucial loss mechanism in many low-cost, thin-film devices.

Up to now we have considered loss mechanisms that primarily affect the number of photogenerated carriers delivered by the photovoltaic cell, and thus its current output under short-circuit conditions. We now focus on the voltage appearing across the terminals of an illuminated cell connected to a resistive load. Figure 2B shows the energy band diagram under this operating condition. A semiconductor junction has the electrical characteristics of a diode—that is, significant current flows only for one polarity

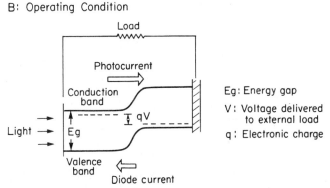

Fig. 1. Schematic exploded view of a generalized solid-state photovoltaic cell.

Front Active Element
• Semiconductor (p-n junction)
• Semitransparent metal (Schottky barrier)

Incident radiation

Front Electrode
• Grid
• Transparent conducting layer

Back Active Element
• Semiconductor

Back Electrode
• Opaque metal

//// High-field region (present only in Back Active Element for Schottky barrier structures)

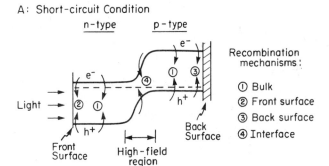

A: Short-circuit Condition

Recombination mechanisms:
① Bulk
② Front surface
③ Back surface
④ Interface

Fig. 2. Electron energy diagram of a *p-n* homojunction photovoltaic cell: (A) short-circuit condition, (B) operating condition, and (C) maximum open-circuit voltage condition. Electrons and holes are denoted e^- and h^+, respectively.

B: Operating Condition

Eg: Energy gap
V: Voltage delivered to external load
q: Electronic charge

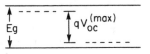

C: Maximum Open-circuit Voltage Condition

97

of applied voltage. This forward polarity is precisely the one appearing across an illuminated solar cell connected to a load. The current thus produced, the diode current, flows in the direction opposite to the photogenerated current discussed above. The net current flow is therefore reduced from that under short-circuit conditions by the magnitude of this diode current, and becomes zero when the external load resistance becomes infinite. This is called the open-circuit condition, and the voltage measured under this condition is the one at which the diode current exactly balances out the photocurrent. High open-circuit voltages, therefore, are obtained for small diode currents.

The magnitude of the diode current for a given voltage bias depends on the current-carrying mechanisms across the junction. The minimum diode current, and thus the maximum open-circuit voltage, occurs when current is carried only by diffusion of energetic carriers over the top of the potential barrier at the junction. Voltage of forward polarity reduces the height of this barrier and increases the diode current. If this voltage is large enough, the barrier vanishes and the diode current can balance any photocurrent level produced by the cell. This condition, therefore, defines the maximum open-circuit voltage of a photovoltaic cell; it is illustrated in Fig. 2C for a homojunction. Inspection of this energy band diagram shows that the maximum open-circuit voltage achievable is slightly smaller than the band gap of the semiconductor, and consequently the use of large-band-gap semiconductors produces high voltages. Such semiconductors, however, produce small currents because they can absorb less radiation. Another consequence of the voltage-output limitation is the occurrence of significant energy losses from the high-energy portion of the solar spectrum. Since the maximum cell voltage is smaller than the band gap, carriers generated by light of energy larger than the band gap can deliver to the load only a fraction of the absorbed energy. The excess energy is dissipated as heat. These factors limit the choice of useful semiconductors to those having band gaps in the range 1.0 to 2.0 electron volts, and favor those with values near 1.5 electron volts (2).

In practice, because the diode current due to diffusion increases exponentially with temperature and additional mechanisms are present for generating diode current, the current-balancing condition at room temperature occurs at lower voltages than that illustrated in Fig. 2C. The prime route to enhanced solar cell performance, therefore, is minimization of the diode current. In particular, note that all the recombination channels mentioned earlier also provide routes for diode current flow. Bulk, interfacial, and surface defects thus decrease not only the current but also the voltage output of the cell.

The power delivered by a solar cell to a load is equal to the product of the current and the voltage across the cell. A typical current-voltage curve for a solar cell is illustrated in Fig. 3. The shape of the curve reflects the voltage dependence of the diode current, since the photogenerated current is nearly independent of voltage. The power delivered by the cell is strongly dependent on the load. When the load is zero (short circuit) or infinite (open circuit), the power output is zero. Between these two extremes the power reaches a maximum at some finite load resistance. The efficiency of a solar cell is usually reported as the ratio of maximum electrical power output to total incident solar energy.

Since cell efficiency depends not only on the values of short-circuit current and open-circuit voltage but also on the shape of the current-voltage curve, it is important to optimize this shape. A quantitative description is given by a parameter called the fill factor, which is defined as the ratio of the maximum output power to the product of short-circuit current and open-circuit voltage. The fill factor depends on diode current mechanisms, but more important, it is influenced by electrode design, contact fabrication, and electrical shorts across the active elements. These effects can be described in terms of an added series resistance between the cell and the load or a shunt resistance across the cell terminals. The former must be minimized and the latter maximized for optimum cell operation. The front electrode and front active element are particularly important in this regard. To maximize the amount of solar radiation reaching the active elements, the front electrode is either a transparent continuous layer or a fine grid structure with minimum shading. Electrode light transmission is usually improved at the expense of increased resistance, leading to a reduction of the fill factor. Increasing the conductivity of the transparent electrode or the front active element, therefore, improves device performance.

The fill factor is also improved by lowering the electrode-semiconductor contact resistance. Semiconductor contact technology, therefore, is by no means trivial, especially where novel semiconductors are involved. Low fill factors also result from materials having very short diffusion lengths or very large interfacial recombination rates. In these cases the photocurrent, normally voltage-insensitive, becomes voltage-dependent and reduces the current output of the cell under load, lowering the overall efficiency.

Equally important to cell efficiency is cell stability over many years of exposure to sunlight and other environmental stresses. Identification of the causes of cell degradation and achievement of adequate lifetime are central issues in photovoltaic research, especially when new materials show encouraging performance.

Silicon

Single-crystal silicon solar cells are the most advanced of all photovoltaic devices. Nearly all the commercial cells sold to date have been silicon cells. This dominance has resulted from the availability of techniques for growing the large, defect-free, silicon single crystals required by the electronics industry. Strictly speaking, silicon does not have the most desirable physical properties for a solar cell device. Its band gap of 1.1 eV is not optimum, and its optical absorption is rather weak; thus active silicon layers must be thicker than 100 micrometers to adequately absorb the solar radiation. This demands carrier diffusion lengths of the same magnitude. High-quality single-crystal silicon with a moderate resistivity (about 1 to 10 ohm-centimeters) and an adequate diffusion length is readily available, and p-n homojunction cells with efficiencies higher than 10 percent have been routinely reported since the late 1950's (3). Progress during the past decade demonstrates that silicon can be fabricated into devices with efficiencies approaching 19 percent (4). Refinements in technology are expected to increase this value.

Improvements in device design account for most of the progress during the decade. Front-surface recombination losses were significantly reduced by decreasing the thickness of the front silicon layer (5), thus reducing the number of carriers generated in the front layer and subject to recombination there. Passivation of the front surface with transparent thin layers, such as SnO_2, also helped to reduce the front-surface recombination loss (6). Current gains from increased light absorption were achieved by using textured surfaces and antireflection coatings, which virtually eliminate light reflection at the front surface (7). Adoption of a back-surface-field structure decreased back-surface recombination, resulting in reduced diode current and increased voltage output (8).

The real challenge in producing solar cells for terrestrial applications is to reduce the overall cost so that they can compete economically with other energy sources. The present high cost of silicon cells includes significant contributions from all fabrication steps—preparing high-purity raw materials, growing single crystals, slicing and polishing the single crystals into thin wafers, fabricating devices, and packaging. Efforts are being made to reduce costs in all these areas. High-capacity, efficient processes for less costly silicon production and purification are being explored, and cells of 15 percent efficiency have recently been fabricated on slices of cut polycrystalline silicon ingots (9). Several processes have provided essentially single-crystal silicon ribbons directly from the melt. These processes can potentially reduce the silicon cell manufacturing costs by eliminating the bulk crystal growth and slicing steps. Cell efficiencies around 15 percent have been obtained with ribbons made by a dendritic web process (10). The next steps in ribbon-growth processes are improving the material quality and increasing the throughput. Alternative methods of junction formation are being investigated; metal-insulator-semiconductor configurations and ion implantation show promise of achieving high-efficiency cells at reduced fabrication costs.

Polycrystalline thin films of silicon deposited on low-cost substrates promise further cost reductions and higher throughput than the ribbon processes. An efficiency approaching 10 percent was recently reported (11) for films prepared on lower-purity silicon bulk substrates, but deposition onto other substrates produces cells with much lower efficiencies. The current output of thin-film silicon cells is much lower than that of single-crystal cells, even though the voltage output is comparable. It is believed that grain boundaries provide efficient recombination centers, where a large fraction of photogenerated carriers are lost. Methods for increasing grain size or passivating grain boundaries are being investigated.

Gallium Arsenide

Gallium arsenide has a much higher optical absorption than silicon, and its ~ 1.4 eV band gap is nearly an ideal match for the solar spectrum. The high optical absorption reduces the thickness

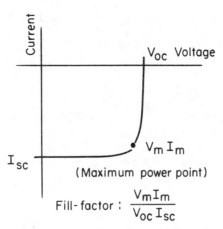

Fig. 3. Current-voltage characteristics of a photovoltaic cell.

of material required for adequate light absorption to a few micrometers, and hence reduces the required photocarrier diffusion length to ~ 2 μm, compared to $\geq$ 100 μm for silicon.

However, GaAs has a set of dominant loss mechanisms which, until the early 1970's, limited the efficiency of single-crystal GaAs cells to about 12 percent (12). For example, as a compound semiconductor, GaAs is more prone to imperfections than silicon, making it more difficult to obtain the crystal and material quality required for even modest carrier diffusion lengths. Moreover, in homojunctions the high optical absorption of GaAs produces a large fraction of photocarriers in the front side of the junction, thus increasing front-surface recombination losses. This enhanced carrier recombination reduces both current and voltage and accounts for the disappointing performance of early GaAs cells.

Advances in materials and device technology have solved many of these problems. Improvements in liquid- and vapor-phase growth techniques have produced GaAs layers with diffusion lengths adequate for high-efficiency cells. These layers are grown epitaxially on low-quality single-crystal substrates, yielding a high-quality single-crystal region where carrier generation, separation, and collection take place. Front-surface recombination losses have been practically eliminated by use of a so-called heterostructure, where p-GaAlAs, which has a wider band gap and low optical absorption, is grown over the p layer of a p-n junction GaAs single-crystal cell. An efficiency of 22 percent has been achieved in such devices (13). Other approaches that have been effective in reducing front-surface recombination and improving cell efficiency include the use of metal-insulator-semiconductor structures (14), AlAs/p-GaAs (15) heterostructures (12), shallow homojunctions (16), and heterostructures where InGaP (17) replaces GaAlAs.

Although GaAs cells are very efficient, fabrication costs and unavailability of materials have limited their use. Two approaches have been used to address these problems. The first uses solar concentrators in which sunlight incident over a large area is optically concentrated onto a small-area solar cell. The electric current produced is roughly proportional to the concentration ratio, which can be very high. Electrode design and contact quality are of prime importance in order to minimize series-resistance losses. Well-designed systems with efficiencies of ~ 22 percent at a concentration ratio of 900 have been reported (18).

The second approach employs thin-film technology. A recent report describes a 21 percent efficient, shallow homojunction GaAs cell, which was fabricated with 4 μm of GaAs epitaxially grown onto a single-crystal Ge substrate (16). This demonstrates that highly efficient cells can be made with single-crystal, thin-film GaAs. Polycrystalline thin-film GaAs cells on various substrates, however, have efficiencies of only 5 to 6 percent without antireflection coatings (19). High photocurrents are usually obtainable but voltage outputs are low—just the opposite of the situation with thin-film Si cells. Since the techniques for growing thin films are similar to those for making high-efficiency epitaxial layer cells, it appears that the poorer performance of polycrystalline cells must result from grain boundaries.

Cadmium Sulfide/Copper Sulfide

Thin-film Cu_2S/CdS cells have followed a different development path and thus have faced different problems. These cells were discovered in 1954, and by the early 1960's the basic structure and manufacturing techniques for a 5 percent efficient thin-film cell were established (20). An understanding of the properties of the cell materials and the device operation, however, took longer to develop and even today is incomplete. The cell preparation process summarized below illustrates some unique features of this cell.

Fabrication normally begins with a conductive substrate onto which a layer of CdS 20 to 30 μm thick is vacuum-evaporated. The Cu_2S layer is prepared by dipping the CdS-coated substrate into a hot, aqueous solution of CuCl. A fine-grid collecting electrode is then applied to the Cu_2S layer. In most cells illumination is from the Cu_2S side; it is generally agreed that Cu_2S absorbs most of the light and is the locus of most photocarrier generation.

Cuprous sulfide prepared by methods other than that outlined above usually yields much poorer results. One exception is a dry process in which a thin layer of CuCl is evaporated onto the CdS layer; this is followed by a heat treatment which, in an ion-exchange reaction, converts the thin surface layer of CdS into Cu_2S (21). Most workers believe that Cu_2S formation occurs through interdiffusion of Cu and Cd without disturbing the sulfur sublattice. Although this so-called topotaxial process (21) produces fewer interfacial defects than other methods, interfacial recombination still plays a major role in limiting cell efficiency (22).

The Cu_2S layer is normally heavily doped and the carrier diffusion length is extremely small (~ 0.2 μm). As a result, the Cu_2S layer in the more efficient cells must be extremely thin (0.1 to 0.3 μm) in order to reduce bulk recombination losses. In spite of the high optical absorption coefficient of Cu_2S, the thin layer does not absorb enough light in one pass to account for the high current observed in the best cells. The unexpectedly high current results primarily from an increase in the area of light-absorbing Cu_2S to several times the area of the cell. This is accomplished by using an etched, textured CdS surface and forming Cu_2S along CdS grain boundaries. Reflection by the substrate of transmitted light back to the Cu_2S also increases current output.

Current gains obtained in this way are accompanied by voltage losses due to differences between junction and cell areas. Since the diode current increases linearly with junction area while the photogenerated current is determined by cell area, a cell with a textured surface has a larger diode current than a planar cell for a given photogenerated current, and hence a lower open-circuit voltage. If penetration of Cu_2S along grain boundaries reaches the back electrode, the cell is shorted. To avoid this problem, relatively thick CdS layers (20 to 30 μm) must be used, which increases materials and fabrication costs.

In spite of these difficult materials limitations, an efficiency of 9.15 percent has been obtained by optimizing the collecting grid structure and reducing light reflection losses from the front surface (23). Further improvements may require more drastic modifications in cell design. One approach is to replace CdS with $Cd_xZn_{1-x}S$, giving a better match with Cu_2S in energy of the conduction band edge and lattice parameter. This increases the voltage, but the efficiency is not improved because the current output is lower (24).

Cell degradation has seriously affected the performance of Cu_2S/CdS cells, and encapsulation has been necessary to increase their lifetime. Stoichiometric changes due to oxidation of the Cu_2S layer are believed to be primarily responsible for the deterioration.

Amorphous Silicon

In addition to the polycrystalline thin films discussed earlier, silicon can be prepared in the form of thin amorphous films. Because of the differences in structure, amorphous thin films have many physical properties that are distinctly different from those of their crystalline counterparts. Optical absorption is much stronger in amorphous films, since lack of crystal symmetry relaxes the selection rules governing the optical transitions; amorphous films are also free of the grain boundaries that plague polycrystalline silicon thin films. Both characteristics make amorphous silicon (a-Si) films attractive candidates for low-cost solar cells. The key limitations of conventional a-Si are its high density of structural defects and, more important, its unsatisfied chemical bonds, which introduce electronic states within the energy gap. Since the density of these states is high (10^{17} to 10^{20} per cubic centimeter per electron volt), they essentially dominate all electronic processes by providing a large number of recombination sites and preventing any observable doping effects; thus conventional a-Si is totally unsuitable for device applications.

It was recently discovered, however, that the density of the gap states is greatly reduced in a-Si films prepared by glow discharge of silane (SiH_4) (25) or reactive sputtering of silicon in a hydrogen-containing ambient (26). These films respond to p- and n-type doping, thus allowing the preparation of a-Si solar cells having either p-n, Schottky barrier, or metal-insulator-semiconductor structures (27). An efficiency of 5.5 percent has been reported (28), although most cells are in the 2 to 3 percent range.

The low efficiency of these cells is evidently due to diffusion lengths so small that only carriers generated within the high-field region near the junction are collected. Since the high-field region is narrow, only a fraction of the incident light is absorbed and the current output of the cells is low. The high-field region can be broadened by placing a relatively insulating a-Si layer between the p and n regions (28). Major improvements in cell efficiency, however, require significantly longer diffusion lengths.

The short diffusion length observed in these materials is apparently related to the band-gap states, which, even in the best a-Si materials, have a much higher density than in crystalline silicon. It has been shown that the reduction in gap states in these new a-Si materials is due to the incorporation of a large amount of hydrogen, as much as 50 atomic percent, during the deposition process (29). The mechanism by which hydrogen incorporation affects the density of gap states is complicated and not well understood. Hydrogen could satisfy all dangling bonds, but this does not seem to account for the large amounts of hydrogen incorporated into these films—roughly 100 times the amount of gap states identified by electron spin resonance spectroscopy (30). Infrared spectroscopy shows the existence of SiH, SiH_2, and SiH_3 bonds, and the distribution of hydrogen among these bonds depends strongly on the method and conditions of preparation. Once the role of hydrogen in a-Si is better understood, a further reduction of gap states is likely. Simultaneous incorporation of fluorine and hydrogen into a-Si reduces the gap states even further (31), and beneficial effects of oxygen have also been reported (32). It seems likely that the quality of a-Si will continue to improve with further chemical modification.

Other Semiconductors

Solar cells have been prepared with many other semiconductors, but efficiencies have been poor, even though many semiconductors have nearly optimum band gaps and are good light absorbers. Part of the difficulty is that little is known about the properties of most of these semiconductors. Materials of good quality are not available, and junction- and contact-forming techniques have not been established. This is especially true for thin-film cells, in which grain boundaries and film-substrate interactions complicate matters further. A few materials, however, show promise.

Indium phosphide is a III-V semiconductor, as is GaAs, but its materials technology is much less advanced. Heterojunction cells made with CdS films evaporated onto single-crystal InP have efficiencies of about 15 percent, although the efficiency of analogous thin-film cells is less than 6 percent (33). Cadmium telluride is another compound that has the appropriate band gap and very high optical absorption, and it can be prepared in p and n types. Effective use of this material, however, seems hindered by materials technology, and the best efficiency reported for single-crystal cells is about 12 percent (34). This material can be readily prepared in thin-film form, and efficiencies as high as 8.7 percent have been observed in thin-film CdS/CdTe heterojunction cells (35). An interesting ternary material, $CuInSe_2$, produces heterojunction cells with CdS yielding 12 and 6 percent efficiencies in the single-crystal (36) and thin-film (37) forms, respectively.

Conclusions

Photovoltaic solar energy conversion has been considered from the viewpoint of materials, with particular emphasis on cell efficiencies. Promising solar cells are obtainable from a variety of materials, provided cell design and fabrication processes are matched with key materials properties. A weakly light-absorbing material such as silicon is useful because current fabrication procedures yield nearly perfect materials with very large diffusion lengths. Strongly absorbing materials such as GaAs yield efficient devices because surface recombination is effectively reduced by proper device design and improved crystal-growth techniques. Heterojunction Cu_2S/CdS cells with moderate efficiencies, in spite of their thin-film polycrystalline format and mismatch at the interface, are fabricated by processes that minimize interface defects and enhance light capture. Finally, noncrystalline silicon shows promise, as a result of chemical modifications that remove defects inherent in amorphous systems.

Photovoltaic research and development is a relatively young, rapidly moving discipline, but the simultaneous achievement of adequate cell efficiency, low materials and fabrication costs, and long-term stability are yet to be demonstrated. Some existing systems are promising, but further advances must be made in materials properties and in automated, large-volume, cost-effective manufacturing processes before photovoltaic conversion can compete with conventional sources for large-scale electrical power generation.

The rapid growth of innovative research and the beginning of significant industrial production of photovoltaic systems are positive and encouraging. We look forward with optimism to a decade in which this most desirable process of energy conversion may come to fruition and help meet the world's changing energy needs.

References and Notes

1. B. T. Debney and J. R. Knight, *Contemp. Phys.* **19**, 25 (1978); H. J. Hovel, *Semiconductors and Semimetals* (Academic Press, New York, 1975), vol. 11; paper presented at the International Electron Devices Conference, Washington, D.C., 1979; F. A. Shirland and P. Rai-Choudhury, *Rep. Prog. Phys.* **41**, 1839 (1978); H. Kelly, *Science* **199**, 634 (1978); E. Bucher, *Appl. Phys.* **17**, 1 (1978); R. M. Moore, *Sol. Energy* **18**, 225 (1976); H. Ehrenreich and J. H. Martin, *Phys. Today* **32**, 25 (September 1979); J. Javetski, *Electronics* **52**, 105 (July 1979).
2. J. J. Loferski, *J. Appl. Phys.* **27**, 777 (1956).
3. M. Wolf, in *Proceedings of the 25th Power Sources Symposium* (PSC, Red Bank, N.J., 1972), p. 120.
4. J. G. Fossum and E. L. Burgess, *Appl. Phys. Lett.* **33**, 238 (1978).
5. J. Lindmayer and J. Allison, in *9th IEEE Photovoltaic Specialists Conference* (IEEE, New York, 1972), p. 83.
6. J. Chevalier and F. Duenas, in *2nd E.C. Photovoltaic Solar Energy Conference* (Reidel, Dordrecht, 1979), p. 817; I. Chambouleyron, E. Saucedo, J. Montoya, in *ibid.*, p. 647.
7. R. A. Arndt, J. F. Allison, J. G. Haynes, A. Meulenberg, Jr., in *11th IEEE Photovoltaic Specialists Conference* (IEEE, New York, 1973), p. 40.
8. P. A. Iles, in *8th IEEE Photovoltaic Specialists Conference* (IEEE, New York, 1970), p. 345; H. Fischer and W. Pschunder, in *ibid.*, p. 70.
9. J. Lindmayer and Z. C. Putney, paper presented at the 14th IEEE Photovoltaic Specialists Conference, San Diego, 7–11 January 1980.
10. R. B. Campbell, A. Rohatgi, E. J. Seman, J. R. Davis, P. Rai-Choudhury, B. D. Gallagher, paper presented at the 14th IEEE Photovoltaic Specialists Conference, San Diego.
11. T. L. Chu, S. S. Chu, C. L. Lin, R. Abderrassoul, *J. Appl. Phys.* **50**, 919 (1979).
12. J. M. Woodall and H. J. Hovel, *J. Vac. Sci. Technol.* **12**, 1000 (1975).
13. _____, *Appl. Phys. Lett.* **30**, 492 (1977).
14. Y. C. M. Yeh and R. J. Stirn, *ibid.* **33**, 401 (1978).
15. W. D. Johnston, Jr., and W. M. Callahan, *ibid.* **28**, 150 (1976).
16. J. C. C. Fan and C. O. Bozler, in *2nd E.C. Photovoltaic Solar Energy Conference* (Reidel, Dordrecht, 1979), p. 938.
17. G. H. Olsen, M. Ettenberg, R. V. D'Aiello, *Appl. Phys. Lett.* **33**, 606 (1978).
18. R. Sahai, D. D. Edwall, J. S. Harris, Jr., *ibid.* **34**, 147 (1979).
19. S. S. Chu, T. L. Chu, Y. T. Lee, paper presented at the 14th IEEE Photovoltaic Specialists Conference, San Diego, 7–11 January 1980.
20. F. A. Shirland, *Adv. Energy Convers.* **6**, 201 (1966).
21. T. S. teVelde and J. Dieleman, *Philips Res. Rep.* **28**, 573 (1973).
22. A. Rothwarf, J. Phillips, N. Convers-Wyeth, in *13th IEEE Photovoltaic Specialists Conference* (IEEE, New York, 1978), p. 399.
23. A. M. Barnett, J. A. Bragagnolo, R. B. Hall, J. E. Phillips, J. D. Meakin, *ibid.*, p. 419.
24. T. L. Hench and R. B. Hall, in *2nd E.C. Photovoltaic Solar Energy Conference* (Reidel, Dordrecht, 1979), p. 379.
25. W. F. Spear and P. E. LeComber, *Solid State Commun.* **17**, 1193 (1975); *Philos. Mag.* **33**, 935 (1976).
26. W. Paul, A. J. Lewis, G. A. N. Connell, T. D. Moustakas, *Solid State Commun.* **20**, 969 (1976).
27. D. E. Carlson and C. R. Wronski, *Appl. Phys. Lett.* **28**, 671 (1976).
28. D. E. Carlson, *IEEE Trans. Electron Devices* **ED-24**, 449 (1977).
29. _____, C. W. Magee, A. R. Triano, *J. Electrochem. Soc.* **126**, 688 (1979).
30. H. Fritzsche, C. C. Tsai, P. Persans, *Solid State Technol.* **21**, 55 (Jan. 1978); M. H. Brodsky, *Thin Solid Films* **50**, 57 (1978).
31. S. R. Ovshinsky, *Nature (London)* **276**, 482 (1978).
32. K. Ishii, M. Naoe, S. Yamanaka, S. Okano, M. Suzuki, *Jpn. J. Appl. Phys.* **18**, 1395 (1979); M. A. Paesler et al., *Phys. Rev. Lett.* **41**, 1492 (1978).
33. J. L. Shay, S. Wagner, M. Bettini, K. J. Bachmann, E. Buehler, *IEEE Trans. Electron Devices* **ED-24**, 483 (1977).
34. K. Yamaguchi, H. Matsumoto, N. Nakayama, S. Ikegami, *Jpn. J. Appl. Phys.* **15**, 1575 (1976).
35. H. Uda, H. Taniguchi, M. Yoshida, T. Yamashita, *ibid.* **17**, 585 (1978).
36. J. L. Shay, S. Wagner, H. M. Kasper, *Appl. Phys. Lett.* **27**, 89 (1975).
37. L. L. Kazmerski, F. R. White, G. A. Sanborn, A. J. Merrill, M. S. Ayyagari, S. D. Mittleman, G. K. Morgan, in *12th IEEE Photovoltaic Specialists Conference* (IEEE, New York, 1976), p. 534.
38. Critical reviews of the manuscript and useful suggestions by B. W. Rossiter, D. J. Trevoy, L. C. Isett, and C. W. Tang are gratefully acknowledged.

III-V Compounds and Alloys: An Update

Jerry M. Woodall

Nowadays, meaningful discussions of solid-state materials are usually coupled to their function, either for the study of solid-state phenomena or for use in the realization of a device. This is particularly true for semiconductor materials. Nearly all research on the preparation of semiconductor materials is motivated by anticipated advantages from their use in advanced device technology, and it is rare to find such research in this field for its own sake. A semiconductor research

several that have been demonstrated using III-V's, for only a few—microwave devices and light-emitting diodes (LED's)—have economic processing procedures been developed.

To focus in more detail on this issue, I compare in Table 1 some relevant properties of silicon with those of the III-V compound GaAs, a material widely researched and currently used in commercial devices. At first glance, those who recognize the significance of these

rust than GaAs and can be more reliably processed into intricate devices.

Nevertheless, in the research and development laboratory, III-V materials work is very visible and appears to be increasing. In view of the commercial position of silicon, it may well be asked (particularly by board members of companies researching III-V materials) why. Several compelling reasons justify this research. First, some of the III-V materials have electrical properties that should lead to transistor devices with significantly higher performance than those made with silicon. Second, most III-V materials have optoelectronic properties that silicon does not have, and hence can be made into devices with unique or exclusive functions—for example, injection electroluminescence and injection lasing. Third, the III-V compounds can be alloyed or mixed together. This allows the formation of material with continuously variable properties, such as band-gap energy, charge carrier mobility, and lattice constant, and adds several new dimensions in the design and function of new device structures. For example, lattice-matched heterojunctions formed between III-V compounds or alloys with different band gaps can confer optical, electrical, and chemical advantages over silicon devices and circuits. When these advantages are considered for transistor devices and circuits, it is quite likely that this will lead to improved processing methods with better reliability, lower cost, and higher yields. The implications of this possibility have been partly responsible for a resurgence of research on III-V materials applied to data-processing devices and circuits. This article highlights some of the advances in III-V materials and devices that have led to this situation. As this discussion cannot be comprehensive, the material presented will reflect to some degree the biases of the author.

Summary. The III-V compounds and alloys have been studied for three decades. Until recently, these materials have been commercialized for only a few specialized optoelectronic devices and microwave devices. Advances in thin-film epitaxy techniques, such as liquid phase epitaxy and chemical vapor deposition, are now providing the ability to form good quality lattice-matched heterojunctions with III-V materials. New optoelectronic devices, such as room-temperature continuous-wave injection lasers, have already resulted. This newfound ability may also affect the field of high-speed integrated circuits.

effort usually considers the material, the device, and processing as a unit. It is within this framework that III-V (1) compound semiconductor research has evolved.

From a commercial point of view, the current economic impact of III-V materials is barely noticeable. As is well known, the semiconductor world is dominated by silicon devices. This is because silicon devices and integrated circuits are the result of an optimal combination of material properties and reliable low-cost processing methods. Optimal combination is the key consideration (2), for no matter how exciting the transport or optical properties of some newly made material are, it will never see the marketplace unless an economically competitive device, circuit, or electronic system can be made from it. The III-V materials have not yet effectively competed with silicon, for of the many devices that have been conceived and the

properties in terms of potential device performance might wonder why silicon is used at all (3). For instance, the direct band gap implies high radiative recombination efficiencies (efficient light-emitting devices). The larger band-gap energy generally means higher device operating temperatures. Higher carrier mobilities and larger electron saturation drift velocities suggest transistors with higher speed or lower power dissipation. However, there are factors not shown in Table 1 that are equally important. First, silicon is a very abundant element and is easily refined into a high-purity form. Silicon can be easily converted into bulk single crystals of almost any desired size and shape and then sliced into wafers with dimensions that facilitate economic device processing. The thermally grown oxide of silicon, SiO_2, used for passivation and several device structures is both electronically and chemically superior to the oxides grown on GaAs. The control of the electronic properties of silicon during processing is superior to that of GaAs. Simply stated, silicon has a better

Structures and Devices

To put into proper context the discussion of the advances in materials preparation techniques, it is useful to describe some generic structures and phenomena related to a broad class of devices, particularly optoelectronic devices based on heterojunction structures.

It is not overstating the case to say that over the past 15 years the major impact of III-V materials has been in the area of optoelectronic devices and that one of the most important reasons for this has been the development of the

The author is a research staff member at IBM T. J. Watson Research Center, Yorktown Heights, New York 10598.

102

ability to epitaxially form good quality latticed-matched heterojunctions between dissimilar materials in a controllable and reproducible fashion. In my opinion, this accomplishment represents the single most important advance to date in the field of III-V materials science and technology.

Most of the beneficial aspects of lattice-matched heterojunctions can be generically understood by considering Fig. 1. A somewhat schematic representation of an energy band diagram for a typical III-V material is shown in Fig. 1a. The ordinate direction represents electron energy and the abscissa direction represents a spatial coordinate. The lines marked E_c and E_v show the energies of electrons in the bottom of the conduction band and the top of the valence band, respectively, as a function of position in the crystal. The termination points of these lines are the crystal surfaces. The line marked E_f is the Fermi energy level and is uniform throughout, indicating no net current flow. In this example, E_f is near the bottom of the conduction band in the interior of the crystal, indicating that the crystal is n-type and hence electrons are the majority carrier. Notice that the lines E_c and E_v are parallel throughout and are bent upward at the surface of the crystal. This is due to a large density of surface states associated with the native oxides present on III-V material surfaces. These states are distributed between E_c and E_v and can trap both electrons and holes near the surface. If the crystal is n-type, the states will trap electrons, producing a negative charge on the surface. This negative charge is balanced by the positive charges, due to the ionized donors stripped of their electrons, distributed near the surface. The density of surface states is so large that E_f tends to be "pinned" over a relatively small energy range. This pinning phenomenon is not well understood and is a topic of great interest in the field of surface science. The consequence of the "band bending" due to E_f pinning will become apparent.

Let us introduce a hole—a positive mobile charge—into the crystal near the surface. In a real device, this is accomplished by a hole-injecting junction (a p-n junction) or by absorption of a photon. In n-type material, holes are minority carriers that greatly perturb the equilibrium state of the crystal, and hence the crystal is eager to eliminate them. One method is direct recombination with electrons, which results in the emission of photons. This occurs, for example, in GaAs and is the mechanism responsible for photoluminescence and injection

electroluminescence. Another way to eliminate holes is to have them cross a nearby p-n junction, producing a photocurrent. This process occurs in photodetector and solar cell devices. For the situation shown in Fig. 1a, it is most likely that holes near the surface will recombine through the surface states. This represents a loss in most optoelectronic devices and was a very important problem before the development of lattice-matched heterojunctions.

Next consider the lattice-matched heterojunction shown in Fig. 1b. This is similar to Fig. 1a, except that the surface of the n-type material, A, has a thin epitaxial n-type layer of a lattice-matched material, B, having a larger band gap. In this case, material B was chosen to have the same E_c as material A. As before, let us introduce some holes into material A. In this case, the hole will not be lost by surface recombination. The surface of material A in Fig. 1a is replaced by an interface in Fig. 1b. When a hole approaches the interface, it meets an ener-

gy barrier. This barrier is a consequence of material B having a larger band gap than material A. Elimination of the holes will now most likely occur by radiative recombination, unless, of course, the material contains other impurities or defects that act as traps. If we now replace one of the layers of material B with a p-n junction, we have one of the basic configurations of currently important optoelectronic devices. If light is absorbed by material A, it acts as an efficient photodetector or solar cell. If the p-n junction is electrically forward-biased, it can operate as either an LED or an injection laser.

Finally, if the lattice mismatch between material A and material B is sufficiently large (> 0.5 percent), the diagram in Fig. 1c will result. In this example, the surface recombination of Fig. 1a is replaced by interface recombination due to midgap states associated with a density of "misfit" dislocations, which have formed to accommodate the lattice mismatch.

Let us now examine the "menu" of III-V materials available to form heterojunctions. The rich variety of available combinations of band gaps, materials, and lattice constants are shown in Fig. 2. This variety has encouraged the study of ternary and quaternary III-V materials, and the development of epitaxy techniques for these materials has been responsible for many recent advances in optoelectronic devices. The chemical rule that determines the allowable compositions is that the sum of the atom fractions of the group III elements must equal the sum of the atom fractions of the group V elements in the crystal. For example, the ternary alloy of Ga, Al, and As can be expressed as $Ga_{1-x}Al_xAs$, where $0 \leq x \leq 1$, and x is arbitrarily assigned to either of the group III elements. Similarly, the quaternary alloy of Ga, In, As, and P can be expressed as $Ga_{1-x}In_xAs_{1-y}P_y$. The points in Fig. 2 designate "pure" III-V compounds; the lines are ternary alloys that connect two different III-V compounds, whose compositions are expressed as band gaps and lattice constants.

A menu of some widely researched de-

Table 1. Properties of silicon and GaAs relevant to high-speed and optoelectronic device performance.

Property	Si	GaAs
Band-gap energy (eV)	1.1	1.4
Band-gap type	Indirect	Direct
Intrinsic electron mobility (cm²/V-sec)	1,450	10,000
Electron mobility for 10^{17} cm^{-3} (cm²/V-sec)	700	4,500
Electron saturation drift velocity (cm/sec)	1×10^7	2×10^7
Practical p-type conductivity for n-p-n transistors (ohm-cm)$^{-1}$	16	65

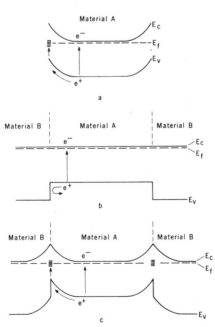

Fig. 1. Energy band diagram of a typical III-V material for three different cases: (a) surfaces with native oxides, (b) surfaces covered with a lattice-matched epitaxial material with a larger band gap, and (c) surfaces covered with a lattice-mismatched material with a larger band gap.

103

vice structures is shown in Fig. 3. This figure is a highly schematic representation of generic cross sections of some currently important devices that utilize III-V compounds and alloys. Several of these structures will be discussed in more detail below (4). In Fig. 3, E_1 and E_2 refer to materials having different energy gaps, and, except for Fig. 3g, E_1 and E_2 have nearly the same lattice constants. For example, E_1 and E_2 could be either $Ga_{1-x}Al_xAs$ having different x values, or $Ga_{1-x}In_xAs_{1-y}P_y$ having different x or y values, or both (see Fig. 2). Figure 3g refers to a special type of LED made of either $GaAs_{1-x}P_x$ or GaP in which the lattice constant of the substrate, E_1, does not closely match the LED p-n junction material, E_2. A layer "graded" in composition from E_1 to E_2 is necessary to prevent the formation of defects that degrade the performance of the LED. It can be seen that except for $GaAs_{1-x}P_x$ and GaP LED's (both of which have significantly affected the small-character-density display business) and GaAs metal-semiconductor field-effect transistors (MESFET's), most of the other devices employ heterojunctions. Of these, an injection laser having two heterojunctions and referred to as the double-heterostructure (DH) laser is perhaps the most unique. Its principles of operation can be understood with reference to Fig. 1b. If one of the material B layers in Fig. 1b is made p-type and the other made n-type, and the material A layer is made n- or p-type and about 0.2 micrometer thick, this would describe the most common form of the DH laser. Minority carriers are injected

into material A across the p-n heterojunction from material B. The other heterojunction "confines" the minority carrier to material A. When the density of the minority carriers exceeds a threshold value, stimulated emission or lasing occurs. The emission is guided out of material A because material B has a lower index of refraction than material A. As a result of the carrier confinement and waveguide properties of the structure, the laser can be made to operate in the continuous-wave mode at room temperature and be modulated at frequencies above 10^9 hertz. This accounts for the development of the use of this type of laser in large-bandwidth optical fiber communication links.

Currently, the most widely studied lasers are those having either GaAs-$Ga_{1-x}Al_xAs$ or InP-$Ga_{1-x}In_xAs_{1-y}P_y$ heterojunction structures. On the other hand, Fig. 2 indicates that it should be possible to use a variety of materials to make devices that lase at wavelengths covering a large portion of the visible and infrared (IR) spectrum. In fact, the wavelengths of interest are those that have the best transmission characteristics in the currently available optical fiber materials. Most of the current research activity is concerned with developing lasers that operate at the desired wavelengths.

Two other optoelectronic devices should be briefly noted: photodetectors and solar cells. As discussed above, these devices produce current or voltage as the result of minority carriers generated by the absorption of light crossing a p-n junction. Research on photodetec-

tors is very active, and most of it is directed toward IR devices for military applications. In this field III-V materials do not always have significant advantages over other semiconductors such as II-VI materials. For solar cells the situation is somewhat different. Theoretically, materials with a direct band gap of 1.4 to 1.6 electron volts are the most interesting, and III-V materials of this type have produced solar-to-electrical conversion efficiencies that are equal to and even better than those for silicon. Three of the most notable device structures include p-p-n $Ga_{1-x}Al_xAs$-GaAs-GaAs, n-p GaAs, and p-n InP-CdS cells.

Special mention must be made of GaP and $GaAs_{1-x}P_x$ LED's. Since the early 1970's the visible-wavelength LED has become one of the more technologically important additions to the field of optoelectronic devices. Because of its inherently large energy gap, the first LED's were realized in the GaAs-GaP ($GaAs_{1-x}P_x$) alloy system (see Fig. 2). The most familiar devices have been those formed from $GaAs_{0.6}P_{0.4}$, where the energy gap is direct and strong room-temperature electroluminescence is easily obtained through homojunction formation. These have been used in many alphanumeric display applications and have attained external quantum efficiencies of about 0.1 percent; they emit radiation in the red portion of the spectrum.

Larger gap compositions of $GaAs_{1-x}P_x$ have also been used for LED's. However, because the band gap is indirect for $x > 0.45$ and therefore inherently inefficient for intrinsic radiative recombination, selected "deep-level" impurities must be used for these compositions to assist the recombination process. Historically, GaP was the first indirect material for which selected impurities were used in the fabrication of LED's. The use of zinc and oxygen produced efficient red-emitting diodes, whereas the use of nitrogen (N), a so-called isoelectronic trap, produced efficient green-emitting diodes. These impurities act to confine carriers in their neighborhood. For indirect materials this relaxes the usual crystal momentum selection rules forbidding direct electron-hole recombination and makes the material appear quasi-direct.

These results were then extended to the $GaAs_{1-x}P_x$ system, allowing the fabrication of LED's with a continuously variable emission wavelength. There are commercially available $GaAs_{1-x}P_x$:N LED's that emit yellow and red light. The discovery and utilization of radiation-enhancing impurities in indirect

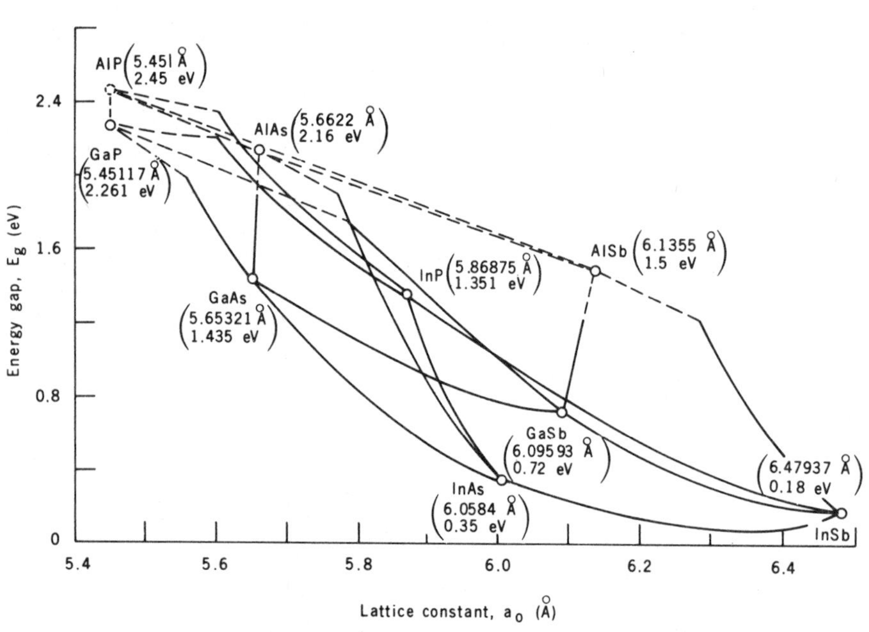

Fig. 2. Diagram representing energy gaps and lattice constants of III-V compounds and ternary alloys.

band-gap materials has been an important milestone in the development of optoelectronic devices.

So far, this discussion of structure and devices has focused on optoelectronic applications. In the past 3 or 4 years, however, there has been a large increase in research on III-V materials applied to high-speed devices such as field-effect and bipolar transistors. Recently, many integrated circuit functions previously limited to silicon-based structures have been realized in GaAs-based structures. In addition, the performance of most, if not all, of the GaAs circuits—as measured by the product of the dissipated power and circuit delay time, a figure of merit for logic circuits—greatly exceeds that of equivalent silicon circuits. Currently, the most widely researched material and device in this field is the GaAs MESFET, which has a relatively simple structure compared to an injection laser. The basic device consists of a thin n-type layer epitaxially grown on or ion-implanted into a semi-insulating substrate. There are two ohmic contacts, the "source" and the "drain," separated by a metal Schottky barrier "gate." An in-depth discussion of the various MESFET structures and their many circuit functions is beyond the scope of this article. It is sufficient to note that varying the voltage on the gate electrode produces a varying resistance to current flow between the source and drain electrodes. Consequently, the materials parameters of primary interest are electron mobility, saturation drift velocity, band gap, and Schottky barrier height. Some of the most important device processing considerations are (i) the reliable production of large, high-purity, semi-insulating GaAs crystals with thermal stability during processing; (ii) a controllable ion implantation technology including thermal activation of dopants; and (iii) Schottky barrier height control.

This discussion of structures and devices has been exemplary rather than comprehensive in nature. It is intended to help put the advances in materials technology described below into proper perspective.

Chemistry and Bulk Crystal Growth

Currently, most semiconductor devices are fabricated on single-crystal substrates of the same material. As a result, there is a demand for reliable bulk crystal growth methods that produce high-quality single-crystal boules of sizes and shapes such that they are easily formed into wafers useful for device pro-cessing. For silicon, bulk crystal growth is not considered a problem area. The same is not true for the widely used bulk crystals of III-V materials. The crystal growth problems of these materials can be understood by considering Table 2, which lists some properties of III-V compounds relevant to bulk crystal growth. It can be seen that the advantages of a wide range of band gaps and lattice constants are somewhat offset by the problems presented by the large variations in melting point, dissociation pressure, and chemical reactivity among the compounds. This means that a universal technique or apparatus is not likely to produce the best material for all the III-V compounds. Rather specialized procedures, apparatus, and containment materials must be used for each compound.

To illustrate this point, consider what would happen if the Czochralski method and apparatus for the growth of bulk silicon crystals were used unmodified to grow GaAs. First, how do we get GaAs? For the purpose of bulk crystal growth, the III-V compounds are usually synthe-sized from the elements. Rapid and efficient synthesis of GaAs occurs when As is added to a Ga melt held at a temperature slightly above the freezing point of the stoichiometric GaAs liquid, $\approx 1240°C$. The stoichiometric liquid has an As vapor pressure of about 1 atmosphere in equilibrium with it. Freezing this melt would produce single-phase solid GaAs. Freezing melts with higher or lower As vapor pressures would result in nonstoichiometric GaAs and, in some cases, two-phase material. The second phase could be either occluded Ga or voids due to trapped As vapor or oc-cluded As. Thus an attempt to use the unmodified silicon apparatus for GaAs would result in As evaporating away from the melt and condensing on the colder parts of the apparatus, leaving behind nearly pure Ga. Even if As evapora-tion were prevented, the melt would become contaminated with silicon because of reaction of Ga with the fused SiO_2 crucible normally used to contain silicon melts. This silicon contamination leads to uncontrolled n-type doping in GaAs

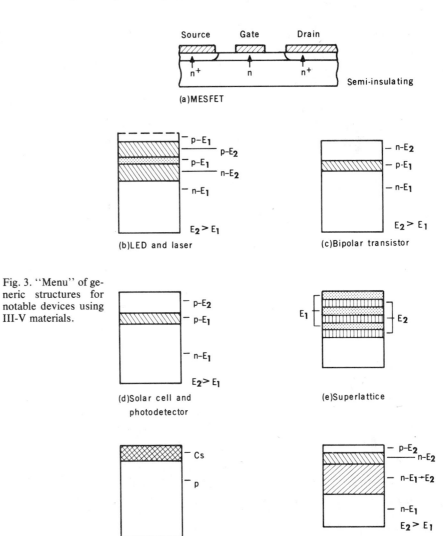

Fig. 3. "Menu" of generic structures for notable devices using III-V materials.

(a) MESFET

(b) LED and laser

(c) Bipolar transistor

(d) Solar cell and photodetector

(e) Superlattice

(f) Negative electron affinity device

(g) LED

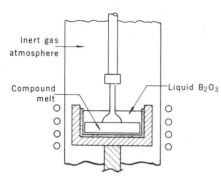

Fig. 4. Schematic of the liquid encapsulation method for bulk crystal growth illustrating the containment of the volatile group V element in the melt.

crystal and is still a topic of research. Thus the thermochemistry of III-V materials plays an important role in the development of bulk crystal growth methods.

In the past, the dissociation problem has been handled with moderate success in the laboratory. The basic solution was to provide a condensing surface for the volatile element (As from GaAs) whose temperature would produce a solid-gas equilibrium vapor pressure equal to that over the stoichiometric liquid—615°C for As solid → 1/4As$_4$ gas (1 atm). The remaining internal surfaces of the apparatus were maintained above the condensing temperature. This was accomplished in two widely used devices. One was the modified Czochralski apparatus, which contained internally glass-sealed magnetic lifting and rotating devices. The other was a modified horizontal Bridgman apparatus, which was a sealed fused SiO$_2$ device with an As "cold" finger at 615°C and a boat, usually made of fused SiO$_2$, containing the melt at 1240°C.

These two methods adequately served the research community in the past. As the need for device circuit-type substrates has increased, however, so has the need for better crystal growth methods. Several developments have helped to solve this problem, one of which is illustrated in Fig. 4. The technique, called the liquid encapsulation (LEC) method, is essentially the Czochralski method with two important modifications. First, the melt is completely encapsulated by a viscous glass, B$_2$O$_3$, which prevents melt dissociation when the dissociation vapor pressure is matched or exceeded by the pressure of an inert gas applied in the apparatus. Second, the apparatus has features common to pressure vessels that contain lifting and rotating feed-through. This apparatus is now routinely used to grow GaAs, InP, and GaP single crystals in sizes useful for device fabrication. The problems that remain for this method ap-

plied to GaAs include uncontrolled background impurity contamination and problems with crystal perfection and functional reproducibility. Another technique recently modified to improve size and shape control is the vertical Bridgman or Stockbarger method, shown in Fig. 5. In this method the melt, which is GaP in this example, fills the vessel made of pyrolytically formed BN (Boralloy); B$_2$O$_3$ is also used, as it is for the LEC method. The advantage of this method is that the crystal assumes the cross section of the vessel, which can be made circular. Hence the crystal can be sliced into circular wafers, which is very desirable for device processing. This method has been used to grow single crystals of GaP, InP, and, with suitable temperature profiles, Ga$_{1-x}$In$_x$P.

In addition to method modifications, significant progress has been made in reducing background contamination of the melt and improving crystalline perfection. Use of the more exotic refractory materials such as AlN and BN has reduced silicon contamination of the melt, especially in the LEC and Stockbarger methods. The silicon contamination from SiO$_2$ vessels, especially in the horizontal Bridgman method, has been reduced by a chemical trick. Silicon contamination in GaAs occurs because of the reaction of Ga with SiO$_2$ to produce Ga$_2$O vapor and Si. Even though SiO$_2$ is much more stable than the oxides of Ga and As, the reaction still produces silicon contamination and unwanted n-type doping of about 5×10^{16} to 50×10^{16} Si atoms per cubic centimeter of melt. This exceeds the desired controlled doping levels for several applications. The chemical trick to reduce silicon contamination is to add Ga$_2$O$_3$ to the Ga melt prior to synthesis. The Ga$_2$O$_3$ reacts

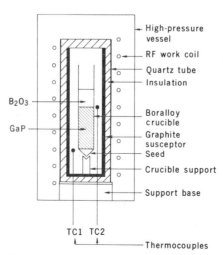

Fig. 5. Schematic of modified Stockbarger method for bulk crystal growth.

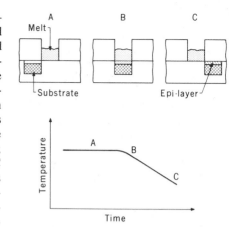

Fig. 6. Simplified representation of the liquid phase epitaxy method.

readily with Ga to produce Ga$_2$O vapor before the Ga reacts with the SiO$_2$ during heat-up. The initial presence of Ga$_2$O vapor greatly suppresses further reaction of the Ga with SiO$_2$, according to the mass action laws of chemistry. This technique has been used to form the highest-purity bulk GaAs reported to date. The semi-insulating form of this material contains $< 10^{15}$ electrically active impurities. Until recently, most semi-insulating GaAs was of lower purity and was made by overcompensating the n-type silicon contamination with Cr during growth. The Cr acts as an efficient donor compensating dopant in GaAs, but does not significantly dope the crystal p-type. Unfortunately, this type of semi-insulating GaAs produced some problems in the fabrication of GaAs MESFET structures, and therefore methods that produce high-purity semi-insulating GaAs are very desirable. Semi-insulating GaAs of intermediate purity was made recently by using BN vessels and the LEC method.

Improvements in the control of crystal perfection have been made on a somewhat limited scale. For example, the horizontal Bridgman method has been modified to allow the seeding of GaAs crystals on fixed orientations. It has been found that dislocation-free crystals can be grown with the [031] direction parallel to the crystal axis. So far, this has been limited to relatively small ingots about 200 grams in weight and 1.5 square centimeters in cross section. Dislocation-free GaAs has also been grown sporadically by the older modified Czochralski methods, but it has not yet been prepared by the LEC method.

To summarize the state of the art, GaAs crystals 5.0 to 7.0 cm in diameter and of moderate purity can be grown commercially by the LEC method. These crystals can be doped n-type

$(10^{16} < n < 5 \times 10^{18}$ cm^{-3}), p-type $(5 \times 10^{16} < p < 2 \times 10^{19}$ cm^{-3}), and semi-insulating (resistivity $>10^6$ ohm-cm). Currently, the most widely used semi-insulating material is doped with Cr. Commercial GaAs crystals of lower purity grown by the horizontal Bridgman method are also available with about the same doping ranges. The horizontal Bridgman material usually has a lower dislocation density and produces noncircular wafers. In the laboratory, it is possible to grow high-purity (ionized impurities $<10^{15}$ cm^{-3}), dislocation-free GaAs with a 1.5-cm^2 cross section.

Two other widely used III-V compounds, InP and GaP, are commercially available as bulk crystals and substrates. These are grown primarily by the LEC method. Except for AlP and AlAs, which are difficult to form into bulk crystals even by special techniques, the other materials in Table 2 with low dissociation pressures ($<10^{-3}$ atm) are usually available on special order and do not pose special problems for crystal growth in the research laboratory.

Thin-Film Fabrication

Most III-V device fabrication begins with deposition of one or more active epitaxial layers. Epitaxial layers are those which conform to the crystal structure and orientation of the material on which they are deposited. There are three principal epitaxial techniques for III-V materials: liquid phase epitaxy (LPE), chemical vapor deposition (CVD), and molecular beam epitaxy (MBE). The MBE method is discussed in another article in this issue (5).

The LPE method has been the most widely used one for thin-film III-V materials and device research. It was first developed in the late 1950's and applied to the formation of p-n junctions in germanium and GaAs. In the LPE method a solid solute phase is epitaxially grown by supercooling a melt or solvent containing the solute. For III-V materials, this generally means that the melt composition is rich with respect to one of the group III element in the solid phase. For example, LPE growth of Ga$_{1-x}$Al$_x$As is generally performed by using a melt of more than 90 percent Ga, the rest being Al, As, and some dopant. The term LPE originally referred to supercooling melts—lowering the temperature of melts that were in solid-liquid equilibrium. More recently the technique has been expanded to include growth from melts that are supersaturated by methods other than supercooling. Over the years many LPE ap-

paratuses and procedures have been developed, and they cannot be adequately discussed here. It is more useful to examine the nature of the LPE method in terms of its wide use as a research tool and to a lesser extent its use in production lines, and also to discuss some recent innovations that attempt to overcome some disadvantages of the method.

The "typical" LPE method is shown schematically in Fig. 6. The basic features are a melt chamber of fixed geometry and a method for moving a substrate in and out of the melt. Usually, operation begins by establishing a melt at solid-liquid equilibrium with the substrate removed from the melt region. The melt is usually supercooled by lowering the temperature at a constant rate. The substrate is then brought into contact with the melt while maintaining the cooling. After a prescribed change in temperature or elapsed time, the substrate is removed from the melt. For small cooling rates and small melt volumes, nearly equilibrium growth occurs since the composition

of the melt remains near the liquidus line of the phase diagram. The composition and layer thickness are then mainly a function of the temperature change during substrate-melt contact. For large cooling rates and large melt volumes, the composition of the melt at the solid-liquid interface is very close to the liquidus composition, while the bulk of the melt is of a different composition. This results in diffusion-controlled growth, and hence for small temperature changes the layer thickness and composition are mainly a function of time of substrate-melt contact.

Table 3 shows notable advantages and disadvantages of the LPE method. Perhaps the most important advantage of the technique, especially in the exploratory research environment, is the beneficial role of the solid-liquid interface in terms of quality of grown material. This is because melts of group III elements can be made with high purity. Also, as growth of the solid layer proceeds from LPE melts, most impurities in the melt

Table 2. Some physical and thermochemical properties of III-V compounds relevant to bulk crystal growth.

Compound	Lattice constant (Å)	Band gap (eV) and type (D, I)*	Dissociation pressure (atm)	Melting point (°C)
AlP	5.45	2.45 I	8	1840
GaP	5.45	2.24 I	35	1465
AlAs	5.66	2.16 I	1.4	1760
AlSb	6.14	1.5 I	$< 10^{-3}$	1065
GaAs	5.65	1.4 D	1	1240
InP	5.87	1.35 D	25	1065
GaSb	6.10	0.72 D	$< 10^{-3}$	706
InAs	6.06	0.35 D	0.3	947
InSb	6.48	0.18 D	$< 10^{-3}$	525

*Direct or indirect.

Table 3. Comparison of the LPE and CVD methods.

LPE	CVD
Advantages	
Near-equilibrium growth conditions	In situ substrate cleaning or etching
Growth of material with good optical and electrical properties	Good morphology and thickness control
	Good doping control
Liquid-solid interface lowers contamination of layer during growth	Low-temperature growth (600° to 800°C)
	Multilayer and multijunctions easily grown
High-purity melts	Growth start and termination well defined
Nonreactive melt containers	Large-scale production
Low-temperature growth (600° to 800°C)	In situ process monitoring; for example,
Low defect densities	ellipsometry and IR spectrometry
Melt getters impurities from solid layer	
Disadvantages	
Imprecise thickness and morphology control due to variable growth mechanism, terracing, and meniscus line effects	Ultrapure chemicals required
	Leak-free systems required for high-purity materials with good optical properties
Growth termination a fundamental problem	Vapor phase reactants may produce defects, traps, or unwanted doping in epilayer
Unfavorable phase diagrams for growth of some III-V alloys	Reactivity of chemicals with containment vessels
Cross-doping effects of volatile dopants in multimelt systems	Autodoping of epilayer by substrate etching during growth
Large batch or continuous processing of multilayer structures very difficult	

107

tend to be rejected rather than incorporated into the layer. These effects together mean that useful materials can be grown by "quick and dirty" experiments. An example of this was the LPE growth of $Ga_{1-x}Al_xAs$ layers. High-efficiency $Ga_{1-x}Al_xAs$ LED devices were first made around 1967 by this method. It has taken more than 10 years of intensive research to duplicate the quality of this material by the CVD and MBE methods.

There are two principal disadvantages of this method applied to III-V materials.

To terminate growth by the LPE method, the melt must be physically separated from the substrate at the end of the growth period. Failure to do this would result in unwanted growth during the return of the apparatus to room temperature, because the solubility of the solute species decreases with decreasing temperature. In the early development of LPE methods, much effort was devoted to designing apparatus with features that would cleanly shear away the melt from the substrate. In addition, much effort

was spent on multichamber fixtures to contain several different melts in order to grow multiple p-n and heterojunction structures. Consequently, most of the currently interesting optoelectronic heterostructure devices were pioneered by the LPE method, and the method is still the preferred technique for many of these devices.

Another disadvantage of the LPE method is the rough surface morphology associated with the near-equilibrium growth kinetics and melt separation dynamics. Terrace-like surfaces and interfaces occur as a result of a slight misorientation of the substrate surface with respect to a low-index crystallographic plane. Meniscus line motion during melt separation produces trails on surfaces and interfaces. Recently, several advances have been made in the LPE method to improve surface morphology. Studies of melt supersaturation and substrate orientation have led to an understanding of the conditions necessary for LPE growth of smooth surfaces and planar layers and interfaces. It has been found that smooth surfaces can be grown under conditions of large melt supersaturations and critical substrate orientation. Specialized techniques have been developed for achieving controlled melt supersaturation. They include electroepitaxy, in which electric current flowing across the solid-liquid interface causes localized cooling and melt supersaturation, and isothermal melt mixing techniques, in which equilibrium melts of different compositions are mixed at constant temperature to produce a supersaturated melt. The principles of critically oriented substrates for achieving microscopically smooth epitaxial layers are based on the interplay of the variation of interface free energy with crystallographic direction, the interface area, and the interface configuration that minimizes its total energy during growth.

A device-material-method menu illustrating the use of the LPE method is shown in Table 4. Much of the LPE research is focused on the following devices and materials: MESFET's of GaAs, InP, and $In_{1-x}Ga_xAs$; injection lasers of GaAs-$Ga_{1-x}Al_xAs$ and InP-$Ga_{1-x}In_xAs_{1-y}P_y$; photodetectors and solar cells of InAs, InSb, $Ga_{1-x}In_xSb$, and GaAs-$Ga_{1-x}Al_xAs$; and LED's of GaP, GaAs, and $Ga_{1-x}Al_xAs$. I believe that as the demand for greater device dimension control increases, the role of the LPE method in the laboratory will diminish in favor of the CVD and MBE methods. There is evidence that this trend has already started, and that ulti-

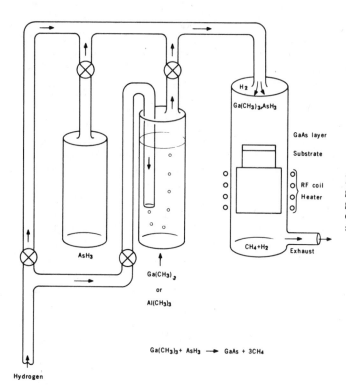

Fig. 7. Simplified diagram of the metal-organic method for chemical vapor deposition.

Table 4. Exemplary III-V materials and devices and epitaxial methods used for fabrication.

Device	Material	Method
LED	GaAs: Si-doped	LPE
	$Ga_{1-x}Al_xAs$-GaAs	LPE, CVD
	$GaAs_{1-x}P_x$	CVD
	$Ga_{1-x}In_xAs_{1-y}P_y$-InP	LPE, CVD
	GaP	LPE, CVD
DH laser	GaAs-$Ga_{1-x}Al_xAs$	LPE, CVD
	InP-$Ga_{1-x}In_xAs_{1-y}P_y$	LPE, CVD
Solar cell	GaAs-$Ga_{1-x}Al_xAs$	LPE, CVD
	n-p-GaAs	CVD
	InP-CdS	CVD
IR photodetector	InSb	Bulk
	$Ga_{1-x}In_xAs_{1-y}P_y$	LPE, CVD
	$GaAs_{1-x}Sb_x$	LPE, CVD
	InAs	Bulk
Heterojunction transistor	GaAs-$Ga_{1-x}Al_xAs$	LPE, CVD
	InP-$Ga_{1-x}In_xAs_{1-y}P_y$	LPE, CVD
Negative electron affinity device	GaAs	LPE, CVD
	GaP	LPE, CVD
	$Ga_{1-x}In_xAs_{1-y}P_y$	LPE, CVD
MESFET	GaAs	LPE, CVD
	InP	LPE, CVD

mately the LPE method will be used for specialized applications in which it offers some advantage for materials quality or device structure.

The CVD method is sometimes referred to as vapor phase epitaxy (VPE) to distinguish it from the more generalized chemical vapor deposition method used to form both epitaxial and nonepitaxial films by thermochemical vapor phase reactions. Most of the research on CVD applied to III-V materials has concentrated on two different chemistries: (i) the III-halogens and V-halogens or V-hydrogen, such as $GaCl_3$ and $AsCl_3$ or AsH_3, and (ii) the III metal-organics and V-hydrogen, such as $Ga(CH_3)_3$ and AsH_3. The thermochemistries of these systems are very different. The halogen reactions are usually "hot" to "cold" ones, in which the III-halogen is generated in a hot zone by reaction of the III element with HCl and then diffuses to the cold zone, where it reacts with the V species to form the III-V material. The metal-organic (MO) chemistry is a "hot wall" process in which the III-R compound "cracks" or pyrolyzes away the organic (R) group and the remaining III and VH_3 react to form III-V. A schematic of the MO methods applied to GaAs is shown in Fig. 7.

Both of these chemistries have been widely developed and used successfully for GaAs devices. For other III-V compound and alloys, the situation is somewhat confusing and is changing rapidly. The halogen techniques were more intensely studied than the MO method early in the development of III-V materials. During this period several important devices were developed with the halogen method, most notably the $GaAs_{1-x}P_x$ LED and the GaAs Gunn effect device. Others include $In_{1-x}Ga_xP$ LED's and lasers. Until very recently, the MO technique was not developed to the point of fabricating some of the important devices. When success did occur, it was on GaAs-$Ga_{1-x}Al_xAs$-based devices,

which at the time were practically inaccessible by the halogen method, primarily because of unfavorable thermochemistry and reactivity of containing vessels. Since the success of the MO method with $Ga_{1-x}Al_xAs$, there have been increased efforts to extend it to other III-V materials. Many workers consider this method to have long-term advantages including better thickness, purity, and composition control and the ability to grow a wide range of III-V compounds and alloys.

Referring again to Table 3, it is useful to consider some of the advantages and disadvantages of the CVD method. The principal advantages are (i) good dimension control, (ii) good doping and composition control, and (iii) straightforward scale-up for wafer processing in a production environment. The principal disadvantages are (i) the requirement for ultrapure chemicals, (ii) high reactivity of the vapor with containment vessels, and (iii) possible introduction of defects or unwanted doping by chemical reactants from the formation of the III-V material. A reexamination of Table 4 shows that the CVD method has been successfully applied to the most currently important devices.

Conclusions

At present, silicon-based devices dominate the semiconductor device industry. The impact of III-V materials so far has been on specialty optoelectronic devices, including lasers, LED's, photodetectors, and solar cells. This has been due mainly to the large variety of band-gap energies available with alloying of III-V compounds, the formation of lattice-matched heterojunctions, special impurities that enhance radiative recombination, bulk crystal growth techniques that produce useful substrates, and the development of the LPE and CVD methods for producing dimensionally con-

trolled epitaxial layers of good optical and electrical quality. As integrated circuits become more dense and complicated, research in this field must consider material, device, and processing as a unit. The future role of III-V materials is likely to be significant because of their processing advantages associated with the chemical and functional aspects of structures similar to those found in optoelectronic devices.

References and Notes

1. The III and V refer to the group IIIA and group VA elements of the periodic table, respectively.
2. It is even more appropriate that material, device, processing, and packaging be considered as a unit for high-speed data-processing applications.
3. It was not the first choice of the pioneers of the transistor industry. Shockley and Bardeen worked unsuccessfully for a decade on CuO, which was attractive for integration with copper wires. They finally succeeded in showing bipolar transistor action after switching to germanium. It was only after the U.S. government's insistence on high-temperature operation for military purposes forced the development of silicon that the great advantages of that material and its uniquely high-quality passivating oxide were discovered.
4. For a more complete description, the interested reader is encouraged to consult C. J. Nuese, *J. Electron Mater.* **6**, 253 (1977); A. A. Bergh and P. J. Dean, *Proc. IEEE* **60**, 156 (1972); C. B. Duke and N. Holonyak, Jr., *Phys. Today* (Dec. 1973), p. 23; M. G. Craford and W. O. Groves, *Proc. IEEE* **61**, 862 (1973); S. M. Sze, *Physics of Semiconductor Devices* (Wiley, New York, 1969); A. G. Milnes and D. L. Feucht, *Heterojunctions and Metal-Semiconductor Junctions* (Academic Press, New York, 1972); H. Hovel, *Solar Cells*, vol. 11, *Semiconductors and Semimetals* (Academic Press, New York, 1975); J. W. Matthews, *Materials Science and Technology—Epitaxial Growth* (Academic Press, New York, 1975); P. Hartman, *Crystal Growth: An Introduction* (North-Holland, Amsterdam, 1973); G. M. Blom, Ed., *J. Cryst. Growth* **27** (1974); H. Nelson, *RCA Rev.* **24**, 603 (1963); J. M. Woodall, *J. Electrochem. Soc.* **118**, 150 (1971); H. Rupprecht, J. M. Woodall, G. D. Pettit, *Appl. Phys. Lett.* **11**, 81 (1967); J. M. Woodall, *Trans. Metall. Soc. AIME* **239**, 378 (1967); S. E. Blum and R. J. Chicotka, *J. Electrochem. Soc.* **120**, 588 (1973); M. B. Small, K. H. Bachem, R. M. Potemski, *J. Cryst. Growth* **39**, 216 (1977); J. C. C. Fan, C. O. Bozler, B. J. Palm, *Appl. Phys. Lett.* **35**, 875 (1979); Y. Imamura, L. Jastrzebski, H. C. Gatos, *J. Electrochem. Soc.* **126**, 1381 (1979); M. G. Craford and N. Holonyak, Jr., in *Optical Properties of Solids*, B. O. Seraphin, Ed. (Gordon & Breach, New York, 1976), p. 188; D. J. Wolford, B. G. Streetman, R. J. Nelson, N. Holonyak, Jr., *Solid State Commun.* **19**, 741 (1976); D. J. Wolford, B. G. Streetman, W. Y. Hsu, J. D. Dow, R. J. Nelson, H. Holonyak, Jr., *Phys. Rev. Lett.* **36**, 1400 (1976).
5. M. Panish, *Science* **208**, 916 (1980).

Molecular Beam Epitaxy

M. B. Panish

The advances in solid-state device technology that have taken place since the invention of the transistor have required constant improvements in methods for preparing and processing semiconductors with precise dimensional and compositional constraints. In addition to spectacular growth in the technology of silicon for integrated circuits with increasingly larger densities of devices and functions, there has been steady progress in electronics technology into opti-

Summary. Molecular beam epitaxy is an ultrahigh vacuum technique for growing very thin epitaxial layers of semiconductor crystals. Because it is inherently a slow growth process, extreme dimensional control over both major compositional variations and impurity incorporation can be achieved. The result is that it has been possible, with one combination of lattice-matched semiconductors, GaAs and $Al_xGa_{1-x}As$, to demonstrate a large variety of novel single-crystal structures. These results have important implications for fundamental studies of the physics of thin-layered structures and for the development of new semiconductor electronic and optoelectronic devices.

GaAs and $Al_xGa_{1-x}As$, or InP and $Ga_xIn_{1-x}P_yAs_{1-y}$ with the correct ratio of x to y, the size of the crystal lattice can be kept virtually unchanged in spite of the compositional variations, so that changes in important electrical and optical properties can be achieved in very small regions of the crystal without seriously perturbing the crystal structure.

Most of the binary III-V compounds are manufactured as large single-crystal ingots that are sliced into wafers to be

cal and microwave frequencies that has directed interest to several compound semiconductors. These are usually either binary compounds of one of the group III elements Al, Ga, and In and one of the group V elements P, As, and Sb, or crystalline solid solutions of the binary compounds such as $Al_xGa_{1-x}As$ or $Ga_xIn_{1-x}P_yAs_{1-y}$ with $0 \le x,y \le 1$.

The III-V semiconductors have the same gross electronic bonding structure as the better known semiconductors Si and Ge; but because of differences in the detailed nature of that structure, they are sometimes more useful for microwave devices and always more useful as light-emitting devices. In addition, the ability of III-V compounds to form solid solutions means that abrupt transitions in such properties as the energy of the forbidden energy gap and the refractive index can be achieved by compositional changes. For some combinations of these semiconductors, most notably

used for device fabrication. Frequently, in order to capitalize on the versatility of the III-V semiconductors for solid-state devices, these single-crystal wafers are used as substrates for the subsequent growth of very thin layers of the same or other III-V compounds having the desired electronic or optical properties. This must be done in such a way as to continue, in the grown layer, the crystal structure of the substrate. Such crystal growth, in which the substrate determines the crystallinity and orientation of the grown layer, is called epitaxy, and a variety of epitaxial growth techniques have been developed. The most common of these are vapor phase epitaxy (VPE) and liquid phase epitaxy (LPE). The former utilizes a heated stream of gaseous elements or compounds that interact at the surface of the substrate to form the crystalline layer. In the latter the same end is accomplished by cooling a heated metallic solution saturated with the components needed to grow the layer, while that solution is in contact with the substrate.

In addition to VPE and LPE, vacuum epitaxy has been studied. Günther (*1, 2*) showed in 1958 that thin films of polycrystalline III-V compounds could be grown when the elements comprising the semiconductor were evaporated in a vacuum system with heated walls onto a heated substrate material. In 1966 Steinberg and Scruggs (*3*), using a glass bell jar vacuum system without hot walls, grew epitaxial GaAs onto a heated NaCl single-crystal substrate. Then in 1968 Davey and Pankey (*4*) showed that by using a heated vacuum system with a sufficiently high ambient arsenic pressure, epitaxial GaAs could be grown when a beam of gallium atoms impinged on a heated single-crystal GaAs substrate surface. In the same year Arthur (*5*), using a metal vacuum system that did not have heated walls, impinged beams of both gallium atoms and arsenic molecules onto a heated GaAs substrate wafer and obtained epitaxial growth. The technique that is now generally called molecular beam epitaxy (MBE) derives directly from Arthur's work. His use of a cool-wall, all-metal vacuum system; impinging beams of all components; cryopaneling to yield some beam collimation and to reduce radiative heating of the vacuum system walls by the heated beam sources; and fast pumping techniques to maintain ultrahigh vacuum eventually yielded the cleanliness of the ultrahigh vacuum that now permits vacuum epitaxy of high-quality semiconductor material.

There has been a vigorous effort over the past decade to demonstrate MBE of semiconductor materials useful for semiconductor devices. Much of the credit for bringing this technology to its present state of maturity must be given to Cho (*6, 7*), who used GaAs and $Al_xGa_{1-x}As$ as prototype materials in most of his work. There is now a rather extensive literature on MBE as it has become the subject of research at many laboratories; detailed discussions and bibliographies are given in (*6–9*).

The work of the past few years has clearly demonstrated that MBE is an extraordinarily versatile epitaxy technique that is applicable to a variety of conventional microwave and optoelectronic devices. In addition, because of the extreme dimensional control with MBE, it is possible to build essentially new crystals with periodicities not available in nature and to prepare structures whose properties depend on confinement of holes and electrons to crystalline regions so small that quantum confinement effects become important.

The author is head of the Materials Science Research Department, Bell Laboratories, Murray Hill, New Jersey 07974.

Molecular beam epitaxy is, in principle, applicable to the growth of epitaxial layers of a variety of compound semiconductors, and experimental studies of MBE of many different materials have been done or are under way in a number of laboratories. Particularly notable are the studies of InP on InP (*10, 11*), $Ga_x In_{1-x}As$ on InP (*12*), InAs and GaSb ultrathin layered structures (*13*), Si on Si (*14, 15*) and on sapphire (*16*), and IV-VI compounds (*17*). The MBE efforts on InP and $Ga_x In_{1-x}P$ on InP and other III-V compounds yield light-emitting devices and detectors at a variety of wavelengths, mostly in the near infrared. The studies of IV-VI compounds do the same for devices for the far infrared. The work with silicon provides an added degree of freedom in doping control and shows promise for higher quality Si on sapphire (SOS) than is now commonly achieved with chemical vapor deposition. In spite of the great variety of MBE work on other materials, most MBE studies to date have been done with GaAs and $Al_x Ga_{1-x}As$. For that reason the discussions presented in this article deal only with those semiconductors, which may be considered prototypes, at least for epitaxy of structures of other III-V compounds.

The Molecular Beam Epitaxy Process

Reduced to its essentials, a system for MBE of GaAs consists of an ultrahigh vacuum system containing sources for atomic or molecular beams of Ga and As and a heated substrate wafer, as illustrated very schematically in Fig. 1. The beam sources are usually containers for the liquid Ga or solid As. They have an orifice that faces the substrate wafer. When the container, or effusion oven as it is usually called, is heated, atoms of Ga or molecules of arsenic effuse from the orifice. The effusing species constitute a beam in which the mean free path is large compared to the distance between the oven orifice and the substrate wafer. If the orifice diameter is small compared to the mean free path of the gaseous components inside the effusion oven, the flux of Ga or As_4 at the target wafer may readily be shown to depend on the partial pressure of the species within the oven, the distance from orifice to substrate, the temperature, the species molecular weight, and the orifice area. Additional ovens, not shown in Fig. 1, may be used to generate a beam of Al, for the growth of $Al_x Ga_{1-x}As$, and to generate beams of impurity elements

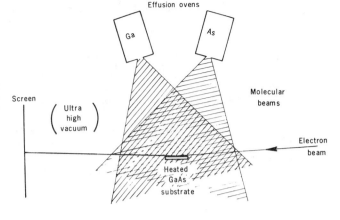

Fig. 1. Schematic representation of a molecular beam epitaxy system.

that can be used to make the epitaxial semiconductor *n* or *p* type. Most current MBE systems have about six effusion ovens. The beams may be shut off with shutters interposed between the substrate and the oven orifice, or the beam intensity may be varied by varying the oven temperature. Also illustrated in Fig. 1 is an electron beam that impinges, at a glancing angle, on the growing surface of the crystal for in situ evaluation of surface morphology.

The successful use of MBE for epitaxy of GaAs, $Al_x Ga_{1-x}As$, and other III-V compounds is a direct consequence of the behavior of group III atoms and group V molecules on striking the heated substrate surface. Arthur (*5*) showed for GaAs—and it is true for other III-V compounds—that there is a range of substrate temperatures over which virtually all of the group III element adsorbs on the surface. This holds for the entire usual temperature range of 450° to 650°C for the growth of GaAs. The surface lifetime of Ga on GaAs is greater than about 10 seconds, while the arsenic molecules desorb rapidly from a heated GaAs surface unless adsorbed Ga is present. In the latter case the surface lifetime of As increases as it bonds to the Ga. It decreases again when the excess Ga is consumed. The result of this is that one As atom remains on the surface for each Ga atom provided in the Ga beam. For the growth of GaAs the 1 : 1 ratio of Ga to As is maintained in the growing layer simply by having the As_4 flux be greater than the Ga flux. For epitaxy of $Al_x Ga_{1-x}As$ the ratio of Al to Ga atoms in the solid is simply the ratio of the atom flux of each during growth, while the ratio of total group III (Al plus Ga) to As atoms in the solid is unity.

Achievement of high crystalline and semiconductor quality of the epitaxial layers also requires that clean ultrahigh vacuum conditions be maintained and that the substrate temperature be suffi-

ciently high that the atoms adsorbing on the surface are mobile enough to migrate to the proper crystal sites. For GaAs, growth usually takes place with a substrate temperature above 450°C, and for $Al_x Ga_{1-x}As$ the temperature is usually above 550° or even 600°C.

Electron Diffraction and Surface Morphology

Molecular beam epitaxy is unique among crystal growth techniques in that it is possible to examine the crystal surface in some detail during the growth process. The electron beam shown in Fig. 1 is diffracted by the regular array of atoms that constitute the crystal structure near or at the surface of the crystal much as light is diffracted by a grating. The diffraction pattern of the electrons yields information about the arrangement of the atoms on the growing crystal surface. If the surface is microscopically rough, the diffraction pattern will be characteristic of the three-dimensional crystal since the beam must penetrate protuberances on the surface. If it is almost smooth on an atomic scale, the diffraction pattern will show the characteristic two-dimensional spacing of the atoms on that surface. This is illustrated in Fig. 2 from a study by Cho (*18*). A set of electron diffraction patterns of a {100} GaAs surface is shown in various stages of MBE growth, with associated electron micrographs showing the morphology of the surface. In Fig. 2a the spotted pattern is for the highly polished, but still microscopically rough starting crystal. The spots result from diffraction from the three-dimensional crystal lattice as the electron beam penetrates the protuberances on the atomically rough surface. Figure 2b shows that after average growth of 150 angstroms the surface is smoother and the diffraction pattern is streaking. The streaked pattern shows the relaxation of one dimension in the diffraction as the electron beam is diffracted more by atoms on the surface than by atoms in the bulk. A new diffraction streak halfway between the main peaks shows that the surface atoms have a unit cell size twice that of the atoms in the bulk. Figure 2c shows that after growth of 1 micrometer the surface is smooth and the diffraction pattern fully streaked. Thus the examination of the electron diffraction pattern during growth of the epitaxial layer yields information about the microscopic smoothness of the crystal surface, and demonstrates that during MBE the microscopic smoothness can be improved over that obtainable with mechanical and chemical polishing.

Impurity Incorporation and Profiling

When some impurity elements (dopants) that have either more or fewer valence electrons than are needed for bonding are incorporated into the semiconductor, it becomes either n or p type. The impurity atoms in the n-type semiconductor have donated electrons that are mobile current carriers and occupy energy states in the conduction band. The impurity atoms in the p-type semiconductor have removed electrons from the valence band, leaving behind mobile, positively charged "holes." For the semiconductor to be useful for most solid-state devices, the concentrations of carriers in the n- and p-type regions must usually be precisely controlled. The quantity of impurity is typically much less than 10^{-5} atom fraction, and in a given structure it is often useful to be able to vary the impurity profile as a function of depth into the crystal so that the conductivity type and profile may be tailored to the device requirements.

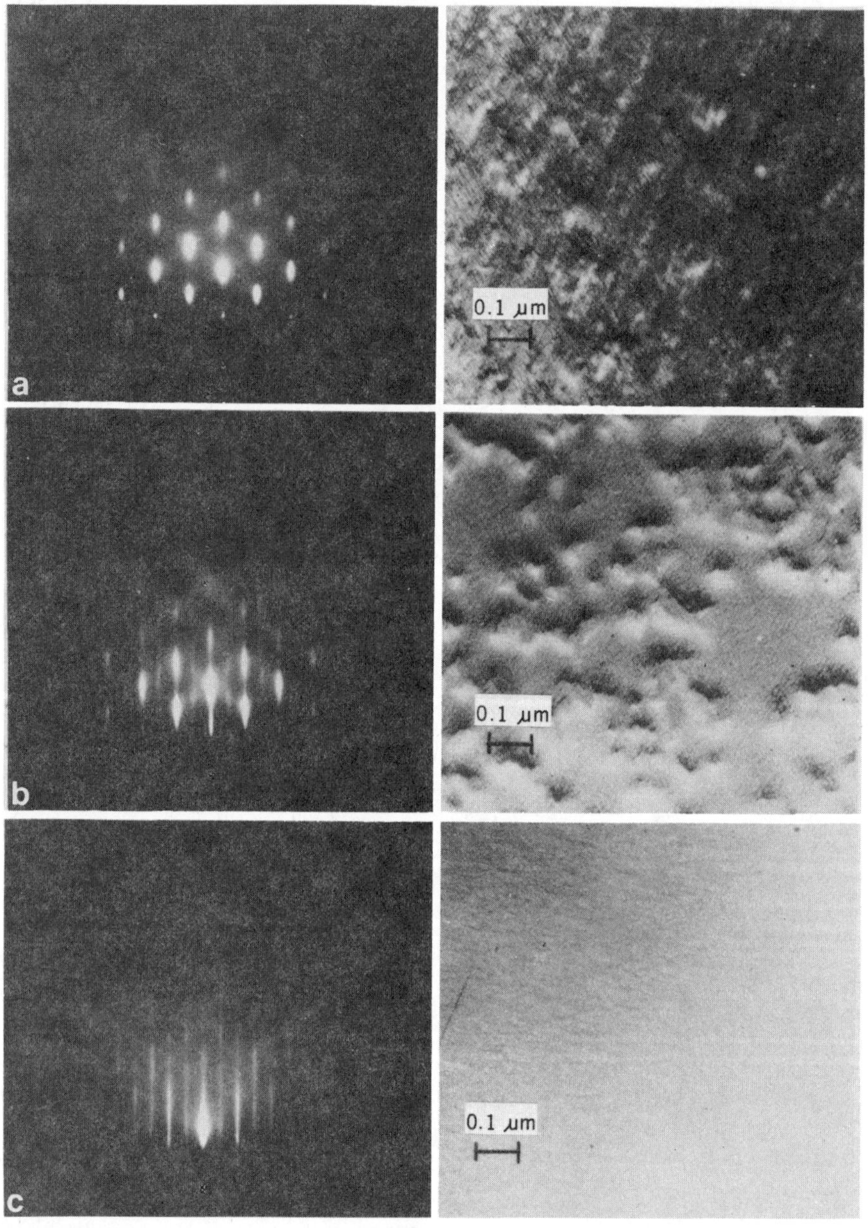

Conventional epitaxy techniques provide only limited control over doping profiles, usually permitting only a fixed or a slowly varying level of impurity in a given layer. Frequently, impurity atoms are diffused or ion-implanted into the surface of a semiconductor. These methods, which are technologically very important, yield a very restricted range of impurity profiles, usually involving either a concentration that decreases monotonically into the crystal or a single peaked distribution.

Because MBE is a slow growth process, layer thicknesses typically increase by 1 μm per hour, and it is possible by varying impurity effusion oven temperatures or by the use of shutters to arbitrarily vary the impurity concentration as a function of depth. This is illustrated (19) for Ge added to GaAs in Fig. 3. Three rectangular pulses of Ge in the 1-μm-thick GaAs epitaxial layer yield electron concentrations shown by the dashed curve. The electron concentration profile is, in fact, what would be expected for rectangular impurity profiles. Thus MBE can yield the most abrupt electron concentration profile possible and also all gradations away from that limit. The ability to obtain such arbitrary doping profile control with MBE has been used for the preparation of a number of solid-state devices. These include hyperabrupt varactor diodes that require an exponential doping profile with depth (20); IMPATT (impact ionization avalanche transit time) diodes of the so-called low-high-low variety (21) that require a 1000-Å-thick doping spike in which the impurity concentration increases by a factor of 10; state-of-the-art field-effect transistors requiring precision doping control in a layer only several thousand angstroms thick (22); and state-of-the-art microwave mixer diodes for low-temperature, low-noise operation (23). The latter require a very abrupt transition from highly to lightly doped material and are used mostly for radio astronomy.

All of the applications of MBE described above utilize conventional doping levels in the range of about 10^{16} to 10^{18} impurity atoms per cubic centimeter. In addition, it is possible with MBE to achieve doping levels around 5×10^{19} cm^{-3} for both n- and p-type dopants (Sn and Be) in GaAs and to use oxygen as a dopant for Al$_x$Ga$_{1-x}$As to render it semi-

Fig. 2. Reflection electron diffraction patterns (40 kiloelectron volts, $\bar{1}\bar{1}0$ azimuth) and electron micrographs of replicas of the same GaAs surface (18). (a) Polish-etched GaAs after heating in a vacuum for 5 minutes. (b) Same after 150 Å of GaAs was deposited by MBE. (c) Same after 1 μm of GaAs was deposited by MBE.

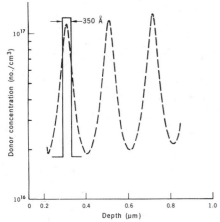

Fig. 3. A periodic doping profile of Ge in GaAs (*19*). The solid curve shows the Ge profile for one peak as estimated from the growth conditions. The dashed curve shows the measured electron distribution in the sample containing three such Ge pulses above a background level.

insulating. The former permits the application of high-quality metal ohmic contacts either in situ by evaporation (*24, 25*) or out of the MBE system by electroplating, without subsequent heating. The oxygen-doped $Al_xGa_{1-x}As$ has been found to prevent surface leakage of current due to recombination of carriers at GaAs surfaces (*26*). Such passivation provides hope for a metal-insulator-semiconductor technology with GaAs. In addition, since it is an epitaxial part of the crystal structure, the oxygen-doped $Al_xGa_{1-x}As$ provides the solid-state device designer with a new degree of freedom, the possibility of designing devices incorporating lattice-matched semi-insulating regions into the single-crystal device structure.

Heterostructures

Single-crystal multilayered structures having component layers that differ in composition but are lattice-matched form the basis for semiconductor devices in which both light and current carriers (holes and electrons) can be manipulated. The double-heterostructure (DH) laser (*27*) is perhaps one of the best illustrations of these devices. It also provided one of the earliest motivations for MBE studies at Bell Laboratories.

The GaAs-$Al_xGa_{1-x}As$ DH laser in its simplest version is a small rectangular single-crystal parallelepiped consisting of an *n*-type GaAs substrate with at least three layers—*n*-$Al_xGa_{1-x}As$, *p*-GaAs, and *p*-$Al_xGa_{1-x}As$—grown epitaxially onto it as illustrated in Fig. 4, a and b. The alignment of the conduction and valence bands of the composite structure,

when forward-biased (*n*-side negative) with a voltage of about the width of the GaAs energy gap, is shown schematically in Fig. 4c. As the result of forward bias, electrons are injected into the conduction band of the *p*-GaAs layer, where they recombine with the majority holes and emit radiation with approximately the energy of the GaAs energy gap, $E_{g_{GaAs}}$. Note in Fig. 4c that at the heterojunction there are potential barriers that prevent holes (e^+) and electrons (e^-) from diffusing beyond the GaAs region. The injected electrical carriers are then confined to the GaAs layer. In addition, because GaAs has a higher refractive index than $Al_xGa_{1-x}As$, the $Al_xGa_{1-x}As$-GaAs-$Al_xGa_{1-x}As$ three-layer sandwich is a waveguide so that the generated light tends to be confined to the GaAs layer. The cleaved ends of the parallelepiped act as partial mirrors. Thus, light of energy approximately $E_{g_{GaAs}}$ is generated by an electronic transition in a waveguide within a Fabry-Perot cavity formed by the mirrors. With a sufficiently high current through the device, stimulated emission and lasing result.

The DH laser provides an excellent illustration of how heterostructures are used to manipulate light and electrical carriers in a single solid-state device. A variety of other heterostructure devices have also been fabricated. These include optical modulators, optical switches, waveguides, and couplers, all utilizing combinations of GaAs and $Al_xGa_{1-x}As$ layers. There is a growing technology, mostly in optical communications, that utilizes heterostructure lasers as discrete devices. The multilayered wafers used for fabrication of these lasers are at present most often grown by LPE. Because of difficulties in obtaining growth sufficiently free of contamination, the MBE technique has only recently been used to demonstrate high-quality lasers. These lasers have had current densities for the onset of lasing that were comparable to or even lower than those of comparable lasers made with LPE wafers (*28, 29*). In addition, the properties of many such lasers selected from various areas of several large wafers demonstrated a degree of uniformity that is clearly much better than has been achieved by liquid epitaxy. This achievement is important, not only for the eventual fabrication of discrete devices more reproducibly than has been possible with LPE, but also because the use of integrated electrooptic circuits may become desirable.

Integrated electrooptic circuits made possible by MBE would incorporate not

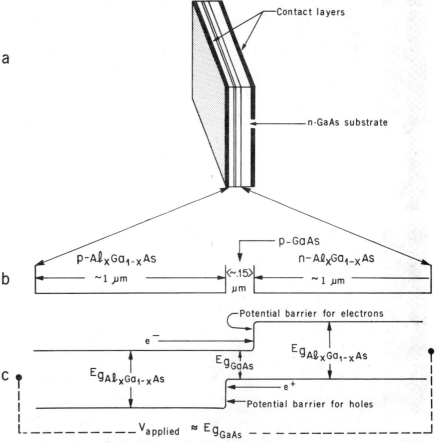

Fig. 4. (a) A double-heterostructure laser. (b) The epitaxial layers. (c) Conduction and valence band edges under forward bias from an applied voltage $V_{applied} \approx E_{g_{GaAs}}$.

113

only the devices mentioned above but also optical detector and microwave devices such as field-effect transistors, all on a single semiconductor chip. The MBE technique may be particularly useful for such complex structures because it permits lateral dimensional control by means of shadow masking. Studies of shadow masking during MBE (30) are illustrated in Fig. 5a, which shows how mesas and tapers are grown. A three-layer mesa is shown in Fig. 5b. A similar technique has already been used to prepare a laser-taper coupler combination (31).

Monolayers and Quantum Wells

Given the ability to obtain extremely smooth surfaces, a slow and precise growth rate, controlled impurity incorporation as a function of depth, and composition variation as a function of depth, several workers have elected to study a range of multilayered structures with extraordinarily small dimensions in layer thickness. The structures with the thinnest layers are the so-called monolayer structures described by Gossard et al. (32).

An approach to the ultimate in such single-crystal structures is illustrated in Fig. 6, which is a cross-sectional transmission electron micrograph of a stack of alternating GaAs and AlAs layers. To visualize the scale of these individual layers, the GaAs or AlAs crystal is shown schematically as a stack of alternating planes of gallium (or aluminum) and arsenic atoms. For the structure shown in Fig. 6, referred to as an alternate bilayer structure, each dark or light band consists of four atomic planes, two mostly of gallium atoms or mostly of aluminum atoms, interleaved with two planes of arsenic atoms, as illustrated on the right in Fig. 6. The group III element layers are, to some degree, mixtures of Al and Ga, hence the terms mostly Ga and mostly Al. Each nominally GaAs or AlAs layer is 5.6 Å thick, and stacks of such layers that are 10,000 Å thick have been grown on GaAs substrates. Stacks of alternating GaAs and AlAs layers consisting of single 2.8-Å monolayers were also grown and, in x-ray and electron diffraction, have shown the alternate single-monolayer composition. All these structures are new crystals that have periodicities not available in nature. They are grown by maintaining an unvarying arsenic beam on the substrate while alternately exposing it to the Ga and Al beams with precisely timed opening and closing of shutters interposed between the effusion ovens and the wafer. So far, such structures have been used to provide microscopic information on the MBE crystal growth process, to demonstrate the limits in dimensional precision of the MBE technique, and to correlate structural and physical properties of crystals in the limit of atomic dimensions.

Quantum well structures consist of somewhat thicker alternating layers of GaAs and, in this case, $Al_xGa_{1-x}As$, with each layer in the range of 50 to ~ 400 Å thick. One such heterostructure is illustrated in Fig. 7a. The band edge energy diagram corresponding to the structure of Fig. 7a is illustrated schematically in Fig. 7b. The important characteristic of these structures is that the steps in the conduction and valence band edges at the heterojunctions form the boundaries of potential wells for

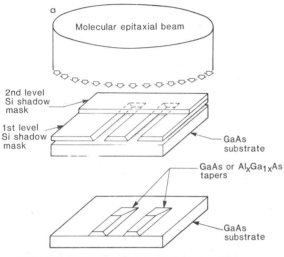

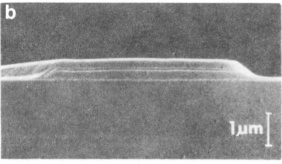

Fig. 5. (a) Schematic representation of shadow masking with MBE. (b) Cross section of a three-layer mesa grown by MBE through a shadow mask (30).

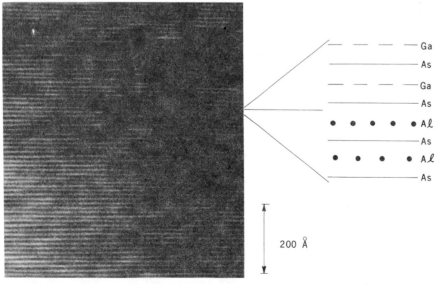

Fig. 6. Cross-sectional transmission electron micrograph of a "monolayer structure" consisting of interleaved bilayers nominally of GaAs and AlAs. The layer interfaces are in the 100 plane. [Micrograph provided by P. M. Petroff]

electrons in the conduction band and holes in the valence band. When such quantum wells are present they can modify the macroscopic properties of the components of multilayer structures so that they differ from the equivalent properties of the bulk semiconductor. Keldysh (33) predicted effects due to quantization of carriers in artificial periodic potential structures in 1963. Esaki and Tsu (34) suggested in 1970 that such a structure, which they called a superlattice, would result from a one-dimensional periodic variation of solid solution composition in a III-V semiconductor. Particularly striking examples of the effects of quantum wells on properties of MBE layers of GaAs and $Al_xGa_{1-x}As$ have been demonstrated with optical absorption and electron mobility.

The individual quantum well is essentially a one-dimensional container for electrons, and quantum theory requires that for a sufficiently small container the electron moving in the confining direction can have only fixed energies. Thus,

the quantum wells of Fig. 7a contain sets of energy subbands—the dashed lines in Fig. 7b—for electrons. Such "confined carrier quantum states" are clearly observable by optical absorption in GaAs quantum wells less than about 400 Å thick.

Bulk GaAs is essentially transparent to light of energy less than that of the width of the forbidden energy gap of GaAs (0.9 μm), but is strongly absorbing for light of higher energy because the higher energy photons are absorbed in the excitation of electrons from the valence band into the conduction band. This is illustrated in the top curve of Fig. 8, which shows the absorption spectrum of a 4000-Å-thick GaAs layer. The peak at the leading edge represents absorption by electrons that are excited into the conduction band, but are still bound by the Coulomb interaction to the holes in the valence band (exciton effects). When light is passed through a quantum well layer of GaAs the absorption of photons results from excitation of electrons from

energy subbands in the valence band quantum well to subbands in the conduction band quantum well. These transitions occur at energies greater than the bulk GaAs band gap and provide the major optical absorption mechanism. As a result, the absorption spectrum of the quantum well shows bands characteristic of such transitions. These are illustrated in the bottom three curves in Fig. 8. Dingle et al. (35) showed that the transition energies can be quite accurately predicted with the well-known particle-in-a-box quantum theory calculation.

The structures used to obtain absorption data such as those shown in Fig. 8 are stacks of GaAs quantum wells with 20 or more GaAs layers. The substrate has been removed by chemical etching. Since the measurements are usually made by passing light through the structure in a direction orthogonal to the layer surface, the large number of GaAs layers is needed to provide sufficient material thickness for a useful absorption measurement. For these measurements the

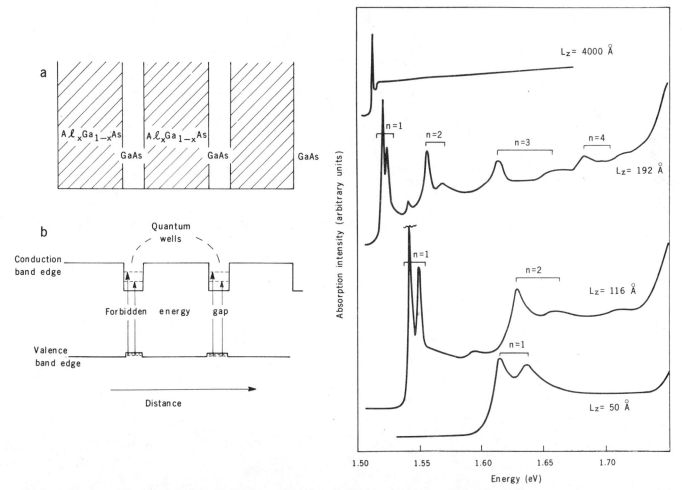

Fig. 7 (left). Schematic representation of (a) a cross section of a GaAs-$Al_xGa_{1-x}As$ quantum well structure and (b) band edges in the quantum structure of (a).   Fig. 8 (right). Absorption spectra of GaAs layers 4000, 192, 116, and 50 Å thick between $Al_xGa_{1-x}As$ barriers. The quantum well structures consist of a sufficient number of GaAs layers to yield approximately the same absorption as the single GaAs layer used for the upper curve. The peaking at the transition energies is the result of exciton effects. Otherwise, the quantum well subbands would have given a more stepped absorption spectrum. The splitting of the various levels, n, results from the resolution of quantized levels in the hole quantum well. [Figure provided by R. Dingle]

energy range of the light used is varied between that of the GaAs band gap and that of the $Al_xGa_{1-x}As$ band gap. Since the energy of the levels within the quantum well depends on its depth and width, both the Al content and the layer thickness must be highly reproducible for the absorption measurements to yield well-resolved data for the transition energies. In Fig. 8 the transitions from the confined hole subbands are clearly resolved. This requires a layer-to-layer reproducibility in thickness of less than 5 Å, a dramatic illustration of the precision in dimensional reproducibility that MBE permits. Large-angle x-ray studies of quantum well structures grown both at IBM (36) and at Bell Laboratories (37) have confirmed that the individual layers are flat and uniform with abrupt interfaces in which the entire composition change is achieved within 5 Å in the growth direction.

The possibility of modifying the transport characteristics of electrical carriers in the material comprising a quantum well results from the ability of the quantum well structure to provide a separation in space between the part of the structure occupied by dopant elements and the carriers they contribute to the semiconductor. The mobility of carriers in semiconductors is affected by many factors, including scattering by the ionized impurity atoms that donated the electrons in the first place. If, during the MBE growth of a quantum well structure, donor atoms are added to the $Al_xGa_{1-x}As$ layer but not to the GaAs layer, the electrons in the conduction band, which seek the lowest possible energy states, fall into the GaAs quantum well. They then occupy quantum well energy subbands. Thus it is possible, by using MBE to grow such "modulated-doping" quantum well structures (38), to have GaAs that contains donated electrons without having the donor atoms present in the semiconducting layer. The reduction in scattering by impurity atoms is reflected in the greater mobility of the electrons in the quantum well material than in equivalent doped bulk GaAs.

The MBE-grown quantum well structures and other heterostructures that require extremely thin epitaxial layers are under investigation from a number of different points of view. Heterostructure lasers that incorporate several quantum wells in the center (active) layer have better longitudinal mode stability and a smaller temperature dependence of threshold current than the equivalent more conventional structures (39). The higher mobility of electrons in GaAs modulated-doping quantum wells has led to studies, now under way, of the possibility for improvement in the speed of GaAs microwave devices. Quantum well structures have also been used to investigate the fundamental properties of the two-dimensional electron gas (40, 41), the possibility of fabricating energy-selective phonon mirrors (42), and, in InAs-GaSb structures, the creation of artificial semimetals (13). In addition, GaAs-Al_xGa_{1-x}GaAs heterostructures in which the $Al_xGa_{1-x}As$ layer is very thin (≤ 500 Å) and has a graded band gap obtained by varying x have been shown to form very simple current rectifiers that should be totally compatible with other optical or electronic devices that may be integrated onto the same chip (43).

Conclusion

A rather brief and general description of some of the developments in molecular beam epitaxy, using only the semiconductors GaAs and $Al_xGa_{1-x}As$ as prototypes, has been presented. Because it is possible to maintain high semiconductor quality at very low growth rates, MBE permits extensive dimensional control in epitaxial layer thickness and dopant element profiles. Unlike other crystal growth methods, it also permits growth of complex epitaxial structures with controlled lateral dimensions because of the possibility for shadow masking. These developments have resulted in a burgeoning of MBE studies of fundamental properties of very thin structures and of the use of MBE for optoelectronic and microwave devices.

References and Notes

1. K. G. Günther, *Naturwissenschaften* **45**, 415 (1958).
2. _____, in *Compound Semiconductors*, R. K. Willardson and H. L. Goering, Eds. (Reinhold, New York, 1961), vol. 1, p. 313.
3. R. F. Steinberg and D. M. Scruggs, *J. Appl. Phys.* **37**, 4586 (1966).
4. J. E. Davey and T. Pankey, *ibid.* **39**, 1941 (1968).
5. J. R. Arthur, *ibid.*, p. 4032.
6. A. Y. Cho and J. R. Arthur, *Prog. Solid State Chem.* **10**, 157 (1975).
7. A. Y. Cho, *J. Vac. Sci. Technol.* **16**, 275 (1979).
8. L. L. Chang and R. Ludeke, in *Epitaxial Growth*, J. W. Mathews, Ed. (Academic Press, New York, 1975), part A, p. 37.
9. R. Ploog, in *Crystal Growth, Properties and Applications*, L. F. Boschke, Ed. (Springer-Verlag, Heidelberg, 1979).
10. R. C. Farrow, *J. Phys. D* **7**, 121 (1974).
11. J. H. Miller, B. I. Miller, K. M. Bachmann, *J. Electrochem. Soc.* **124**, 259 (1977).
12. B. I. Miller and J. H. McFee, *ibid.* **125**, 1311 (1978).
13. L. L. Chang, G. A. Sai-Halsz, N. J. Kawai, I. Esaki, *J. Vac. Sci. Technol.* **16**, 1504 (1979).
14. Y. Ota, *J. Electrochem. Soc.* **124**, 1795 (1977).
15. G. E. Becker and J. C. Bean, *J. Appl. Phys.* **48**, 3395 (1977).
16. J. C. Bean, *Appl. Phys. Lett.* **36**, 741 (1980).
17. H. Holloway and J. N. Walpole, in *Progress in Crystal Growth and Characterization*, B. R. Pamplin, Ed. (Pergamon, New York, 1979), vol. 2, pp. 49–94.
18. A. Y. Cho, *J. Vac. Sci. Technol.* **8**, S31 (1971).
19. _____, *J. Appl. Phys.* **46**, 1733 (1975).
20. _____ and F. K. Reinhart, *ibid.* **45**, 1812 (1974).
21. A. Y. Cho, C. N. Dunn, R. L. Kuvas, W. E. Schroeder, *Appl. Phys. Lett.* **25**, 224 (1974).
22. S. G. Bandy, D. M. Collins, C. K. Nishimoto, *Electron. Lett.* **15**, 218 (1979).
23. M. V. Schneider, R. A. Linke, A. Y. Cho, *Appl. Phys. Lett.* **31**, 219 (1977).
24. J. V. DiLorenzo, W. C. Niehaus, A. Y. Cho, *J. Appl. Phys.* **50**, 951 (1979).
25. W. T. Tsang, *Appl. Phys. Lett.* **33**, 426 (1978).
26. H. C. Casey, Jr., A. Y. Cho, P. W. Foy, *ibid.* **34**, 594 (1979).
27. H. C. Casey, Jr., and M. B. Panish, *Heterostructure Lasers* (Academic Press, New York, 1978).
28. W. T. Tsang, *Appl. Phys. Lett.* **34**, 473 (1979).
29. A. Y. Cho, H. C. Casey, Jr., C. Radice, P. W. Foy, *Electron. Lett.* **16** (No. 2), 72 (1980).
30. W. T. Tsang and M. Ilegems, *Appl. Phys. Lett.* **31**, 301 (1977).
31. F. K. Reinhart and A. Y. Cho, *ibid.*, p. 457.
32. A. C. Gossard, P. M. Petroff, W. Wiegmann, R. Dingle, *ibid.* **29**, 323 (1976).
33. L. V. Keldysh, *Sov. Phys. Solid State* **4**, 1658 (1963).
34. L. Esaki and R. Tsu, *IBM J. Res. Dev.* **14**, 61 (1970).
35. R. Dingle, W. Wiegmann, C. H. Henry, *Phys. Rev. Lett.* **33**, 827 (1974).
36. L. L. Chang, A. Segmuller, L. Esaki, *Appl. Phys. Lett.* **28**, 39 (1976).
37. R. M. Fleming *et al.*, *J. Appl. Phys.* **51**, 357 (1980).
38. R. Dingle, H. L. Störmer, A. C. Gossard, W. Wiegmann, *Appl. Phys. Lett.* **33**, 665 (1978).
39. W. T. Tsang, C. Weisbuch, R. C. Miller, R. Dingle, *ibid.* **35**, 673 (1979).
40. D. C. Tsui, H. Störmer, A. C. Gossard, W. Wiegmann, *Phys. Rev B* **21**, 1589 (1980).
41. L. L. Chang, H. Sakaki, C. H. Change, L. Esaki, *Phys. Rev. Lett.* **38**, 1489 (1977).
42. V. Narayanamurti, H. L. Störmer, M. A. Chin, A. C. Gossard, W. Wiegmann, *ibid.* **43**, 2012 (1979).
43. C. L. Allyn, A. C. Gossard, W. Wiegmann, *Appl. Phys. Lett.* **36**, 373 (1980).
44. W. T. Tsang and M. Ilegems, *ibid.* **35**, 792 (1979).

New Methods of Processing Silicon Slices

Thomas Clifton Penn

Within the semiconductor component factories for the last decade there has been a proliferation of fabrication methods generally classed as dry processing. To appreciate the advantages and limitations of such processes requires some basic knowledge of the classical fabrication of silicon integrated circuits (IC's). This knowledge will also be applicable to

Summary. Through the use of room-temperature, radio-frequency plasma ionization of gases, the insulating, conducting, and semiconducting materials associated with the fabrication of silicon integrated circuits can be patterned to submicrometer dimensions. A tutorial description is presented of the fabrication techniques used in the past with an overview of where plasma processing has made noteworthy improvements in the lithography of materials.

some of the other articles in this issue dealing with the fabrication of microscopic geometries.

A silicon IC typically starts with a slice of single-crystalline material about 0.5 millimeter thick and 100 millimeters in diameter whose mechanical and electrical characteristics are held to critical specifications. Many IC's are fabricated on this slice simultaneously by (i) growing or depositing specific layers of materials, (ii) partially removing these materials to define microscopic geometries, and (iii) altering the electrical properties of these materials. In this way a three-dimensional structure can be constructed by varying the geometries in two dimensions while the thickness is essentially constant throughout a particular material layer. The first layer on the slice is usually epitaxially grown single-crystal silicon or oxidized silicon.

Thousands of metal-oxide-semiconductor (MOS) transistors are used in a single pocket calculator, so I will choose this device as a typical semiconductor part. The completed structure of such a device is sketched in Fig. 1. Table 1 shows the materials and typical dimensions of such a structure. In other semiconductor products such as linear circuits, bipolar logic, Schottky and so on,

different materials and processes are used, but the techniques are similar enough for descriptive purposes. In this case we start with an initial oxidation of the slice of the order of 1000 angstroms and deposit silicon nitride of the order of 1000 angstroms. (Note that neither of these layers is evident in the finished product.) Throughout the fabrication process the manipulation of layer dimensions and properties is effected by the use of secondary materials that protect or "resist" the etching, oxidation, deposition, or ion implantation to which other parts of the slice are exposed. In this case the nitride layer protects the area the transistor will later occupy from thick oxidation, as in the LOCOS (local oxidation of silicon) process (*1*).

To pattern this material we make use of another secondary material—photoresist. The photoresist pattern is determined by yet another secondary material, chrome, with a photomask pattern which itself was patterned with photoresist. A master mask was used to expose this "working plate" photomask. The master mask itself was made by photoreduction of a "reticle," which replicates what we wish chip geometries to be. This is a rather tortuous path for such a simple result, and I have made the explanation complete to show how important the proper execution of material lithography is to the manufacturing process. One blemish in the pattern of the reticle for just one material layer can destroy every chip on every slice in the manufacturing line. Because of its great importance, lithography has been studied with intensity, and this has led to several advanced techniques that are discussed later in this article.

Continuing with the description of the MOS fabrication process, Fig. 2 shows the stages of producing the "moat" area in which the MOS transistor will reside. The nitride surface is coated with a polymeric material that is photosensitive to ultraviolet light. In this case the photoresist is assumed to be negative, which means that it will be polymerized where the ultraviolet light strikes it. Thus the mask that produces this pattern is clear where the pattern will become opaque in the resist. Now the photoresist layer is "developed" in a solvent solution, which washes away all the unpolymerized resist and leaves the resist that was exposed to ultraviolet, as shown in Fig. 2b. A batch of several resist-patterned slices can now be etched simultaneously by placing them in a hot phosphoric acid retort. The edges of the resist may tend to lift during this operation and degrade the geometric definition, as in most wet etching processes. Plasma (dry) etching is now fairly common at this process step and avoids the photoresist lifting problem. The resulting slice pattern is shown in Fig. 2c. Many of these patterned slices are now placed in a furnace tube and subjected to a hot oxidizing environment for several hours. The nitride layer oxidizes very little, while the surrounding silicon is oxidized to a thickness of about 10,000 Å. This thick "field" oxide is used to space the MOS conductors away from the silicon substrate so that parasitic MOS transistors will not be created. The finished oxidation pattern is shown in Fig. 2d.

It is not my purpose here to require the reader to become a process expert, and I have omitted many steps (such as channel stops and cleaning steps) while trying to retain enough detail to highlight areas where dry processing can be used to advantage. To continue with the fabrication process, the moat area is stripped to the bare silicon. A very clean gate oxide is now grown to a thickness of nearly 1000 Å. Polysilicon is deposited and patterned in a lithographic manner similar to that described above. The wet etching of polysilicon MOS gate electrodes has traditionally been more art than science. Dry etching allows so much improvement that narrower gate dimensions have been readily achieved. This is very desirable for very large scale integrated circuits (VLSI). The previous moat patterning step did not require any alignment with an underlying pattern. The

The author is a Fellow at Texas Instruments Incorporated, Dallas 75265.

117

silicon gate pattern must, however, be placed within the moat area. After several steps we now have the cross section shown in Fig. 3a. The source and drain areas may now be created in the silicon substrate without additional lithographic techniques by using the thick field oxidation and the silicon gate electrode to protect the material they cover. Either furnace tube diffusion of dopant materials (such as phosphorus or boron) or ion implantation may be used to create the source and drain areas. In the more modern ion implantation method, ions of the dopant are accelerated in an ion gun structure and a beam of these ions is focused and scanned across the slice surface in a uniform, controlled-dosage manner. Subsequent furnace tube processes drive these ions to the proper crystallographic locations to create an active transistor, as shown in Fig 3b.

This should be enough exposure to the basic problems of semiconductor fabrication. High-quality, low defect density

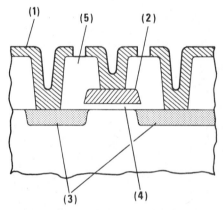

Fig. 1. A metal-oxide-semiconductor transistor structure with material and dimensions outlined in Table 1. Shown are (1) aluminum leads, (2) silicon gate, (3) source and drain, (4) gate oxide, and (5) field oxide.

Table 1. Semiconductor materials and common dimensions for the MOS transistor shown in Fig. 1.

Material	Thickness (Å)	Minimum dimension (μm)
Primary		
Silicon oxide (gate)	1,000	
Polysilicon (gate)	5,000	5
Doped silicon (source, drain)	7,000	5
Silicon oxide (field)	10,000	5
Aluminum	10,000	7
Secondary		
Photoresist	5,000	5
Silicon oxide (initial)	1,000	10
Silicon nitride (moat)	1,000	10
Chrome	1,000	5

118

material lithography must be accomplished. To do this, resist patterning and alignment to previous patterns must be precise to less than 1 micrometer over a linear dimension of 100 mm. While one layer of material is being modified, the layers of material exposed beside it and underlying it must not be harmed. For example, when the polysilicon gate was formed above, polysilicon was removed over oxide. The etchant for this silicon must not attack the oxide at a high removal rate. The advantage of dry processing is its ability to define very small (< 1 μm) widths. Removal of a specific material without affecting others is more difficult.

Masks and Pattern Generation

The patterning sequence outlined in Table 2 will be used to trace how the ideas of the circuit designer are reduced to the material patterns desired. Through computer simulation and modeling techniques, the designer defines the pattern requirements for every layer of the desired device. A pattern generation program is written and stored on a medium such as computer magnetic tape. This tape is used to manipulate an x, y stage, which holds a plate coated with photographic emulsion beneath a tiny spot of light; the exposure can run to hours. The photosensitive plate is developed by traditional photographic methods. This makes an oversized image (ten times the typical size) of one layer of one chip. The oversizing makes the effect of dirt and emulsion defects less important.

The reticle is mounted on a repeating printer and reduced with lenses to the desired image. One exposure is made of one chip, then the resist-covered chrome master is very accurately stepped to an adjacent location and again exposed. All levels must register from top to bottom at every x, y location. This mask level is developed with solvent and etched to describe the geometry accurately. To preserve these master masks, a set of several submasters may be contact-printed and used to produce working plates.

Although each step can be performed at a relatively low defect level, there are about 100 material levels in a simple device where something can go wrong. If the level integrity is as good as 0.995, one still loses 40 percent of the devices; if it is as poor as 0.98, one loses 97 percent.

One of the first uses of electron beam exposure machines has been in making photomasks. The patterning can be used

Table 2. Traditional photomask fabrication.

Operation	Materials affected
1. Reticle pattern generation	Emulsion
2. Master photomask	Resist, chrome
3. Submaster photomasks	Resist, chrome
4. Working photomasks	Resist, chrome
5. Semiconductor fabrication	See Table 1

to make masters or working plates, depending on the cost and complexity of the final product. Unfortunately, many electron-sensitive materials tend to have adhesion problems when wet-etched. Once again, dry processing can be used advantageously.

Although the mask system discussed above virtually eliminates mask-associated defects, the slow throughput makes

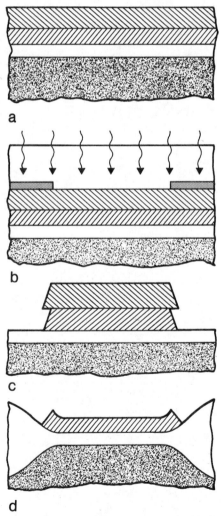

Fig. 2. How material layers are manipulated by overlaying other materials for protection: (a) unpatterned sandwich of silicon, oxide, nitride, and photoresist; (b) exposing the resist pattern to ultraviolet light; (c) after etching nitride; and (d) after furnace tube oxidation.

the process expensive. Another modern approach uses electron beam techniques to define a chrome oversize reticle. Now, instead of stepping a reduced image on a mask, one directly steps this reduced image onto the resist-coated silicon slice. Geometries as small as 1 to 2 μm can be defined in this manner on resist, and with dry processing the resist survives the subsequent patterning operation.

What might be considered the ultimate short circuit of the mask process is the approach of "writing" on the material desired. Although there is activity in this area, maturity is several years away. In the meantime we can come close to this approach by coating the slice itself with electron beam–sensitive resist and generating patterns by computer software directly on the slice itself. This has the great advantage of chip-by-chip alignment and extremely fine geometry generation. The disadvantage is increased pattern generation time.

Plasma Machine Evolution

The use of cold discharge tube reactions in the semiconductor industry is by no means new. The first major use of plasma processing was in removing common organic photoresists (2). A schematic representation of a typical apparatus in shown in Fig. 4. In this apparatus the slices are held upright in a carrier, much as in a furnace tube. The chamber is evacuated to a pressure of a few torrs and a radio-frequency generator is connected to the terminals shown. A mixture of argon and oxygen is introduced into the chamber and generates a glow discharge of relatively low magnitude in the classical plasma discharge sense. This creates an oxidizing environment, which rapidly converts the organic photoresist to the volatile products carbon dioxide and water vapor. Here we have the underlying principle of all semiconductor dry etching processes. A solid material is converted to gaseous byproducts, which are removed by the vacuum pumping system.

The active species generated in a tube reactor must have a sufficient lifetime to reach the surfaces of interest. Some etches and depositions can be made only by generating the active species near the surface of the slice. Sterling and Swann (3) showed how chemical vapor deposition could be promoted between two capacitor plates. The extension of this idea to a plasma reactor of the radial flow type was demonstrated by Reinberg (4),

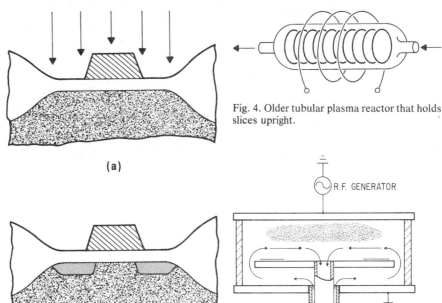

(a)

(b)

Fig. 3. Placing the source and drain automatically by aligning the gate within the moat area: (a) as patterned and (b) after ion implantation and furnace drive.

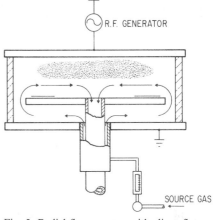

Fig. 4. Older tubular plasma reactor that holds slices upright.

Fig. 5. Radial flow reactor with slices flat on lower surface.

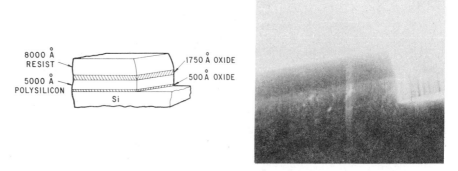

Fig. 6. Plasma etching of multiple layers.

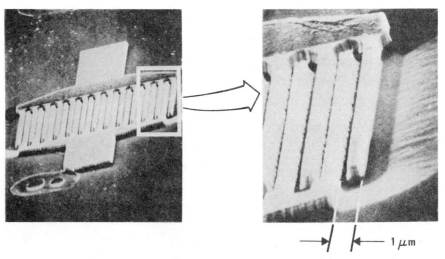

Fig. 7. Reactive ion–milled microwave bipolar transistor.

119

Table 3. Solid materials deposited and removed with gases used for the plasma process.

Solid	Gases
Depositions	
Silicon nitride ($Si_xN_yH_z$)	Silane (SiH_4) and ammonia (NH_3)
Silicon oxide ($Si_xO_yH_z$)	Nitrous oxide (N_2O) and SiH_4
Amorphous silicon	SiH_4 and argon
Material removal	
Silicon oxide (SiO_2)	Silicon tetrafluoride (SiF_4)
	Carbon tetrafluoride (CF_4)
	C_3F_8, C_2F_6, C_5F_{12}, CHF_3
Silicon	CF_4 and O_2
	Carbon tetrachloride (CCl_4) and hydrogen chloride (HCl)
Silicon nitride (Si_3N_4)	CF_4
Vanadium, titanium, tantalum, molybdenum, tungsten	CF_4
Chrome and chrome oxide	CCl_4
Aluminum	CCl_4, boron trichloride (BCl_3)
Photoresist	Argon and O_2

and a sketch of such a reactor is shown in Fig. 5. Such reactors have been in extensive production use at Texas Instruments since 1972, and there are now several manufacturers of similar equipment.

Development of Dry Processing

The scientific understanding of plasma processes was very sketchy in the early 1970's and certain areas are still not well understood. The world's first all-dry-processed IC was produced in 1975 by plasma development of photoresist; plasma etching of nitride, oxide, silicon, and aluminum; and ozone resist removal (5). Nevertheless, virtually all the methods were worked out by determining the volatile products of the solids and choosing likely combinations of gas mixtures to produce them. More recently, scientists at IBM (6) and Bell Laboratories (7) have studied the action of a common fluorocarbon etchant, CF_4, in the etching of silicon in enough detail to explain in general what is observed. However, predicting an outcome before the fact is still a low-probability exercise.

With the parallel-plate configuration there tends to be a continuum of materi-al removal processes as pressure is changed. At pressures of 0.1 to 10 torrs the plasma etching performed is sometimes referred to as chemical plasma etching. In this region one can obtain fairly high specific etch ratios of silicon to oxide by using a fluorinated gas. The etch characteristics tend to be isotropic, although when silicon is etched with a chlorinated species the profiles obtained are quite vertical, and this process was considered novel enough to patent (8). Others have observed anisotropic profiles in plasma etching of silicon, silicon oxide, silicon nitride, and aluminum (9). This more anisotropic etch is usually considered the proper domain of what has been called reactive ion etching (RIE) for pressures low enough to mix ion bombardment with chemical action. In RIE the etch contour is more vertical, but the etch is less specific to the material and erodes the resist faster than higher pressure etches do. As the pressure is lowered still further we enter the sputter etching or ion milling regime, where energetic ions of argon "sandblast" away resist and other materials with equal vigor. The removal rate is lower than that in plasma etching.

There is one other plasma etching method that is different from those dis-cussed above. A high-power-density microwave discharge can produce metastable atoms with long lifetimes which etch material almost as if it were immersed in acid—that is, isotropically. There is some evidence that the etching species produced in this reaction is not the same as the species in the reactions described above.

Applications of Dry Processing

A plethora of plasma etch gases and combinations can now be used to remove almost any material employed in the fabrication of silicon IC's, photomasks, and magnetic bubble structures. The emphasis has been on the small dimensions obtainable at low defect densities. Shown in Table 3 is a list of materials and the etch gases used for their removal. The list is by no means complete, as many manufacturers consider their gas recipes trade secrets. The vertical section in Fig. 6 illustrates some of the exciting possibilities for structures that can be realized by plasma etching with nearly zero undercut. In Fig. 7 the close spacing and narrow geometries of a microwave transistor structure show what can be achieved by direct writing with an electron beam on the resist over the material to be patterned. These are but a few of the many advances that can be expected in material lithography when plasma etching is combined with modern resist techniques.

References

1. J. A. Appels, K. Kooi, M. M. Paffen, J. J. H. Schatorje, W. H. C. G. Verkuylen, *Philips Res. Rep.* **25**, 118 (1970).
2. C. E. Gleit and W. D. Holland, *Anal. Chem.* **34**, 1454 (1962).
3. H. F. Sterling and R. C. G. Swann, *Solid-State Electron.* **8**, 653 (1965).
4. A. R. Reinberg, U.S. Patent 3,757,733 (1973).
5. T. C. Penn, *IEEE Trans. Electron Devices* **ED-26**, 640 (1979).
6. J. W. Coburn, H. F. Winters, T. J. Chuang, *J. Appl. Phys.* **48**, 3532 (1977).
7. W. R. Harshbarger, T. A. Miller, P. Norton, R. A. Porter, *Appl. Spectrosc.* **31**, 201 (1977).
8. A. R. Reinberg and R. Rao, U.S. Patent 4,069,096 (1978).
9. C. J. Mogab and W. R. Harshbarger, *Electronics* **115**, 117 (1978).

Materials Aspects of Display Devices

B. Kazan

With the rapid growth in versatility and complexity of electronic systems, display devices are becoming an increasingly crucial element in the efficient transfer of information from these systems to the human operator. Because of the variety of demands placed on displays to satisfy existing and future requirements in industrial, consumer, and

as the ubiquitous device for displaying electrical information in computer terminals and word-processing and related information systems.

Although the cathode-ray tube is still dominant in these areas, intense effort is now under way to develop new display technologies and associated materials. This interest can be traced to the grow-

Summary. For many years interest in display devices stemmed almost entirely from the needs of television receivers, oscilloscopes, and radar systems whose requirements were almost ideally satisfied by the cathode-ray tube. However, during the past two decades the proliferation of small electronic instruments with digital output has stimulated the development of a number of new display media which, because of their compact nature and low power consumption, are better suited for such applications. In view of the success achieved with these technologies a continuing effort is being directed toward increasing their information capacity to the point where they may ultimately be useful in applications now requiring cathode-ray tubes.

military applications, the field of display technology is today in a state of considerably flux, with numerous devices, both old and new, competing with each other to capture portions of a large and rapidly growing market.

For many years the dominant display medium has been the cathode-ray tube. In 1897 (some months before the publication of J. J. Thomson that established with certainty that "cathode-rays" consisted of electrons) the cathode-ray tube was invented by F. Braun as an instrument for the visualization of rapid time-varying electrical signals. Since then the successive improvements made in it have resulted in today's highly sophisticated oscilloscopes. By about 1930 it became clear that the cathode-ray tube was almost ideally suited for producing television images, resulting in the virtual cessation of work on all other display technologies. During World War II, the cathode-ray tube found an important new role as the display medium in radar systems, whose development could hardly have been possible without such tubes. In more recent years the cathode-ray tube has emerged in still another role

ing need during the 1950's for small devices capable of displaying numerical information in applications such as high-speed pulse-counting. This need was satisfied by the Nixie gas-discharge tube which could display any one of ten digits in a fixed format. With the subsequent growth of semiconductor and digital technology a great variety of small, low-power instruments emerged for which it was desirable to have even more compact numeric or character displays capable of operating at lower voltage and with reduced power consumption—application areas for which the conventional bulky, high-voltage cathode-ray tube was hardly appropriate. These instrument requirements stimulated work on other technologies, particularly light-emitting diodes and liquid crystals as well as improved types of gas-discharge devices.

The success achieved in producing simple displays with these new technologies soon stimulated efforts to develop devices with larger capacity, capable, for example, of displaying tens or hundreds of numerals or characters. As commercial devices emerged containing

an increasing number of display elements, the realization arose that these technologies might be even further extended to satisfy application areas previously considered only the province of the cathode-ray tube. Although the development of display devices capable of competing with the cathode-ray tube (that is, having comparable numbers of picture elements, speed, gray scale, and color) remains a formidable task, major development programs are now under way with this objective. Motivation for this research is in no small measure attributable to the increasing saturation of the television market in many countries and the resultant intense rivalry between companies to produce new television displays in panel form without the size limitations of the cathode-ray tube.

As in the case of other areas of electronics, display systems have profited greatly from the development of semiconductor technology, especially where local information storage and complex addressing schemes are involved. However, the advances in semiconductor technology, characterized by the ever-increasing number of low-power circuit elements which can be packed onto a small semiconductor chip, have been of little direct aid in the development of most display devices. For display purposes optically active materials extending over large areas are required (difficult to achieve with semiconductor crystals of limited size), and the sequential switching of substantial power to a very large number of elements is frequently involved.

The different display technologies now under investigation make use of a bewildering array of physical and chemical phenomena, encompassing materials in the solid, liquid, and gaseous states. In the sections that follow, however, the discussion will be focused primarily on the major technologies and associated materials, particularly those which have attained some degree of commercial acceptance.

Cathode-Ray Tubes

Since their inception, cathode-ray tubes have depended on luminescent screens of powder material (referred to as "phosphors") to convert the energy of the scanning electron beam to light. Although naturally occurring minerals were used in early tubes, these materials have long since been supplanted by syn-

The author is principal scientist at the Palo Alto Research Center, Xerox Corporation, Palo Alto, California 94304.

thetic phosphors with far superior characteristics. Because of the requirements of high stability under electron bombardment, low vapor pressure, and the ability to withstand tube bakeout temperatures in excess of 400°C, only selected inorganic compounds have been found useful. Among these are the sulfides, selenides, silicates, tungstates, and oxides of zinc, cadmium, calcium, and magnesium (1).

All phosphor powders consist of crystalline particles and generally depend for their luminescence on the incorporation of a small amount of one or more "activators" (usually metallic elements such as copper, silver, manganese, chromium, and bismuth). By controlling the quantity and type of activator (as well as the procedures) used during synthesis the characteristics of the phosphor (for example, efficiency, color of luminescence, and decay time) can be varied (1, 2). Associated with the activator atoms or ions incorporated in the host crystal are a number of discrete energy levels, the lower ones being normally occupied. During bombardment by the high-energy electron beam, the luminescent centers corresponding to the activator atoms are excited, producing light as they relax to their lower energy state. All present phosphors are wide-band-gap (> 3 electron volts) materials, making them transparent to the light internally generated. With few exceptions the materials are highly insulating, any conductivity which does exist generally resulting from electron, rather than hole, mobility.

Of the more than 50 different commercial phosphors available, perhaps a third are group II-VI compounds such as ZnS, (Zn,Cd)S, and Zn(S,Se). These are used in a variety of applications and are of particular importance in color television tubes where, for example, ZnS:Mg (that is, ZnS activated with Mg) is used to produce blue emission and (Zn,Cd)S:Cu,Al is used to produce green emission. Phosphors of these types are characterized by high efficiency, materials such as ZnS:Ag and ZnS:Cu being capable of converting more than 20 percent of the electron-beam energy to light. Such high efficiencies allow brightness levels of more than 100 foot-lamberts (fL) to be readily obtained in commercial television receivers.

Another important phosphor used in many applications is Zn_2SiO_4:Mn whose luminescence is green. Although its efficiency is less than half that of the II-VI phosphors it is extensively used in applications where its ability to withstand intense electron bombardment for long periods with minimal degradation is impor-

tant. Typical of such applications are oscilloscopes, projection systems, and cockpit displays where maximum brightness is required.

The most recently developed types of phosphors are those containing rare-earth activators (3), particularly terbium and europium. The stimulus for the development of these phosphors arose from the need in color tubes for a red-emitting phosphor whose efficiency was comparable to the green and blue phosphors. The first rare-earth phosphor to be developed was YVO_4:Eu, which was introduced in 1964. This has been superseded by the more efficient Y_2O_2S:Eu. It is of interest that because of the high efficiency obtained with red phosphors research is now being redirected toward improvement of the green phosphors. For other applications several additional rare-earth phosphors have been developed (4) such as green-emitting La_2O_2S:Tb and white-emitting Y_2O_2S:Tb.

Aside from their long operating life, rare-earth phosphors are important because of their relative freedom from saturation at high input power levels. Unlike the broad emission spectrum of other phosphors, the emission from rare-earth phosphors is confined to one or more very narrow bands (less than 1 nanometer at half value). This allows the screen to be viewed through a narrow-band filter which blocks the reflection of external light at all other wavelengths, thereby enabling high-contrast images to be produced in high ambient light environments (such as airplane cockpits).

For applications such as television, phosphors are used whose persistence is less than the frame time of 30 to 40 milliseconds. For use in radar, where the time to scan a frame may be a second or longer, special phosphors have been developed such as (Zn,Cd)S:Cu whose yellow-orange persistence is of the order of seconds. Since the persistence of this material is maximized when it is optically excited (instead of electron-beam excited) it is coated with a layer of ZnS:Ag whose blue emission under electron bombardment causes luminescence in the (Zn,Cd)S. At the other extreme, phosphors with very short persistence have been developed for use in flying-spot scanners and photographic applications. A phosphor of this type is calcium-magnesium silicate activated with cerium (emitting in the violet and ultraviolet) which has a decay time of about 0.1 microsecond.

Typical phosphor screens consist of particles 1 to 10 micrometers in diameter deposited on the glass faceplate of the tube with a small amount of binder such

as potassium silicate. To absorb most of the energy of the penetrating electron beam, the layer is usually made several particles thick. However, because of light scattering, the resolution is limited to a few tens of micrometers. Although the resolution can be improved by using smaller particles, this usually results in lowered phosphor efficiency since the "dead layer" which covers the particles then absorbs a greater fraction of the electron-beam energy. For very high resolution, thin phosphor films produced by evaporation or vapor-phase reaction methods have been made with materials similar to those used for powder phosphors. Although their intrinsic luminescent efficiencies are comparable to powder materials, their effective efficiency is reduced to about 10 percent of this because of the high index of refraction of the film which prevents the escape of most of the light from the smooth surfaces.

Much of the present work on phosphors is concerned with the development of materials with special characteristics of color, persistence, and reduction of saturation at high brightness levels as well as increased efficiency and life. Despite the increased understanding of the luminescence of solids resulting from advances in solid-state physics, the complex physical processes occurring in the excitation of electron-bombarded phosphors (particularly powder materials) are in many respects understood only qualitatively. The development of new phosphors with specific characteristics, although guided by theoretical considerations, thus remains to a large extent dependent on experimental procedures.

Electroluminescent-Layer Displays

For many years the remarkably good performance obtained with phosphors excited by electron bombardment was a challenge to workers to develop similar screens that could be made to luminesce by direct application of a voltage across them. As far back as 1937, G. Destriau showed that light could be obtained from specially prepared ZnS powder layers when an a-c electric field was applied. Unfortunately, because of the unavailability at that time of satisfactory transparent electrodes for viewing the phosphor and the use of castor oil as a liquid dielectric to embed the particles, this work was either ignored or not considered seriously by the scientific community. In 1950, however, workers at Sylvania, using similar phosphor powder, were successful in fabricating more prac-

122

tical cells in which the powder was embedded in a solid dielectric and which employed recently developed transparent conductive coatings of tin oxide on glass. Since then considerable research effort has been directed toward improving electroluminescent phosphors and developing display devices based on these materials (5). Unlike the large number of materials which are relatively efficient under electron bombardment, in the case of electroluminescent layers acceptable efficiency has been obtained from few materials other than ZnS or closely related compounds such as Zn(S,Se).

In preparing ZnS powders for use in a-c–excited layers sufficient copper is added so that, in addition to its incorporation in the crystal as an activator, isolated regions of Cu_2S are formed on the surface or interior of the grains, resulting in the formation of Cu_2S-ZnS heterojunctions (6). When a-c voltage is applied across a powder-binder layer most of the voltage drop during a particular half-cycle appears across those junctions which are oriented in the back-biased (high-resistance) direction. Free electrons in these regions are then accelerated, causing ionization of luminescent centers by collision processes, with subsequent capture of the electrons in nearby traps. Upon reversal of the field these electrons are released, some of them recombining with the ionized centers to produce light. During alternate half-cycles similar action occurs at junctions oriented in the opposite direction so that a light pulse is produced each half-cycle.

Electroluminescent layers of the above type are usually about 25 to 50 μm thick, with the powder embedded in an epoxy or other organic binder. In typical operation an a-c voltage in the range of 100 to 200 volts root mean square (rms) is applied at some frequency in the range of 50 to 5000 hertz. At the higher frequency and voltage levels a steady-state brightness can be obtained which is adequate for viewing in moderate room light. Although considerable work has been done to develop practical display devices with such layers, the poor life often obtained (for example, more than 50 percent drop in efficiency after several hundred hours), their limited brightness under pulsed conditions, and the lack of a sharp threshold voltage for turn-on (important for addressing large arrays of elements as discussed below) has discouraged use of these materials.

About 10 years ago a new type of powder-binder electroluminescent layer was developed that can be excited with d-c (or pulsed d-c) voltage and whose char-

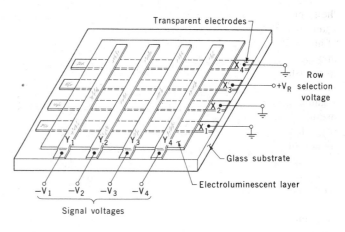

Fig. 1. Electroluminescent-layer display with *X-Y* addressing electrodes.

acteristics are considerably more desirable for display applications (7). As before, ZnS powder is used but with Mn as the activator. During preparation the powder particles are coated with a relatively heavy layer of conductive Cu_2S. When d-c voltage is initially applied across a layer of such particles held together with a limited amount of binder a relatively high current flows because of the particle-to-particle contacts, but no light is produced. If the d-c voltage is gradually increased (up to about 60 V, for example) during a period of about a minute a "forming" process occurs in a thin portion of the layer adjacent to the anode in which copper ions diffuse into the ZnS. This causes a marked increase in resistivity with most of the applied voltage drop now appearing across this portion of the layer. If d-c (or pulsed d-c) voltage of the same polarity whose magnitude is comparable to or higher than the voltage used in forming, is now applied, substantial yellow-orange light emission, characteristic of the Mn activator, is generated in the insulating region. Somewhat similar to the case of a-c powder material, light emission is attributed to excitation of the Mn luminescent centers (either directly or indirectly) by energetic electrons accelerated in the high-field region.

Because of the relatively high light output obtained with short d-c pulses and the sharp threshold voltage, such layers are of considerable interest for display devices containing a large number of picture elements. To reduce the number of electrical connections that would be required if each picture element of a large array had a separate lead, an *X-Y* addressing scheme is used such as shown in Fig. 1. As indicated, the electroluminescent layer is sandwiched between two sets of orthogonal conductors, the individual picture elements being defined by the crossover region of the electrodes. (To permit viewing of the light the *X*-electrodes on the glass surface are made transparent.) In operation a pulse voltage, $+V_R$ is applied to one of the row conductors, for example, X_3, and pulse voltages, $-V_1 \ldots -V_4$, are simultaneously applied to the *Y*-conductors. This causes voltage pulses of varying magnitude above the threshold voltage, V_T, to appear across the elements of row X_3, producing corresponding variations in light output. Negligible light output is produced, however, from the phosphor elements of all other rows since it is assumed that the voltages across them are all below the threshold voltage. By applying the pulse voltage, $+V_R$, to successive rows and cyclically repeating the addressing process at a rapid rate a flicker-free image can be produced. Experimental panels of this type have been developed for displaying both alphanumeric information and television images (8). For the latter purpose panels up to 20 by 27 centimeters in size have been built with 224×224 elements. By using pulses up to about 250 V across the phosphor elements images were obtained with good gray scale and a highlight brightness of 10 fL (about 1/10 of the 100-fL brightness obtainable with commercial television receivers).

As an alternative to powder layers, effort has also been directed toward the development of thin-film electroluminescent layers. Although initial attempts to produce such films resulted in poor light output and short life, significant improvements in such materials were made by workers at the Sigmatron Co. during the period of 1964 to 1970 using manganese-activated ZnS. These films (about 1 μm thick), requiring about 200 to 300 V of a-c for operation, were provided with an additional insulating film on one or both sides to prevent breakdown (5). By coating the rear surface of the phosphor with an additional light-absorbing layer of arsenic selenide the transparent phosphor was made to appear black in the off state (as opposed to the diffuse white appearance of powder layers). Information dis-

123

played with such films could thus be viewed with good contrast even in high ambient illumination.

Subsequent work by the Sharp Company with the same material (9) has resulted in thin-film phosphors capable of higher light output and extremely long life (in excess of 10,000 hours). The ZnS film, in this case 0.5 μm thick, is sandwiched between two thinner insulating films of Y_2O_3. The improved results obtained with such films are in part attributed to the use of better insulator layers as well as a high-temperature annealing of the ZnS film after vacuum deposition.

Similar to d-c powder layers thin-film phosphors exhibit a very sharp voltage threshold. When the threshold field is exceeded electrons trapped in the material are released and accelerated, causing collision, ionization, and excitation of the Mn luminescent centers. As in the case of powder-layer devices, X-Y–addressed multielement panels fabricated from such films have been used to produce television images and to display alphanumeric information. For example, panels with a diagonal of 16 cm containing 240 × 320 elements when operated at a frame rate of 30 Hz have produced television images with good gray scale and a highlight brightness of 20 fL (9).

Arrays of thin-film phosphor elements can also be used as memory devices (9) if the Mn content of the ZnS is increased from the amount normally used (about 1 percent by weight) to about 5 percent. If an a-c sustaining voltage just below the threshold (for example, 250 V rms, 5 kilohertz) is maintained across all the elements they will remain off. However, if the voltage across selected elements is raised during one half-cycle by the addition of a small "write" pulse they will be triggered to an "on" state and continue to emit light for an indefinite period. Since the "stored" level of light emission at different elements can be controlled by the magnitude of the write pulses, gray scale as well as on-off information can be stored. When desired either the entire array or selected elements can be switched off by interrupting for several cycles the sustaining voltage across these elements.

At present the storage mechanism in ZnS films is not clearly understood. However, it appears to be associated with a negative resistance effect produced by the increased Mn content. Somewhat similar to the a-c gas panel (described below) the storage action is attributed to the buildup and retention of charges at the phosphor-insulator interface each half-cycle. Although still in an experimental stage, storage panels of

this type are potentially useful for displaying stationary images with a very large number of picture elements, since writing can be accomplished at arbitrarily slow speeds and problems of flicker are avoided.

Although electroluminescent panels are promising for a number of information display applications they have the disadvantage of requiring high-voltage addressing circuits, particularly in the case of X-Y arrays where many drivers are required. This problem, however, is likely to be ameliorated as suitable low-cost integrated circuits are developed. Another problem is the low efficiency of electroluminescent layers (only a few percent of the efficiency of the best electron-bombarded phosphors) resulting in significant power dissipation in the panel and drive circuits. Aside from this, because of the limitations of present materials, the maximum brightness that can be expected from X-Y addressed panels with full television resolution is about 1/10 of that obtainable from cathode-ray tubes. Although attempts have been made to develop phosphors with colors other than the yellow-orange of ZnS:Mn the efficiency obtained is substantially lower than for this material.

Light-Emitting-Diode Displays

Unlike powder or thin-film electroluminescent layers which are microcrystalline in nature, light-emitting diodes are fabricated from single-crystal material in which p-n junctions are formed by the addition of suitable doping agents (10). Such diodes are a direct outgrowth of developments in semiconductor technology and the understanding gained of the detailed electronic processes occurring in such materials. Early studies of light emission from current flow through crystals were made by Lossew in 1923 using metallic contacts to naturally occurring SiC crystals. However, because of the low efficiency and the difficulty in obtaining reproducible results little interest was generated at the time. In 1962 strong interest in the subject arose as a result of the demonstration that infrared radiation with efficiencies as high as 50 percent could be generated in p-n junctions of gallium arsenide (GaAs). By 1968 diodes capable of emitting visible (red) light were introduced commercially fabricated from crystals of GaAsP. Since then improved materials with higher efficiency have been developed, emitting light with a variety of different colors. Light-emitting diodes have a number of characteristics

making them desirable for display applications. These include: (i) low-voltage operation allowing direct interfacing with semiconductor logic circuits, (ii) small size, (iii) long life (greater than 10,000 hours), (iv) high peak brightness, and (v) rectifying properties useful for X-Y addressing.

When a voltage is applied across a light-emitting diode in the forward or conducting direction electrons and holes present in the n and p sides of the junction, respectively, flow toward each other and, in the recombination processes occurring in the neighborhood of the junction, light is emitted. For the radiation to be in the visible range (as opposed to infrared) the energy difference between the holes and electrons (that is, the band gap of the semiconductor) must exceed about 1.8 electron volts. Although a number of luminescent materials satisfy this criterion (particularly those used for cathode-ray tube phosphors) the inability to fabricate p-n junctions in most of them precludes their use for light-emitting diodes. Materials of interest satisfying the requirements for light-emitting diodes are GaP, GaAsP, GaAlAs, GaN, and SiC. However, at present all commercial diodes are fabricated from crystals of the first two compounds.

The first commercial diodes were produced from $GaAs_{0.6}P_{0.4}$ deposited epitaxially as a thin layer on a GaAs crystal substrate. Using local diffusion techniques similiar to those employed in making silicon diodes, p-n junctions were then formed. Corresponding to the band gap of $GaAs_{0.6}P_{0.4}$ (1.92 eV) an emission band of red light with a peak at about 650 nm is produced as a result of direct recombination of electrons and holes. Since the junction material as well as the GaAs substrate strongly absorb this radiation all of the light emitted in the direction of the substrate is lost. Because of the high index of refraction of the GaAsP, however, only light emitted toward the surface at an angle close to the normal (about 4 percent of the radiation in the forward direction) leaves the crystal, the remainder being reflected back. Frequently the diode is encapsulated in epoxy material shaped in the form of a lens. This concentrates the light in the forward direction and also doubles the light escaping since the index of refraction of the epoxy is intermediate between that of the GaAsP and air. Diodes of this material are particularly useful where a number are fabricated in close proximity on a single-crystal chip since the light produced by each diode is localized and can-

124

not spread to other areas. By taking advantage of this, single chips (for example, 0.2 inch high) with seven bar-shaped diode segments have been developed for numeric displays in pocket calculators.

Although diodes emitting light with shorter wavelengths, for example, yellow or green, can be made by increasing the phosphorus content of GaAsP (thus increasing the band gap), as the phosphorus content is increased beyond about 40 percent, a very rapid drop in efficiency occurs because of changes in the energy band structure which results in an increased number of the injected electrons and holes recombining nonradiatively to produce heat. It has been found, however, that the efficiency can be significantly increased by incorporating nitrogen atoms in the crystal which replace some of the arsenic or phosphorus (11). The nitrogen atoms in this case act as isoelectronic centers that can trap electron-hole pairs in an excited state. In recombining from this state these have a greater probability of producing light than by recombining through other processes. However, since some energy is initially given up by the electron-hole pair in the trapping process, the wavelength of the emitted light is somewhat less than that corresponding to the band gap of the material. The energy conversion efficiency of the nitrogen-doped diodes is, on the whole, comparable with that of red-emitting $GaAs_{0.6}P_{0.4}$ diodes. However, because of the increased sensitivity of the eye to green and yellow light compared to red, the luminous efficiency (lumens per watt) of the nitrogen-doped diodes is generally several times greater.

Three types of diodes with nitrogen doping have become important commercially: $GaAs_{0.65}P_{0.35}$, which emits orange light; $GaAs_{0.85}P_{0.15}$, which emits yellow light; and GaP, which emits green light. In the case of GaP (whose band gap is 2.25 eV), efficient diodes can also be produced if, instead of nitrogen, the material is doped with zinc and oxygen (a Zn atom replacing a Ga atom and an O atom replacing the nearest neighbor P atom). This Zn-O pair also acts to trap an electron-hole pair which may then recombine radiatively. However, because of the increased energy loss during trapping the light emitted is red with a spectral peak at about 700 nm.

In the case of nitrogen-doped materials the junction material is epitaxially grown on a GaP rather than GaAs substrate. Since the bandgap of the substrate as well as that of the GaAsP junction is greater than the energy of the emitted photons, these materials are both transparent to the emitted radiation. By providing a reflecting surface at the rear of the substrate, light emitted into the substrate is reflected back, emerging at the sides and top surface of the diode structure and increasing its effective efficiency.

Except for GaP:Zn-O diodes (whose light output saturates with increasing current) the light output of the remaining diodes increases superlinearly with current. More efficient operation results, therefore, if the diodes are driven with periodic pulses of high current than with constant current. This characteristic, together with the short response time of junction diodes to current pulses (a small fraction of a microsecond), as well as their rectifying property which blocks current flow in the reverse (nonemitting) direction, makes them well suited for use in X-Y addressing arrangements. For example, using a matrix of 7 × 25 individual diodes, five-character displays are commercially available (using 5 × 7 diodes per character) in which successive rows or columns of diodes are simultaneously addressed. Recently display devices (16 by 12 cm in area) have been built which consist of an array of 320 × 240 individually selected green-emitting GaP diodes (each 0.3 by 0.3 millimeters in size). By addressing the diodes, one line at a time, television images with a highlight brightness of 70 fL have been demonstrated (12).

Although light-emitting diodes are useful for on-off indicators, small alphanumeric displays, and X-Y arrays of limited size, the cost of fabrication of large arrays of densely packed diodes as well as their high power dissipation (for example, about 0.1 watt/cm² required to produce a brightness level comparable to television screens) limits their usefulness for applications involving the display of complex images.

Gas-Discharge Displays

The fact that light can be produced by an electric discharge through a gas has been known for more than 200 years. Scientific interest in this phenomenon was stimulated in about 1856 by Geissler who fabricated glass tubes with electrodes sealed in at opposite ends and filled them with commonly available gases such as air, carbon dioxide, and hydrogen at low pressure. Although these tubes had a short operating life they were of technical interest for the study of the radiation spectra of different gases. In the early 1900's, soon after the discovery of neon, elongated tubes containing this gas came into use for advertising signs. In such tubes almost all the light is produced by the "positive column," the luminous region starting a small distance from the cathode and extending almost to the anode. Soon after World War I small neon-filled bulbs became available commercially because of their usefulness as on-off indicators of 110- and 220-V power. However, because of their small electrode spacing (for example, several millimeters) they operate without a positive column, light being produced only from the "negative glow" region (also present in elongated tubes) immediately adjacent to the cathode. Despite the lower efficiency of this light generation (less than 1/10 of that from the positive column) most gas-discharge display devices, because of the small size of their elements, depend on this type of emission.

Gas cells have a number of desirable characteristics for display devices. Aside from their relatively simple structure, they can be made in arbitrary shapes and as small as 0.5 mm or less. Since a relatively sharp threshold voltage exists which must be exceeded to initiate a discharge, gas cells are well suited for use in X-Y addressing schemes. The fact that high peak brightness can be produced with short voltage pulses is also of importance for such addressing.

The gas-discharge cells used in all display devices are of the cold-cathode type, that is, the electron emission from the cathode required to sustain the discharge is not produced by heating but results from the transfer of energy from positive ions of the gas discharge to electrons at the cathode surface. Because of its chemical inertness and relatively high luminescent efficiency compared to other gases neon is almost always used. Generally, however, a small amount of a second inert gas is added to form a "Penning mixture" which allows operation at a reduced voltage (13). In the case of a discharge in pure neon, light is produced by excited atoms during electronic transitions to lower energy states, some of which are metastable. However, relatively few atoms acquire enough energy from the discharge to become ionized. By adding, for example, 0.1 to 1.0 percent of argon, a large fraction of the argon atoms become ionized by transfer of energy to them from the neon metastable atoms (whose energy is slightly higher than the ionization potential of the argon). Because of the increased ion generation the minimum voltage required to establish a discharge in such a mixture may be reduced, for example, to about

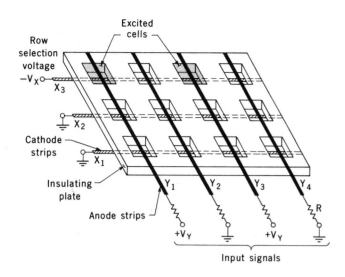

Excited cells

Row selection voltage

$-V_X$

X_3

X_2

X_1

Cathode strips

Insulating plate

Anode strips

Y_1 Y_2 Y_3 Y_4

R

$+V_Y$ $+V_Y$

Input signals

Fig. 2. Array of gas-discharge cells connected for X-Y addressing.

200 V compared to almost 300 V for pure neon.

In simple devices used for displaying a single character or numeral, a common anode may be used with a number of bar-shaped cathodes arranged in a suitable pattern (14). By applying sufficient voltage between the anode and selected combinations of cathodes, localized light emission is produced at these cathodes, allowing different characters or numerals to be displayed. For multicharacter or other displays involving a large number of elements, X-Y addressing is used as in electroluminescent displays to reduce the number of input leads and associated switching circuits (14). An example of such an arrangement is shown in Fig. 2. The individual cells here are defined by the holes in the insulating plate. In registry with the holes is a set of cathode strips on the lower surface of the plate. Running perendicular to these is a set of anode strips on the opposite surface. Although not shown, it is assumed that the entire structure is contained between a pair of glass plates vacuum sealed at the edges and filled with gas at a low pressure, for example, 100 torr. In operation a pulse voltage, $-V_X$, is applied to one of the cathode strips (for example, X_3) while input signals in the form of pulse voltages, $+V_Y$, are applied simultaneously to selected anode strips (for example, Y_1 and Y_3). Since the voltage sum $V_Y + V_X$ is assumed to be greater than the threshold voltage, V_T, for initiating a gas discharge, light emission is produced at the two cells shown. Since the voltages V_Y and V_X are assumed individually to be less than V_T no other cells of the array will fire. In a similar manner other rows of cells corresponding to cathode strips X_2 and X_1 can be addressed repetitively in rapid sequence to produce a flicker-free image. (To avoid excessive current flow through the "on" cells external lim-

iting resistors, R, are placed in series with the anode strips, as shown.)

A somewhat different addressing scheme than indicated in Fig. 2 is also employed in some types of commercial gas panel displays. In these (14a), by means of additional electrodes, an auxiliary gas discharge is generated immediately below the cathode electrode strip whose elements are to be addressed. This auxiliary discharge produces metastable gas atoms, which diffuse into the adjacent display elements, reducing their threshold voltage. The input signal voltages, $+V_Y$, applied to the anode strips are now sufficient to cause the display elements of this particular row to fire without applying a selection voltage, $-V_X$, to the corresponding cathode strip. The advantage of such an addressing system is that a special electrode system with only three external leads can be used to produce an auxiliary discharge at one cathode strip which is then stepped from row to row. (To produce this stepping action, use is also made of the diffusion of metastable atoms from the discharge at each auxiliary electrode to its neighbor.) This avoids the need for the large number of driver circuits required with the addressing scheme of Fig. 2 when a panel with a large number of rows is used.

As in the case of other X-Y-addressed luminescent displays, if the number of rows of a gas panel is increased the fraction of the time that a single row is on is decreased. To compensate for the reduced average brightness it is necessary to raise the addressing voltages to increase the peak currents through the cells. Unfortunately, this results in a rapid rise in cathode sputtering which limits the peak currents which can be used. In the sputtering process a metallic deposit from the cathode builds up on the walls of the cell as a result of bom-

bardment of this electrode by energetic positive ions (14), causing short circuits or blocking the light transmission of the glass viewing surfaces. Although sputtering can be greatly reduced by the addition of a small amount of mercury vapor (a process not well understood) it is not believed practical to build displays with more than about 250 rows if acceptable brightness is to be obtained.

To overcome the brightness limitation experimental panels have been developed in which excited cells remain on continuously after being addressed, thus considerably reducing the deleterious sputtering effects produced by the high peak currents. Use is made here of the fact that a gas discharge, once initiated, can be sustained by a somewhat lower voltage than the threshold voltage (14). If such a d-c sustaining voltage is maintained across all the off cells of an X-Y-connected array, selected cells can then be triggered on by pulse voltages of reduced magnitude superimposed on the sustaining voltage. In a similar manner selected cells can be switched off by superimposing reversed-polarity X-Y pulses which momentarily lower the voltage of these cells below the minimum sustaining voltage. To limit the current flow through individual cells, however, it is generally necessary to provide a high-resistance element in series with each cell, complicating the overall structure. Because of this, as well as addressing problems resulting from variations of firing and erasing voltage from cell to cell, interest in such devices is limited.

An alternative type of storage panel (15) which is in commercial use is shown in Fig. 3. Here the X- and Y-conductors are fabricated on the inner surfaces of two glass plates. These electrodes are then covered with a thin layer of glass (about 25 μm thick) so that they have no direct contact with the gas contained between the two plates. Because of the close spacing between the plates (for example, 0.1 mm) and the relatively high gas pressure (for example, >500 torr) of the neon-argon (or neon-xenon) gas mixtures used, the interaction between discharges at neighboring electrode crossovers is minimized allowing the cellular structure between the plates to be eliminated.

In operation an a-c sustaining voltage (usually in the form of rectangular pulses) is maintained between the sets of X- and Y-conductors. This voltage, capacitively coupled to the gas space is, by itself, insufficient to cause firing of any cell in the off state. However, if during a half-cycle, voltage pulses of suitable magnitude are applied to a selected pair of X-Y conductors a discharge can be initiated. This discharge will then quench itself within about a microsecond because of the buildup of charges on the insulating walls resulting from the current flow. Subsequent to this the cell will continue to fire on successive half-cycles since the voltage (of alternating polarity) built up on the walls during each half-cycle will add to the applied voltage of the next half-cycle. By triggering on selected cells in sequence a complete image can be stored. To switch a selected element off, appropriate X-Y signals are applied to the element (for example, during the off time between two half-cycles) to produce a weak discharge which causes the wall voltage to fall to zero rather than reverse polarity, preventing further firing of the cell by the sustaining voltage. In typical operation, an a-c sustaining voltage of about 100-V peak and 50 kHz is used, resulting in an average brightness of cells in the range of 30 to 75 fL. Commercial panels of this type have been built with 512 × 512 elements and about 21 by 21 cm in size. Experimental panels of larger size have also been built with 1024 × 1024 elements as well as with greater resolution (~ 3.3 lines per millimeter).

An important factor influencing the operation of the panels, aside from the composition of the gas, is the nature of the insulating surfaces in contact with the gas. Although metallic sputtering does not occur, chemical and physical changes produced at the surface by ion bombardment may cause substantial changes in its secondary-emission coefficient which strongly influence the firing and erasing voltages of different elements, making it impossible to address all of the elements with X-Y signals of a given level. To minimize this problem the insulating surfaces are generally coated with a film (several micrometers thick) of evaporated or sputtered MgO which, in addition to being stable under ion bombardment, minimizes the addressing voltages required for the panel.

Advances in gas-discharge technology have also stimulated efforts to develop color displays for television applications. To obtain colors other than the orange-red of neon use is made of different photoluminescent phosphors deposited on the cell walls which can emit red, blue, or green light when excited by ultraviolet radiation. Instead of neon-argon, a gas mixture such as helium with 2 percent of xenon may be used whose discharge is rich in ultraviolet radiation. Although a number of cathode-ray tube phosphors are useful for this purpose, to

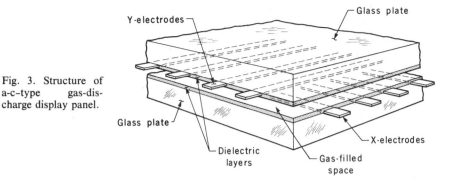

Fig. 3. Structure of a-c-type gas-discharge display panel.

obtain satisfactory life and improved efficiency at the short ultraviolet wavelengths of the gas discharge (100 to 200 nm), special phosphors have been developed such as $BaAl_{12}O_{19}$:Mn for green emission and other europium-activated phosphors for the blue and red (*16*).

One of the limitations of gas panels, particularly for television applications, is their relatively low efficiency. In the case where the neon-argon discharge is viewed directly an efficiency of less than 1 lumen per watt is obtained (comparable to that of electroluminescent-layer and light-emitting diode displays). However, if photoluminescent phosphors are employed and a mixture of light from red, green, and blue cells is used to produce white light an efficiency of less than 1/10 of this is generally obtained. A panel 30 by 30 cm in size operating at the 100 fL brightness of commercial television would thus dissipate considerably more than 100 W.

Liquid-Crystal Displays

Unlike all of the previously discussed display media that generate their own light (referred to as emissive displays), liquid crystals fall into the class of display materials referred to as nonemissive since they control the transmission or reflection of external light (usually ambient illumination). They are thus particularly useful for displays that must be viewed in high ambient light.

Liquid-crystal materials are a class of organic compounds which exhibit, within a certain temperature range of their liquid state, a number of optical and electrical properties characteristic of crystalline solids (*17*). Such materials were already the subject of scientific study before 1900, but it was not until the 1960's that interest in them for display devices arose when it was shown that an electric field applied across a thin layer could produce significant changes in light transmission with little power consumption. Since then, as a result of consid-

erable materials and device research, liquid-crystal displays have come into widespread use in watches, pocket calculators, and a variety of electronic instruments.

The molecules of all liquid-crystal materials are generally elongated in shape. Because of this they are optically anisotropic, that is, the index of refraction is markedly different depending on whether the molecules are oriented parallel or perpendicular to the electric vector of a plane-polarized light wave (*18*). Fluids in the liquid-crystal state also may exhibit a large dielectric anisotropy particularly because of permanent (as well as induced) dipoles associated with the molecules. The latter property allows the alignment of the molecules to be changed by an external field while the former property makes possible a change in light transmission if a suitable optical arrangement is used.

Because of the forces between them the molecules of a material in the liquid-crystal state exhibit some form of long-range order in their alignment. Depending on the specific ordering relationship, different classes of materials exist, namely, nematic, cholesteric, and smectic. For almost all present display applications, however, nematic liquid-crystals are used. In these materials the molecules are aligned essentially parallel to each other but the location of their centers is otherwise random. If the alignment direction is established at the boundary of the liquid (for example, at the surface of the glass plates confining the liquid-crystal material), the remaining molecules tend to assume the same alignment.

The most common displays in present use are the "twisted nematic" type. As shown in Fig. 4a, a layer of nematic material (about 10 μm thick) is contained between the transparent conductive surfaces of two glass plates. This structure is then placed between two polarizers. Before assembly the glass plates are specially treated to make the molecules at the surface align in a particular direction

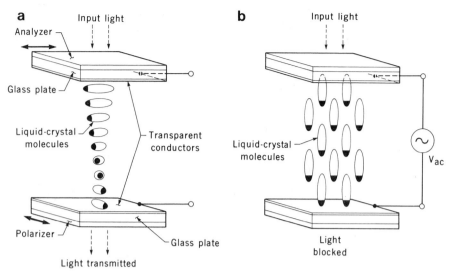

Fig. 4. (a) Twisted-nematic liquid-crystal cell with no voltage applied. (b) Orientation change of liquid-crystal molecules produced by applied a-c voltage.

almost parallel to the surface of the plate. One method for accomplishing this is to evaporate a thin layer (about 100 angstroms thick) of SiO onto the glass surface at an oblique angle. Since the alignment directions of the two plates are set at 90° with respect to each other this causes a gradual twist in direction of the liquid-crystal molecules between the plates as shown. If the input polarizer is oriented, for example, parallel to the alignment of the liquid-crystal molecules at the top plate, the plane of polarization of the light passing through the liquid crystal will be rotated by 90° because of the gradual twist of the liquid-crystal molecules. This light will then pass through the analyzer adjacent to the lower plate whose orientation is set perpendicular to the polarizer.

Since liquid-crystal molecules are used that have a strong dipole moment in the axial direction, if an a-c voltage (for example, 2 to 3 V rms) is applied across the cell, the molecules (except those held by surface forces at the glass plates) will to a large degree orient themselves perpendicular to the glass surfaces as shown in Fig. 4b. In this orientation the liquid-crystal layer can no longer produce a rotation in the polarization plane of the light, causing it to be blocked by the analyzer. In most display devices the ambient light itself is used as the source and, instead of viewing the changes in transmitted light, a reflector is placed at the output to return the light back through the cell, thus permitting viewing from the input side.

Many of the liquid-crystal materials found useful in early device work have a molecular structure whose inner portion consists of two aromatic rings with an additional central group linking them to-

gether (19). One important class, the Schiff base materials (characterized by the −CH=N− central group) was studied extensively. The first compound of this class which could be used at room temperature (commonly known as MBBA) is in the nematic liquid-crystal state over the temperature range of 22° to 48°C. Another compound of this class (known as EBBA) is nematic over the range of 35° to 77°C. However, by mixing the two compounds in the correct proportion eutectic mixtures have been reported (19) whose nematic range extends from 0° to 60°C. For use in twisted-nematic displays, Schiff base and other materials have been developed with an end group (such as −C≡N) attached to one of the aromatic rings which provides a strong axial dipole.

Materials of the above types suffer from decomposition in the presence of small amounts of water or when exposed to ultraviolet radiation, resulting in the breaking of the bonds between the two aromatic rings. In the past few years, however, new materials have been developed such as the biphenyls in which the aromatic rings are directly linked, making the molecules extremely stable in the presence of moisture, air, and light (20). These also may have a CN group attached at one end to form a strong axial dipole and, depending on the particular alkyl or alkoxy group attached at the other end, allow different operating temperature ranges to be obtained. By using two or more compounds of this type, eutectic mixtures have been obtained that are nematic over a temperature range of −10° to 60°C. (Recently, new materials such as phenylcyclohexanes have been developed with comparable stability and temperature range.) It is of interest that

because of the chemical stability and long life of present liquid-crystal materials (which may be considerably greater than 10^4 hours) the life of display devices is frequently limited by the deterioration of the plastic polarizing sheets usually employed which tend to degrade under high temperatures and humidity.

To avoid electrolytic decomposition of liquid-crystal cells, it is usual to apply a-c voltage (for example, 100 Hz) rather than d-c. Because of the high resistivity of the materials (greater than 10^{10} ohm-cm) the power consumption is very small, being less than 1 microwatt per square centimeter (several orders of magnitude lower than for luminescent displays). Liquid-crystal materials are thus ideally suited for portable, battery-operated devices. An important limitation of liquid-crystal displays is their relatively slow response, the turn-on and turn-off times both being of the order of 0.1 second at room temperature (21). Because of the increasing viscosity of the material at low temperature, the turn-off time at −10°C may be about 1 second. At a given temperature the turn-on time varies inversely as the square of the applied voltage. The turn-off time, which depends on the realignment forces between the molecules, increases with the square of the distance between electrodes.

To eliminate the polarizers and provide a wider viewing angle considerable present effort is being directed toward liquid-crystal materials in which a pleochroic dye is dissolved (20). The elongated molecules of such a dye absorb light over a broad spectrum if they are oriented parallel to the electric vector of plane-polarized light but are relatively transparent if they are oriented perpendicular to this vector. Since the dye molecules tend to align in the same direction as the liquid-crystal molecules their orientation can be controlled by electric fields which act on the liquid-crystal material. In the preferred arrangement the dye molecules are incorporated in a cholesteric-type of liquid crystal whose molecules are made to align parallel to the cell walls. Unlike twisted-nematic cells, however, the molecules of the cholesteric material have an intrinsic twist angle which may be much greater (for example, several complete cycles instead of 90°) determined by the thickness of the layer rather than the orientation of the glass plates. In this state the dye molecules, oriented in all directions, absorb unpolarized light. If sufficient voltage is applied, as in the case of the twisted-nematic cell, the liquid-crystal molecules (together with the dye molecules) orient themselves perpendicular to the glass surfaces greatly reducing the light absorption. To obtain satisfactory operation, however, it is necessary to use dye molecules that are strongly oriented by the liquid crystal and are chemically stable. Very recently anthraquinone dyes have been reported which satisfy these conditions well (20).

Attempts to use X-Y addressing techniques for liquid-crystal displays have resulted in limited success. Aside from the threshold voltage not being very sharp, it varies with viewing angle and temperature. In addition, associated with their relatively sluggish response, liquid-crystal cells respond to the integrated effect of repetitive voltage pulses (in particular the rms voltage) rather than the peak value (20). In an X-Y-addressed array the response of unselected elements is thus almost as great as for selected elements since unselected elements are repetitively excited with "half-select" voltages (that is, either X or Y voltages alone), resulting in very low image contrast. With twisted-nematic cells X-Y-addressed arrays having good contrast over an acceptable viewing angle have not yet been developed with more than about ten rows. However, if collimated light can be employed, such as in projection systems, other optical effects in liquid crystals can be used whose threshold is much sharper, allowing images to be obtained from arrays with up to several hundred rows.

To overcome the X-Y addressing limitations of liquid crystals other approaches are being investigated in which a highly nonlinear resistive element (22) or a field-effect transistor is incorporated at each picture element to prevent voltage from appearing across unselected elements. In these arrangements a small capacitor is usually incorporated at each picture element as well. This is charged in accordance with the peak value of the X-Y signal voltages and then allowed to discharge through the relatively slow-responding liquid-crystal element over an extended time.

Miscellaneous New Displays

Aside from the primary display media discussed above, several other types of displays are being investigated, particularly those based on electrochromic and electrophoretic phenomena. Both of these are non-emissive, with interest in them stemming from their relatively low power consumption, memory effects,

wide viewing angle, and usefulness at high ambient light levels.

Electrochromic displays make use of a material whose color can be changed reversibly by passing an electric current through it (23). Such coloration processes involve either a valence change of one of the constituent ions or the formation of a color center associated with a lattice defect. One material that has been studied is deheptyl viologen dibromide, which is colorless in an aqueous solution. If voltage is applied across a cell containing this material an insoluble purple compound is formed on the cathode surface. Another material extensively studied is tungsten oxide. If a thin film of this material is coated on the transparent cathode of a cell containing an electrolyte that can supply H^+ ions (for example, H_2SO_4) the film will change from a transparent to a blue-colored state as a result of current flow through the cell. In this process it is believed that a tungsten bronze, H_xWO_3 ($x < 0.5$) is formed as a result of H^+ ions being injected into the film from the electrolyte together with electrons from the cathode. More recently, electrochromic action has been reported in iridium oxide films (24), the coloration here being attributed to the injection of hydroxyl (or other negative) ions from the electrolyte coupled with the extraction of electrons from the film by the anode. In the case of tungsten oxide and iridium oxide films, attempts have been made to replace the liquid electrolyte with a solid electrolyte or superionic conductor. Although some success has been achieved, with present materials limited life or reduced speed of response is obtained.

Electrochromic displays, like liquid-crystal displays, have the advantage of low-voltage operation (about 1 to 2 V). The switching time (dependent on the material and current density used) is of the order of 0.1 second. Although cells have been reported capable of 10^7 switching cycles, one of the problems encountered is deterioration due to unwanted electrochemical side effects, especially if the applied voltage is raised beyond a certain level. Typically an integrated charge transfer of several millicoulombs (or more) per square centimeter is required to produce a change in coloration, the power consumed in switching being several orders of magnitude greater than for liquid crystals. It should be noted that, because of the electrochemical processes occurring, a reverse voltage is built up during the coloration process. If the cell is short circuited the reverse current flow will cause

decoloration (a situation which may be undesirable in X-Y addressing circuits). However, if the cell is maintained in the open-circuit condition after coloration it can remain in this state for hours or longer, thus providing a memory effect.

Electrophoretic displays make use of a thin layer of dyed fluid in which pigment particles of a strongly contrasting color or reflectivity are suspended (25). Depending on the materials used, as well as charge-control agents added, the particles may acquire either a positive or a negative charge with respect to the liquid. If a layer of such fluid is confined between two parallel electrodes (for example, 50 μm apart) and a d-c voltage (for example, 100 V) is applied, the particles will be drawn to one electrode, building up a coating on the surface. If this electrode is transparent, the color observed will be primarily due to the reflectivity of the particles (for example, yellow or white). After removal of the voltage the cell may remain in this state for hours. However, if a reverse voltage is applied the particles will be drawn to the opposite side of the cell and the color observed through the electrode will be that of the dyed fluid (for example, black or blue) which hides the particles.

Because of their high index of refraction and good light-scattering property titanium dioxide particles (which appear white) have frequently been used. To avoid sedimentation from occurring because of their high density compared to available fluids the particles have in some cases been coated with a resin to reduce their average density. Other particles that have been used are organic pigments, such as Hansa yellow and Diarylide yellow, whose density can be matched by a mixture of suitable fluids. The fluids used should have a high resistivity (for example, 10^{12} ohm-cm), should be chemically stable and, to enable high particle mobility, should have a high dielectric constant and low viscosity. Some fluids that have been used are xylene, perchloroethylene, and trichlorotrifluoroethane.

To obtain satisfactory life times it is important to prevent flocculation or agglomeration of the particles, especially when they are compacted at the electrodes. Although to a limited degree this is prevented by the mutual repulsion of the charged particles, generally steric stabilizers are added to the solution. These provide long-chain molecular groups that attach to the particles and protrude outward, thus preventing the particles from approaching too closely.

To produce switching, relatively little

integrated current flow is required (about 0.1 $\mu C/cm^2$), an advantage for low-power applications. The switching time is typically in the range of 10 to 20 msec, for a given material being proportional to the square of the electrode spacing and inversely proportional to the applied voltage. Because of their poor threshold characteristics, X-Y addressing techniques cannot be used effectively with electrophoretic cells unless some additional circuit component is added at each picture element. However, encouraging results have been obtained with cells incorporating an additional control-grid electrode which prevents particle migration unless both X and Y voltages are simultaneously applied (26).

Conclusion

To satisfy the great variety of applications emerging for displays (covering the gamut from large-screen television to small watches) a continually widening circle of diverse phenomena and new materials is being explored. The field of display technology is thus more a collection of somewhat unrelated topics than a single cohesive subject. Although the performance of different display devices can be evaluated in terms of a number of objective criteria, for example, speed of response, power consumption, or brightness, a comparison of displays on this basis alone may lead to simplistic and erroneous conclusions. In many cases other factors such as flicker, uniformity of elements, variation of contrast with viewing angle, or color may be of comparable importance, requiring subjective judgments to assess their effects. Despite the different immediate and specific goals of the various display technologies, in many cases they share a common long-term goal, namely the achievement of a level of performance comparable to that of the cathode-ray tube. Given the present limitations of these technologies, however, this goal is not likely to be achieved in the next few years but rather in an evolutionary manner extending over a much longer period.

References

1. V. K. Zworykin and G. A. Morton, *Television* (Wiley, New York, 1954), pp. 63–93 and 397–400.
2. D. Curie, *Luminescence in Crystals* (Methuen, London, 1963), pp. 288–327.
3. S. Larach and A. E. Hardy, *Proc. IEEE* **61**, 915 (1973).
4. A. Martin, *Electro-Opt. Syst. Des.* **9**, 35 (1977).
5. B. Kazan, *Proc. of the Soc. for Inf. Disp.* **17**, 23 (1976).
6. H. F. Ivey, *Electroluminescence and Related Effects* (Academic Press, New York, 1963).
7. A. Vecht, N. J. Werring, R. Ellis, P. J. F. Smith, *Proc. IEEE* **61**, 902 (1973).
8. H. Kawarada and N. Ohshima, *ibid.* p. 907.
9. C. Suzuki, T. Inoguchi, S. Mito, *Inf. Disp.* **13**, 14 (1977).
10. C. H. Gooch, *Injection Electroluminescent Devices* (Wiley, London, 1973).
11. M. G. Craford, *IEEE Trans. Electron Devices* **ED-24**, 935 (1977).
12. T. Niina, S. Kuroda, H. Yonei, H. Takesada, *ibid.* **ED-26**, 1182 (1979).
13. F. M. Penning, *Electrical Discharges in Gases* (Macmillan, New York, 1957).
14. R. N. Jackson and K. E. Johnson, *Adv. Electron. Electron Phys.* **35**, 191 (1974).
14a. R. Cola, J. Gaur, G. Holz, J. Ogle, J. Siegel, A. Somlyody, *Adv. in Image Pickup Disp.* **3**, 83 (1977).
15. H. G. Slottow, *IEEE Trans. Electron Devices* **ED-23**, 760 (1976).
16. T. Kojima, R. Toyonaga, T. Sakai, T. Tajima, S. Sega, T. Kuriyama, J. Koike, H. Murakami, *Proc. of the Soc. for Inf. Disp.* **20**, 153 (1979).
17. P. G. de Gennes, *The Physics of Liquid Crystals* (Clarendon, Oxford, 1974).
18. W. J. de Jeu and J. van der Veen, *Philips Tech. Rev.* **37**, 131 (1977).
19. L. A. Goodman, *J. Va. Sci. Technol.* **10**, 804 (1973).
20. E. P. Raynes, *IEEE Trans. Electron Devices* **ED-26**, 1116 (1979).
21. P. Smith, *Electronics* (25 May, 1978), p. 113.
22. D. E. Castleberry, *IEEE Trans. Electron Devices* **ED-26**, 1123 (1979).
23. A. R. Kmetz and F. K. von Willisen, Eds., *Nonemissive Electrooptic Displays* (Plenum, New York, 1975).
24. J. L. Shay and G. Beni, *IEEE Trans. Electron Devices* **ED-26**, 1138 (1979).
25. I. Ota, J. Ohnishi, M. Yoshiyama, *Proc. IEEE* **61**, 832 (1973).
26. B. Singer and A. L. Dalisa, *Proc. of the Soc. for Inf. Disp.* **18**, 255 (1977).

Transmission electron micrograph showing the magnetic domain structure (meandering strips) in a thinned samarium thulium iron oxide garnet crystal. Electrons in the microscope are deflected by the Lorenz force due to the magnetization distribution across the domain walls that define the domains. Submicrometer domains (about 0.5 micrometer wide) and actual wall structure, both unresolvable by light optical methods, are revealed. [S. Herd, IBM Thomas J. Watson Research Center, Yorktown Heights, New York]

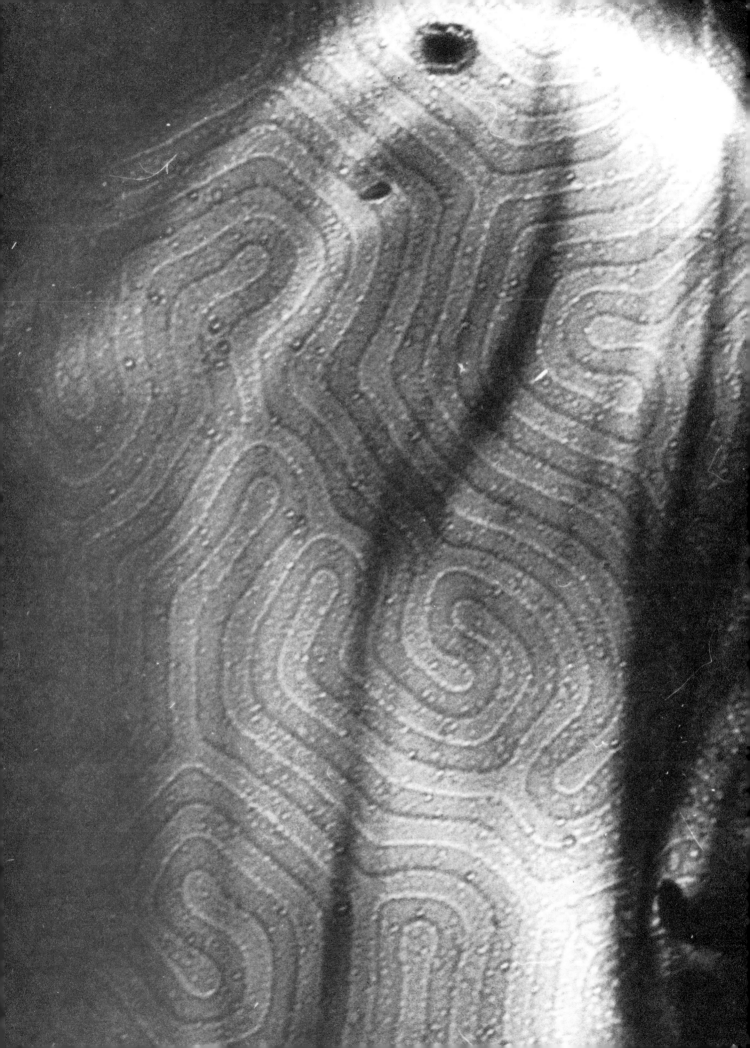

Magnetic Bubble Materials

Edward A. Giess

Bubble domains are small, magnetized, mobile regions within sheets or films of certain magnetic materials. The presence of a bubble can represent a binary "bit" of information, a one, and its absence a zero. Memory and recording devices marketed today store and move a million bubbles, each ≃ 2 micrometers in diameter, within a film ≃ 2 micrometers in thickness and less than 1 square centimeter in area. These solid-state de-

Summary. Physicists, materials scientists, and engineers combined to bring solid-state bubble devices into the computer memory and recording marketplace. Devices with smaller bubbles are being developed for increased data capacity and lower cost. Epitaxial garnet films made by isothermal dipping in molten solutions helped put the technology in place and will probably satisfy the material needs of future devices with bubbles scaled down from 2 to 0.5 micrometer in size.

vices can be made smaller in size and in data capacity; thus smaller, more economical units can be assembled than for the motor-driven disks and tapes they may replace. Devices promise to be especially useful in computer terminals and small data processors. To increase the data capacity of devices and to decrease their cost, the technology is moving toward devices with smaller bubbles in thinner films.

In this article, magnetic bubbles, the devices using them, and the physics of bubbles will be discussed first. Next, the physical forms of bubble materials—platelets and films—that have influenced the evolution of devices will be considered. Finally, materials will be described, especially the magnetic garnet films, which have been the main bubble material and probably will continue to be for quite a while.

Magnetic Bubbles

If a bar magnet were flattened and thereby drastically shortened along its north-south polar axis, into a thin sheet, it would have a large (shape anisotropy) magnetostatic energy from the numerous north poles on one surface opposed by

the south poles on the other surface. In this case the sheet is magnetically uniaxial. There is only one significant axis, the one normal to the sheet. The magnetization lies within the sheet in a direction perpendicular to the plane of the sheet (see arrows in Fig. 1). Sectional magnetic walls could form within the sheet, separating it into magnetic domains where adjacent domains have antiparallel magnetizations. Walls are finite in thickness and have a magnetic energy of their own, but they reduce the net energy and thereby increase the stability of the total magnetic configuration. Neighboring domains have opposite north-south poles at the sheet surfaces and neighboring flux lines can easily close upon each other. Domains are stripe-shaped, and half of them are magnetized north, while the other, intervening half are magnetized south (Fig. 1a).

Stripe domains can be observed directly with a microscope in thin transparent magnetic sheets (such as the garnet films presently used in devices) because the Faraday effect rotates the plane of transmitted polarized light depending on the direction of magnetization in individual domains. This visibility of individual domains has contributed greatly to the rate of development of bubble technology.

A perpendicular bias field—for example, from parallel external permanent magnet sheets—will favor and thereby increase the area of domains magnetized in the same direction, while it will decrease the area of adjacent domains magnetized oppositely. At a certain bias (strip-out) field, the unfavored stripe domains will shrink into right-cylindrical domains called bubbles (*1*). An impor-

tant property of these bubbles is their ability to be moved laterally through the sheet. In commercial devices this propagation is induced by a thin-film array of metal magnets deposited atop the bubble film. The array is controlled by an in-plane magnetic field applied with two external coils.

Bubble Devices

Devices (*2, 3*) must be provided with means for the generation, moving, switching (transfer), sensing, and annihilation of bubbles. Typically, all these device operational functions are achieved with tiny Permalloy (Ni,Fe) soft magnets (which can easily change their direction of magnetization) and electrical conductor lines about 0.2 μm thick deposited on a 0.15-μm SiO_2 layer atop the bubble storage layer (Fig. 2). These metallurgical patterns are shaped by optical lithography techniques (*4*), whose resolution currently limits C-bar (asymmetric chevron) Permalloy magnets to a minimum lateral gap spacing of about 1 μm between adjacent C-bars. The bubble diameter must be about twice this minimum gap. In a propagation and storage track, C-bars that can host one bubble each are aligned along a common axis with a repeat distance of four to five times a bubble diameter. Bubbles rest beneath the C-bars in the gap region, and they propagate when the in-plane drive field rotates and changes the magnetic polarity of the Permalloy C-bars, thereby alternately attracting and repelling the bubbles. Since the stabilizing bias field is supplied by permanent magnets, bubbles are nonvolatile (that is, they continue to exist even after external electrical power is disconnected), and they can be migrated through the device in the solid state without any mechanically moving parts such as are required to move information in conventional disk and tape technologies.

One bubble can be formed and those in the device advance one position with each rotation of the drive field, which determines the device frequency of operation (about 100 kilohertz). An electrically pulsed conductor line chops bubbles from a seed bubble in the generator, and then they are fed into a major track of C-bars (see Fig. 3). A series of orthogonal minor loop tracks face the major track along its length and are connected to it through a series of switching Permalloy patterns actuated by a transfer con-

The author is a research staff member at the IBM T. J. Watson Research Center, Yorktown Heights, New York 10598.

ductor line, so that bubbles can be transferred back and forth between the major track and the minor storage loop. Having minor loops reduces the time needed to access data. At the end of the major track, bubbles are read by a sensor conductor line attached to the ends of a thin-layer Permalloy strip whose resistance decreases in the magnetic field of a bubble. Finally, the bubbles can be eliminated by an annihilator pattern.

Bubble Physics

A magnetic bubble is a physical phenomenon not unique to any one class of chemical compositions. [However, at present practically all bubble devices are made with single-crystal films of multicomponent magnetic rare earth–iron oxides having the garnet structure (5).] Bubble physics (6) considerations dominate device design and materials selection. The stability, size, and speed of bubbles are the key design parameters affecting device reliability, capacity, and data rate, respectively. Stability refers to how well bubbles resist destruction by environmental perturbations, such as the drive fields used to propagate domains. Smaller bubbles are packed more densely and thereby increase data capacity. Since small bubbles are more closely spaced, they move shorter distances and give better data rates than large bubbles with the same velocity. Also, of course, increased bubble velocity improves the data rate in a device. (In present-day devices, the coil drive circuitry limits the data rate.)

Stability of bubble domains is measurable by the in-plane field H_u (called the anisotropy field) required to tip the magnetization M from its easy axis by 90° into the film plane. The product $H_u M/2$ is the uniaxial anisotropy energy constant K_u. (In some bubble materials, this magnetic anisotropy can have three components: growth-induced, strain-induced, and intrinsic.) The stability factor is defined as the ratio of H_u to saturation magnetization, or in energy terms as

$$Q \equiv \frac{K_u}{2\pi M_s^2} \qquad (1)$$

the ratio of the anisotropy energy to the magnetostatic energy. For stable isolated bubbles to exist, Q must be greater than unity. Furthermore, unless Q is appreciable, the in-plane drive field can strip out bubbles into stripes, a failure mode for devices. Stable bubbles exist in a range of bias fields between the (higher) collapse field H_0, where the entire film becomes a single domain, and the

(lower) strip-out field (about $0.7\ H_0$), where bubbles change back into stripe domains. Depending on film material parameters and thickness, H_0 is 0.4 to 0.6 of $4\pi M_s$, the saturation magnetization.

The film thickness is generally made to be just less than the bubble diameter for energetic and stability reasons, and the bubble diameter is about eight times the characteristic length

$$\ell = \frac{\sqrt{A K_u}}{\pi M_s^2} \qquad (2)$$

where A is the magnetic exchange stiffness constant, which is a measure of how strongly neighboring ions are magnetically coupled. The characteristic length is the ratio of the domain wall energy to the demagnetization energy, which is directly related to the magnetostatic energy. Equation 2 shows that bubble size is determined mainly by magnetization. The important thing to remember is that smaller bubbles require larger magnetization.

From a materials standpoint, the data rate of a device and the time required to access data are governed by the speed of bubbles. It will be seen that stability and speed have contrary materials requirements. Stability demands relatively high anisotropy, and speed needs low anisotropy. The speed of a bubble is determined by the product of the drive field across the bubble minus the threshold field for movement (coercivity) and the bubble mobility (velocity per drive field gradient)

$$\mu = \frac{\gamma}{\alpha} \sqrt{\frac{A}{K_u}} \qquad (3)$$

where γ is the gyromagnetic ratio, which has to do with the dynamics of atoms changing their directions of magnetization, and α is the Gilbert damping (magnetic viscosity) parameter. The quantity $\sqrt{A/K_u}$ times π is the domain wall width within which the magnetization reverses direction in any one or a combination of possible wall magnetic structures. For example, simple bubbles have a Bloch wall, wherein the magnetization vector rotates in the plane of the wall in either a left- or right-handed (chirality) direction, but in a Néel wall, which can partly form near the film surfaces, the magnetization vector rotates directly into the wall. Different structures within a given wall meet along Bloch lines, which can meet at Bloch points. These magnetic wall structures provide a mechanism for information coding of a higher order than binary (7).

Walls, whether simple or complex, define bubbles, and defects that lie in the plane of a wall are particularly effective in blocking bubble motion. Accordingly, films for devices must have a high degree of physical perfection and freedom from flaws. The growth techniques developed for making garnet films and the properties of garnets themselves provide products with the high quality required for devices.

Platelets and Films

The early devices had large bubbles and were made with platelets cut and polished from large single crystals. Now competition from other technologies creates the need for devices with smaller bubbles in thinner sheets, which are produced as thin films on flat substrates. Today, magnetic garnet films are grown as single-crystal layers, which are a crystallographic continuation of single-crystal substrates that are nonmagnetic platelets. This process for making films is called epitaxy.

Bubbles have been studied in single-crystal platelets of Fe_2O_3-based compounds with magnetoplumbite, orthofer-

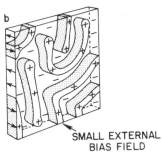
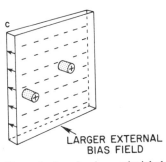

Fig. 1. Magnetic domains in a uniaxial sheet under the influence of an external bias field. [Reprinted from Bobeck and Scovil (22). Copyright © Scientific American, Inc., 1971. All rights reserved]

rite, and garnet crystal structures and in single-crystal films of the latter. Also, films based on metallic $GdCo_5$ alloy compositions (8) with a glasslike amorphous structure have been used. Devices have been fabricated with all but the magnetoplumbites.

Amorphous films were developed late and have bubble properties that are susceptible to temperature change in larger bubble films; hence they could not displace garnet films, which have satisfied design needs to date. Amorphous films could be revived for future devices using very small bubbles, but they would require an electrical insulation layer because they are conductors.

It is possible to slice and polish bulk crystal platelets, 1 cm^2 in area, down to a thickness of about 25 μm; this sets a similar minimum bubble diameter, which in turn limits data storage density. A pref-

erable technology, which considerably relieves both the areal and thickness constraints, is to grow magnetic films epitaxially on thicker nonmagnetic crystal substrate wafers that have the same crystallographic structure and lattice spacing. An epitaxial film has chemically different, but similarly sized, ions in exact positional registration with ions in its supporting substrate lattice. Fortunately, the nonmagnetic gallates (Ga_2O_3-based compounds) have a crystal chemistry related to that of ferrites, and they function well as substrates in the case of garnets. Both Ga and Fe prefer to be trivalent and have similar ionic radii.

Magnetoplumbite, orthoferrite, and garnet crystals (9) can all be obtained from the pseudoternary RE_2O_3-Fe_2O_3-PbO system, where RE is a rare earth or Y (see Fig. 4). In this system the fluxing action of PbO lowers crystallization tem-

peratures below the range where Fe^{3+} tends to be chemically reduced Fe^{2+} unless the oxygen pressure is considerably greater than atmospheric. The latter effect causes magnetoplumbite and iron garnets to melt incongruently—that is, to melt into a liquid, but also into a new solid phase as well, each having a different composition. Consequently, garnet ($RE_3Fe_5O_{12}$) crystals are grown from molten solutions with Fe_2O_3 : RE_2O_3 ratios much greater than the 5:3 stoichiometric ratio.

Bulk crystals and epitaxial films of Fe_2O_3-based compounds grow from PbO-fluxed melts (molten solutions) at linear growth rates around 1 cm per 10^6 seconds; thus it takes at least 1 week to grow a bulk crystal but only several minutes to produce a thin film. The same physical process that is too slow to produce economical bulk crystals allows enough time to make films in a controlled and reproducible way. Crystals of $Gd_3Ga_5O_{12}$ garnet (GGG) (substrates for magnetic garnet epitaxial films), on the other hand, are pulled at almost 1 cm per hour from GGG melts as right cylinder-shaped boules by the Czochralski technique. Substrates of device films are hundreds of times thicker than their films. Fortunately for the sake of economy, substrate boules can be grown hundreds of times faster than films because boules grow from GGG melts with the same composition. Unlike iron garnets, GGG melts congruently. The same Czochralski apparatus geometry is employed with different materials to grow semiconductor silicon boules. Furthermore, the development of garnet epitaxial film technology was accelerated by processing GGG substrate wafers with essentially the same diamond saws and chemical-mechanical polishing equipment that existed for processing Si wafers. Also, the development of yttrium aluminum oxide garnet (YAG) boules for lasers helped pioneer the growth techniques for GGG boules, which are now grown as large as 7.5 cm in diameter routinely.

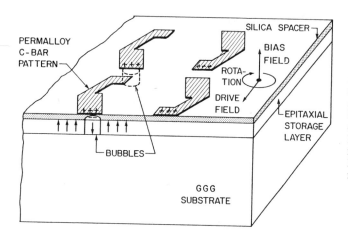

Fig. 2. Conventional field-accessed bubble device chip (schematic section). [Reprinted from George and Reyling (3). Copyright © Mc-Graw-Hill, Inc., 1979. All rights reserved]

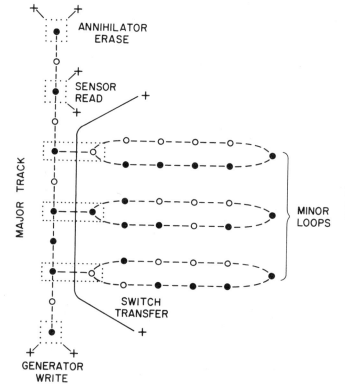

Fig. 3. Conventional device chip organization (schematic).

Materials

Magnetoplumbite is hexagonal (uniaxial) and is one of a class of so-called hexaferrites, wherein Pb can be replaced by Sr or Ba. Isolated bubbles were first observed (9) in 1960 in single-crystal platelets of magnetoplumbite (PbO · $6Fe_2O_3$), although their device potential was not recognized until later. Ten years later it was shown that bubbles in other hexaferrites have limiting velocities too low (< 600 cm/sec) for bubble devices (10). However, it has never been demonstrated that this limitation is intrinsic to hexaferrites, some of which do function in high-frequency microwave devices. It may be possible to make useful submicrometer bubble devices, given the correct hexaferrite chemical composition in the form of a physically perfect epitaxial hexaferrite film. It is not obvious what crystal would be a suitable substrate.

Orthoferrites, with the general formula REFeO$_3$, were the first materials employed in bubble devices (1). Platelets 100 μm thick and containing bubbles 100 μm in diameter were used. They served well as bubble materials for early engineering studies, but were soon replaced by garnets because orthoferrites have a limited range of low 4πM values, resulting in large bubbles.

Garnets have a cubic crystal structure and consequently are not intrinsically uniaxial. They became the dominant bubble materials after it was found that certain magnetic garnets can have growth-induced uniaxial anisotropy K_g when grown from molten solutions (11). Both bulk and epitaxial crystal films exhibit this phenomenon in the absence of macroscopic stress. Despite the importance of anisotropy to the existence of stable bubbles, two other considerations are given at least as much attention in designing garnet compositions for bubble devices. The first is magnetization, which is the principal factor in determining bubble size and is adjusted mainly by partially substituting nonmagnetic Ga or Ge for magnetic Fe. The second is the film lattice parameter, which must closely match that of the substrate and is adjusted mainly by a critical admixture of two or more rare earths.

Magnetic Garnet Crystal Chemistry

Compared to the garnets found in nature, magnetic garnets have the same arrangement of atoms, but different kinds of atoms. Magnetic garnet films and their substrate crystals contain yttrium or rare earths and iron or gallium oxides while natural garnets usually contain calcium, aluminum, and silicon oxides.

The garnet structure belongs to the highest symmetry and most complex space group Ia3d. The unit formula for yttrium iron garnet (YIG) is {Y^{3+}}$_3$-[Fe^{3+}]$_2$(Fe^{3+})$_3$O$_{12}^{-2}$. There are three types of cation sites designated as follows: {dodecahedral} with eight oxygen ion nearest neighbors, [octahedral] with six, and (tetrahedral) with four.

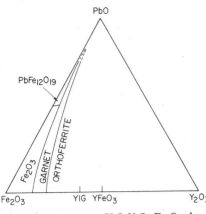

Fig. 4. Pseudoternary PbO-Y$_2$O$_3$-Fe$_2$O$_3$ phase equilibrium diagram showing magnetoplumbite, PbFe$_{12}$O$_{19}$; hematite, Fe$_2$O$_3$; garnet, Y$_3$Fe$_5$O$_{12}$ (YIG); and orthoferrite, YFeO$_3$ primary phase areas. Any possible combination of the three constituents at the apexes of this composition triangle can be represented by a point within the triangle. The areas of this phase diagram are labeled according to the first (primary) crystal phase that will freeze on cooling a molten solution whose composition point is in the area. [Reprinted from Nielsen and Dearborn (23). Copyright © Pergamon Press, Inc. 1958]

Each garnet unit cell contains eight formula units—for example, Y$_{24}$Fe$_{40}$O$_{96}$, which has a 1.2376-nanometer repeat distance. The high symmetry and large lattice spacing probably account for the crystals' reluctance to nucleate from solution spontaneously or form imperfections such as dislocations. Both tendencies aid the controlled growth of substrate crystals and magnetic bubble films with the high physical perfection required for devices.

A very considerable number and range of possible chemical substitutions (12) provides the basis for selecting homogeneous solid solutions useful in bubble devices—for example, {Y, La to Lu, Bi, Ca, Pb}, [Fe, Mn, Sc, Ga, Al], and (Fe, B, Ga, Al, Ge). Each Fe^{3+} ion in the crystal acts as a single magnet with a net magnetic moment. Within each domain, all [Fe^{3+}] have parallel magnetic moments, as do all (Fe^{3+}), but the octahedral and tetrahedral moments have antiparallel alignment, so there is a net magnetization M, the difference between the two sublattices. These magnetic sublattices are coupled by the exchange energy, which decreases with increasing temperature. At the Curie temperature, thermal energy finally breaks the magnetic coupling and M goes to zero. Bubble diameter (Eq. 2) is smallest, about 0.5 μm, for the simple undiluted RE$_3$Fe$_5$O$_{12}$ garnets, where RE is Y or a rare earth, because 4πM is a maximum, about 1800 gauss (= 0.18 tesla). The magnetic {RE = Sm to Yb} ions, when substituted, have their own magnetic sublattice,

which couples antiparallel to the (Fe) and thereby reduces 4πM. However, bubble diameter is principally adjusted (increased) in garnets by nonmagnetic (Ga^{3+}) or (Ge^{4+}) partial substitutions for (Fe^{3+}), which reduce 4πM. When tetravalent (Ge^{4+}) is substituted, its charge must be compensated by an equal amount of {Ca^{2+}} in the otherwise trivalent cation garnet lattice.

The substantial Ca-Ge and especially Ga substitutions necessary to obtain a low 4πM (< 200 gauss) for early ~ 6-μm bubble devices also reduced the Curie temperature sufficiently to make the devices sensitive to ambient temperature fluctuations. This problem is less evident in today's ~ 2-μm bubble devices and will be less so in future devices with smaller bubble garnet compositions. It is not practical to obtain smaller bubbles by increasing 4πM with nonmagnetic (Sc^{3+}) substitutions for octahedral Fe^{3+} because the Curie temperature is concurrently sharply reduced.

The rare earths Sm and Eu are the most effective promoters of K_g, the growth-induced K_u, when the growth direction is [111] or [100]. Either of these large RE ions combined in a 50:50 ratio with a smaller one (Tm, Yb, or Lu) produces the maximum K_g in (111) films (13), but reducing their concentraton diminishes K_g approximately linearly in (100) films (14). Both Sm and Eu dampen domain wall motion (decrease data rate), which means that compositions with more nonmagnetic ions (such as Lu or Y) are to be favored. Also, Sm and Eu enhance the intrinsic magnetocrystalline anisotropy K_1, which favors [111] as the easy direction of magnetization. Mainly because of growth and crystalline anisotropy, nearly all early garnet devices were oriented (111). Crystalline anisotropy contributes proportionally less to the total anisotropy for smaller bubble garnets. Mechanical in-plane strain arising from a lattice spacing mismatch between an epitaxial film and its substrate must be less than ~ 10^{-3} (lattice matching is achieved without difficulty because films are multicomponent). Mismatch strain can be tensile or compressive and can produce a stress-induced anisotropy K_s; this is positive or negative depending on the film magnetostrictive coefficient, which is the summation of contributions from constituent magnetic ions. Bubble stability is principally realized through K_g with K_1 and K_s making at best minor contributions. This is especially true for small-bubble films, which require very high K_u for stability, and consequently the number of useful small-bubble garnet compositions is limited.

Epitaxial Garnet Films

The success of magnetic bubble technology is in no small measure a result of the physical quality, lateral uniformity, and wafer-to-wafer reproducibility of epitaxial garnet films. These films are grown by dipping axially rotated substrate wafers (Fig. 5) into isothermal molten solutions of garnet. A horizontal rotating wafer behaves like a centrifugal fan that draws solution upward in a central plume beneath the wafer and expels the solution radially across the underside of the wafer. This fluid flow leads to a steady-state regime that produces films of the desired quality, mainly because the film growth rate is constant. The film growth process is fairly well understood now and is interesting in its own right (15) because it is possible to both control and measure precisely film growth–related parameters. This and other liquid phase epitaxy processes (such as those used to make GaAs films) are often called LPE processes.

It is appropriate to consider briefly some substrate factors that contribute to the success of garnet LPE. Given the magnetic orientation requirements for garnets, it is fortunately most economical to pull (grow) GGG boules along [111] or [100] axes, the fast growth directions, and then to cut them with a diamond saw normal to their (growth) axes into disk-shaped wafers 0.05 cm thick. This takes advantage of the natural circular shape of boule sections, a shape useful in film growth. It also minimizes the effect of planar striations, which lie in the plane of each disk and arise from slight Gd:Ga stoichiometry fluctuations in the boule growth interface. The fluctuations exist because GGG is a solid solution and the most stable (maximum melting) composition is slightly enriched in Gd_2O_3.

Although much of the crystal chemistry of magnetic garnets was developed for earlier microwave devices with bulk crystals grown from molten salt solutions, the technologically important supercooling tendency of these PbO-fluxed melts was recognized (16) only after epitaxial film growth was attempted for bubbles. Without causing crystals to nucleate spontaneously, the molten solutions can be supercooled as much as 100°C below their saturation temperatures T_s, where garnet crystals would begin to nucleate and to grow if equilibrium conditions were easy to attain. Solutions will remain in this metastable crystal-free state for hours if undisturbed. However, in garnet LPE, the metastable solutions are disturbed and growth is triggered on a GGG wafer (a nucleus) when it is

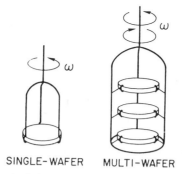

SINGLE-WAFER   MULTI-WAFER

Fig. 5. Substrate assemblies for horizontal plane dipping with rotation ω of single and multiple wafers. These assemblies are lowered into supersaturated garnet solutions in open platinum crucibles.

dipped into the solution. In 30 to 300 seconds epitaxial films will generally achieve sufficient thickness for device applications. Isothermal growth temperatures T_g nearer to T_s—for example, with undercoolings $\Delta T = (T_s - T_g)$ of about 15° to 50°C (15)—are generally used. Isothermal growth in molten solutions supersaturated by supercooling is simpler than growth in conventional systems supersaturated by dynamic cooling, which requires massive heat flow and temperature gradients that depend on apparatus geometry and sophisticated controls.

Growth depletes solute (nutrient) from the solution adjacent to the growth interface, thereby creating a diffusion boundary layer across which solute species subsequently must diffuse to the interface from the bulk of the solution. Control of the boundary layer thickness is the key to a constant growth rate and compositional homogeneity of films. Growth is impeded not only by the growth interface reaction incorporating garnet into the solid but also by transport across the diffusion boundary layer, whose thickness δ can be regulated and made uniform by forced solution flow. The appropriate solution flow pattern is obtained when disk-shaped GGG wafers are rotated singly or oscillated angularly in a multiple coaxial array (Fig. 5) with an angular velocity ω. Without solution flow δ^2 is proportional to growth time t, and the growth rate decreases with time, resulting in nonuniform film compositions.

With forced flow induced by rotation δ^2 is inversely proportional to ω, which is made constant, and growth rate f^+ attains a steady-state value.

$$f^+ \cong (C_L - C_e) \frac{D}{\delta} \frac{1}{1 + r} \qquad (4)$$

where C_L is the garnet concentration in the bulk liquid, C_e is the equilibrium concentration (17), and r is the ratio of D/δ to

K, the interface reaction constant. The critical diffusion constant D is that of the slowest moving species. The diffusion boundary layer is not equivalent to the hydrodynamic one, but they are coupled. (When r is large the interface reaction dominates growth.)

For isothermal growth, this gives a constant f^+, which ensures homogeneous film composition for the multi-constituent (solid solution) garnet systems used in devices. The ratio of constituents in solid solution crystals will change with f^+ and T_g by the so-called segregation effect, which is pronounced in slow-growth processes such as this (18). As growth rate is increased, the crystal composition tends toward that of the solution. Therefore, faster growth leads to more Pb from the solution being incorporated into the crystal.

Both diffusion and reaction are thermally activated, the reaction being especially sensitive to temperature. The diffusive transport process tends to limit growth and therefore predominates at higher growth temperatures ($\gtrsim 900°C$), whereas the interface reaction is rate-determining at lower T_g values. When diffusion is significant, even with forced solution flow, there is a transient epitaxial layer up to about 0.2 μm thick formed during the growth start-up when the boundary layer itself is forming. In this very short time period, the interface reaction essentially alone impedes growth. Therefore the transient layer adjacent to the substrate grows with a very high f^+, giving rise to a different composition than that of the bulk film. The transient layer typically contains more Pb^{2+} (solvent) in solid solution and might adversely affect future small bubble films, which will have to be thinner (< 1 μm thick). Growth conditions favoring reaction over transport—that is, where r in Eq. 4 is large—will suppress the transient. These conditions will be required for the very small bubble films of the future.

Future

Smaller bubbles in thinner garnet films are going to be required. Future epitaxial films made by extending the PbO molten solution technology will probably be grown from more dilute solutions at lower growth temperatures and lower rates, but at larger undercoolings. These garnets will have higher $4\pi M$ and anisotropy, but will be similar to present garnets. Higher magnetization will come from less dilution of the iron lattices with nonmagnetic ions, and probably slightly more Sm (or Eu) will be used for the anisotropy. Conventional Permalloy devices with 2-μm bubbles are approaching bit density limitations imposed by optical lithography and are therefore unlikely to test the limit of available 0.5-μm bubble garnet materials (19). On the other hand, contiguous disk devices (20) have propagation patterns that somewhat relieve lithographic constraints and are taking bubbles into the submicrometer range with conventional field-access technology at hundred-kilohertz data rates. A newly proposed current-access approach (21) has the potential of achieving megahertz data rates. These device engineering considerations can be expected to receive more immediate attention (with only modified garnets) than any search for new materials.

References and Notes

1. A. H. Bobeck, *Bell Syst. Tech. J.* **46**, 1901 (1967).
2. J. A. Rajchman, *Science* **195**, 1223 (1977); R. Kowalchuk, W. S. Chen, T. Mendel, J. Neuhausel, *Circuits Manuf.* **19**, 22 (September 1979); G. Cox, *Electron. Des.* **24**, 154 (22 November 1979).
3. P. K. George and G. Reyling, Jr., *Electronics* **52**, 99 (2 August 1979).
4. T. C. Penn, *Science* **208**, 923 (1980).
5. The following are detailed reviews on magnetic bubble materials: J. E. Davies and E. A. Giess, *J. Mater. Sci.* **10**, 2156 (1975); J. W. Nielsen, *IEEE Trans. Magn.* **MAG-12**, 327 (1976); V. N. Dudorov, V. V. Randoshkin, R. V. Telesnin, *Sov. Phys. Usp.* **20**, 505 (1977); S. L. Blank, in *Crystal Growth: A Tutorial Approach*, W. Bardsley, D. T. J. Hurle, J. B. Mullin, Eds. (North-Holland, Amsterdam, 1979), p. 241; A. H. Eschenfelder, *Magnetic Bubble Technology* (Springer-Verlag, New York, 1980); J. W. Nielsen, in *Annual Review of Materials Science*, R. A. Huggins, R. H. Bube, D. A. Vermilyea, Eds. (Annual Reviews, Palo Alto, Calif., 1979), vol. 9, p. 87; P. Chaudhari *et al.* (8).
6. A. A. Thiele, *Bell Syst. Tech. J.* **48**, 3287 (1969); *J. Appl. Phys.* **41**, 1139 (1970).
7. T. J. Beaulieu, B. R. Brown, B. A. Calhoun, T. L. Hsu, A. P. Malozemoff, *AIP Conf. Proc. No. 34* (1976), p. 138.
8. P. Chaudhari, J. J. Cuomo, R. J. Gambino, E. A. Giess, in *Physics of Thin Films*, G. Hass, M. H. Francombe, R. W. Hoffman, Eds. (Academic Press, New York, 1977), vol. 9, p. 263.
9. C. Kooy and U. Enz, *Philips Res. Rep.* **15**, 7 (1960).
10. A. H. Bobeck, *IEEE Trans. Magn.* **MAG-6**, 445 (1970).
11. _____, E. G. Spencer, L. G. van Uitert, S. C. Abrahams, R. L. Barns, W. H. Grodkiewicz, R. C. Sherwood, P. H. Schmidt, D. M. Smith, E. M. Walters, *Appl. Phys. Lett.* **17**, 131 (1970).
12. S. Geller, *Z. Kristallogr.* **125**, 1 (1967); W. Tolksdorf, in *Landolt-Börnstein: Numerical Data and Functional Relationships in Science and Technology* (Springer-Verlag, New York, 1978), new series, group 3, vol. 12, part a.
13. E. J. Heilner and W. H. Grodkiewicz, *J. Appl. Phys.* **44**, 4218 (1973).
14. T. S. Plaskett, E. Klokholm, D. C. Cronemeyer, P. C. Yin, S. Blum, *Appl. Phys. Lett.* **25**, 357 (1974).
15. E. A. Giess and R. Ghez, in *Epitaxial Growth*, J. W. Matthews, Ed. (Academic Press, New York, 1975), p. 183; W. van Erk, *J. Cryst. Growth* **43**, 446 (1978).
16. H. J. Levinstein, S. Licht, R. W. Landorf, S. L. Blank, *Appl. Phys. Lett.* **19**, 486 (1971); J. E. Davies and E. A. Giess, *J. Cryst. Growth* **30**, 295 (1975).
17. The term $C_L - C_e$ can be given as

$$C_L\left(\frac{1}{T_g} - \frac{1}{T_s}\right) = C_L \frac{\Delta T}{T_g T_s}$$

The equilibrium concentration is determined by T_g.
18. W. van Erk, *J. Cryst. Growth* **46**, 539 (1979).
19. E. A. Giess, R. J. Kobliska, F. Cardone, *J. Appl. Phys.* **50**, 7818 (1979); T. S. Plaskett *et al.*, *ibid.*, p. 7821; E. A. Giess and R. J. Kobliska, *IEEE Trans. Magn.* **MAG-14**, 410 (1978).
20. Y. S. Lin, G. S. Almasi, G. E. Keefe, E. W. Pugh, *IEEE Trans. Magn.* **MAG-15**, 1642 (1979).
21. A. H. Bobeck, S. L. Blank, A. D. Butherus, F. J. Ciak, W. Strauss, *Bell Syst. Tech. J.* **58** (part 2), 1453 (1979).
22. A. H. Bobeck and H. E. D. Scovil, *Sci. Am.* **224**, 78 (June 1971).
23. J. W. Nielsen and E. F. Dearborn, *J. Phys. Chem. Solids* **5**, 202 (1958).
24. The author has been fortunate to have many competent co-workers and to have interacted with others externally over the last decade of bubbles studies. It is a pleasure to acknowledge helpful comments and suggestions from G. S. Almasi, H. Chang, A. H. Eschenfelder, R. Ghez, C. F. Guerci, G. Keefe, R. J. Kobliska, Y. S. Lin, A. P. Malozemoff, E. McCarthy, T. S. Plaskett, E. W. Pugh, L. L. Rosier, M. W. Shafer, and M. B. Small.

Josephson Tunnel-Junction Electrode Materials

C. J. Kircher and M. Murakami

Josephson superconducting devices have exciting potential for use in building ultrahigh-speed computers. These devices, based on the phenomena theoretically predicted by Brian Josephson in 1962, possess two key properties that are essential to building such computers: fast switching speeds and low power dissipation levels (1, 2). The switching speeds of $\sim 10^{-11}$ second obtained for recent Josephson logic devices are faster

ciently dense packaging is already difficult to achieve because the power dissipation of the transistors (> 100 times higher than that of Josephson devices) exceeds the limits for which direct liquid cooling could be used to maintain safe operating temperatures. It is therefore necessary to use bulky heat sinks that result in lower packaging densities. For the potential of Josephson devices to be realized, novel materials and processes must

Summary. Josephson superconducting devices of the tunnel-junction type have exciting potential for use in building ultrahigh-speed computers. We consider the properties of superconducting metals that are needed for such devices that would be used in integrated computer circuits operated at a temperature near absolute zero. Recent advances in lead-alloy thin-film materials are described that have led to substantial improvements in lead-alloy Josephson device reliability. The properties of a $Pb_{0.84}In_{0.12}Au_{0.04}$ alloy, of Nb, and of Nb_3Sn are discussed as examples of three different groups of materials that are of interest. Investigations of lead-alloy and niobium devices have progressed to the point that it is evident that they have good potential for fabricating integrated circuits containing large numbers of devices.

than those of the most advanced semiconductor devices (3). In addition to having fast switching speeds, the devices must be small and closely packed, because the time required for electrical signals to propagate between devices can be the main factor limiting computer performance. (Electrical signals travel only ~ 1.5 millimeters in the device switching time of $\sim 10^{-11}$ second; thus to achieve the highest performance most of the many thousands of logic and fast memory devices in a high-speed computer should be located within a few centimeters of each other.) The low power dissipation of Josephson devices ($< 10^{-6}$ watt per device) should allow them to be packed densely without incurring heat removal problems (1, 4). For very fast semiconductor logic circuitry, suffi-

be developed to enable them to be fabricated with the electrical characteristics desired for logic and memory devices and with the ability to withstand cooling to ~ 4 K above absolute zero (by immersion in liquid helium), the temperature region at which these devices operate.

Josephson Devices

The type of Josephson device most suitable for high-speed computer applications is a SQUID (Superconducting QUantum Interference Device) composed of several superconducting tunneling junctions that share common electrodes (2). A schematic cross section through the center of such a device is shown in Fig. 1. The junction portions of the device consist of two electrodes separated by an ultrathin (5 nanometers) insulating layer. The electrodes are superconductors (metals that have an infinitely small electrical resistivity when

cooled to below a characteristic temperature $T_c \approx 10$ K). The junction regions are defined by windows in a thicker insulating layer. The ultrathin insulating layer is typically an oxide grown on the base electrode. It is sufficiently thin that current flow can pass through it by electron tunneling; hence, it is referred to as a tunneling barrier. The current-voltage characteristic of such a junction has two branches, as shown in Fig. 2. A normal tunneling-current branch (b) for which a voltage occurs across the tunneling barrier, and a superconducting or Josephson tunneling-current branch (a) for which no voltage develops across the barrier. Because the Josephson current is very sensitive to magnetic fields, the junction can be switched from one branch to the other by passing a small current through a thin film wire (the control line in Fig. 1) located in close proximity to the junction. The remaining element of the device is a superconducting ground plane which is used to confine electrical signals to the close proximity of the device, thereby permitting close packing of devices without incurring cross-coupling between them. The control line and ground plane are separated from the junction electrodes by insulating layers not shown in Fig. 1. In addition, a thin-film layer of a nonsuperconducting metal is used to provide load and damping resistors. Integrated logic and memory circuits can be obtained by using the electrode and control-line layers for device interconnections. Such circuits (5) contain up to 14 thin-film layers that range in thickness from 0.02 to 2 micrometers, and are patterned in shapes with minimum dimensions of 2.5 μm by means of photolithography techniques similar to those developed for fabrication of semiconductor circuits. Most of the layers are prepared in vacuum by evaporating a source material and condensing the vapor on a silicon substrate containing any previously deposited patterned layers.

Electrode Materials

The properties desired for Josephson junction electrode materials can be divided into two categories: those which affect the junction characteristics needed for logic and memory device design and those needed to make possible the fabrication of thin-film, multilayer integrated circuits.

Properties desired for junctions. The superconducting property of the electrodes that is of most importance for junctions is the energy gap E_g. (When a

C. J. Kircher, manager of Josephson junction materials and processes, and M. Murakami, a member of the research staff, are at the IBM T. J. Watson Research Center, Yorktown Heights, New York 10598.

metal is cooled below its superconducting transition temperature, T_c, its conduction electrons attain a lower energy state in which they are grouped into pairs. The energy gap can be thought of as the energy required to break up the pairs.) The size of the energy gap influences two features of the normal branch of the junction current-voltage characteristics (see Fig. 2, branch b): the energy gap voltage V_g at which the normal tunneling current rises steeply and the current level I_s below the energy gap. The value of E_g must be large compared to the thermal energy of the electrons in order for the value of I_s to be small compared to the maximum Josephson current I_m and in order for V_g to be well defined, as desired for logic and memory device design. To operate devices at 4.2 K, the boiling temperature of conveniently available liquid helium, a value of $E_g \gtrsim 2.5 \times 10^{-3}$ electron volts is needed. Such E_g values are obtained by using superconducting materials having a $T_c \gtrsim 7$ K. Of the elements, only lead and niobium have high enough T_c values. There are, however, a sizable number of alloys and intermetallic compounds that would also satisfy the $E_g (T_c)$ requirement.

Electrode properties can also affect the tunneling barrier and hence the value of I_m, a junction characteristic that should be controlled to within ± 30 percent for proper circuit operation, as well as the value of the junction capacitance which affects the switching speed of a junction. The I_m value is exponentially dependent on the barrier thickness; as a consequence the barrier thickness typically needs to be controlled to within less than one atomic layer in average thickness. To achieve this degree of control, it is advantageous for several reasons to use an oxide grown on the lower junction electrode as the tunnel barrier. The growth rate of the oxides formed on many metals at low (ambient) temperatures decreases very strongly with increasing oxide thickness, leading to the growth of oxides having a very uniform thickness (6). Moreover, the growth rate becomes slow at oxide thicknesses similar to those desired for tunnel barriers (in part because electron tunneling from the metal to the surface of its oxide is an important factor governing the oxide growth), favoring good control of the oxide thickness. In addition, oxides are among the most chemically stable compounds of metals; thus use of an oxide tunnel barrier favors good stability of junction properties.

The tunnel barrier material will also affect the value of C_j, the junction capacitance per unit area. Since the switching and resetting times required for junctions are governed by the rate at which the capacitance can be charged and discharged (7), small C_j's are desired for devices. The C_j value is determined by the ratio of the dielectric constant of the tunnel barrier to its thickness. Thus, superconductors having metal oxides with small dielectric constants are favored for devices.

Unfortunately, efforts to use metal oxide tunnel barriers have not been successful for all superconductors of interest. This has led to the exploration of tunnel barriers that can be deposited, rather than grown, on the lower electrode (8). Such an approach has the advantages that the barrier material characteristics could be selected independently of those of the electrode—thereby avoiding the high dielectric constant oxides of some potential electrode materials, for example. In addition, the same barrier could be used with different electrodes. However, a problem with this approach is that deposited layers typically grow in a nonuniform manner on the thickness scale of interest for tunnel barriers. They initially form as nuclei at local sites (similar to the water droplets that form when steam condenses on a cool surface) and become a uniform, hole-free film only at larger thicknesses. The difficulty in obtaining pin-hole–free barriers with well-controlled I_m values is thus increased. It is not clear from the experiments that have been tried with this approach whether it could meet the requirements for integrated circuit applications.

Properties desired for integrated circuits. Integrated circuits involve the preparation on a substrate of many superimposed layers of different thin-film materials that are patterned into very small geometries. For Josephson circuits, most of the preparation is carried out at near-ambient temperatures, but

the circuits are operated at 4 K. Thus, a large number of materials and processes must be compatible with each other to allow such circuits to be made successfully. Because of the great diversity of materials, fabrication techniques, and structures possible, as well as the trade-offs that can be made between them, the requirements for this compatibility can really only be considered with any degree of completeness for a rather specific set of the variables. However, a few general constraints that integrated circuit compatibility places on the choice of electrode materials are discernible from the experience we have gained to date.

The first concerns the magnitude of superconducting penetration depth, λ, an exponential decay length that describes the distance that magnetic fields penetrate into a superconductor before they become vanishingly small. For proper electrical operation of circuits the thickness of the superconductors must usually exceed ~ 1.5 × λ to avoid undesirable penetration of magnetic fields (9). However, for small geometries to be patterned on them, the film thicknesses must be small compared to their lateral dimensions. Because each successive layer in a multilayer structure is typically made with a greater thickness than that of those that precede it in order to ensure the coverage of underlying edges, it is important for the first patterned layers to be as thin as possible, for example, ~ 0.1 μm for Josephson devices. Thus, electrode materials with small λ values of ~ 0.1 μm are desirable for use in integrated circuits.

Another superconducting property that can be a constraint on the usefulness of a material for integrated circuits is the superconducting coherence length ξ, a measure of the distance over which a change in the structure or composition of a material will cause a change in its energy gap or transition temperature. In the course of preparing multilayer struc-

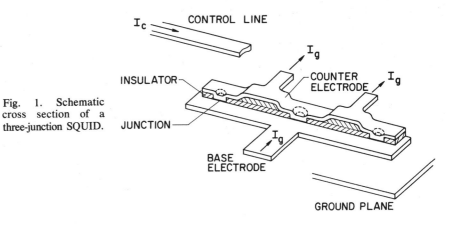

Fig. 1. Schematic cross section of a three-junction SQUID.

tures, materials are necessarily exposed to a variety of ambients, chemical solutions and cleaning processes that can alter the composition or structure of the surfaces of the materials to which they are exposed. For materials with values of ξ that correspond to thicknesses of few atom layers, for example, high-energy gap, high-transition temperature superconductors having the A-15 crystal structure such as Nb_3Sn and Nb_3Ge, these procedures may cause deterioration of the characteristics of junctions and interconnections between superconductors (8).

There are several other constraints that the multilayer nature of integrated circuits places on the materials and processes which can be used to prepare them. The stresses present in thin films that are prepared on substrates held at temperatures less than $\sim 1/4$ of their melting temperature, for example, niobium and other high melting temperature alloys, are typically very high and may exceed the values that underlying thin films or photoresist stencils can withstand. Also, the temperature at which some superconducting materials must be prepared to exhibit desirable super-

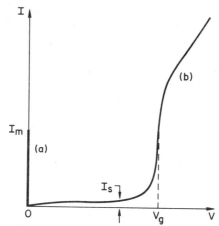

Fig. 2. Current-voltage characteristics of an interferometer similar to that shown in Fig. 1.

conducting properties, for example, above 500°C for the A-15 superconductors, is too high for many other useful thin-film materials to withstand.

Finally, the thermal expansion coefficient of a material can be important. During cooling of devices from ambient to liquid helium temperatures the thin films are constrained to follow the dimensional changes of their relatively massive substrate. A difference in ther-

mal expansion coefficients between film and substrate thus can cause large stresses to develop in the films. These stresses are a problem for superconductors with low melting temperatures, such as those containing lead, which typically have both high thermal expansion coefficients compared to those of useful substrate materials and a lesser ability to support stress without deforming.

Comparing materials. In comparing the properties discussed above with those of various superconducting materials that have been used or have potential for use as Josephson junction electrodes, we find that the materials can be classified into three groups. The properties of a representative material from each group that has been used to prepare successful junctions have been summarized in Table 1. The first group is that of the superconductors with low melting temperatures, exemplified by the $Pb_{0.84}$-$In_{0.12}Au_{0.04}$ alloy and including, for example, lead, tin, indium, and gallium and their various alloys. This group is distinguished from the others primarily by the mechanical properties of its members. They have low values of intrinsic stress σ_i, but larger values of thermal stress σ_T and a greater susceptibility to stress relaxation.

The second group is exemplified by niobium and includes, for example, tantalum and vanadium. The third is exemplified by Nb_3Sn and includes, for example, Nb_3Si, V_3Si, Nb_3Al, and other alloys of niobium, vanadium, or molybdenum with aluminum or silicon. Films of members from both of these groups are typified by their superior mechanical strength and by the large stresses they contain at ambient temperature. The members of the third group are distinguished from those of the second primarily by their much smaller coherence length (ξ) values, by the high substrate temperatures T_s typically required to prepare them, and by the potential they offer for obtaining junctions with reduced capacitance compared to those of the second group because of the elements they contain that form lower dielectric constant oxides.

The present status of efforts to prepare junctions or integrated circuits, or both, from these materials varies. Substantial experience has been gained in the use of lead-alloy Josephson junctions. A process has been developed and successfully used to fabricate experimental integrated circuits containing ~ 100 SQUID devices (5). The materials used and the functions they serve are listed in Table 2. The substrates are oxidized silicon wafers, chosen principally for their

Table 1. The properties of several superconducting materials that have been used as electrodes in Josephson junctions. Each is representative of one of the three groups discussed in the text.

Properties (units)	Pb-In-Au	Nb	Nb_3Sn
Superconducting			
T_c (K)	7	9.2	17
λ (nm)	150	85	170
ξ (nm)	~ 30	~ 30	~ 3
Tunnel barrier			
I_m Control	Good	Good	Unknown
C_j ($\mu F/cm^2$)	4.3	~ 15	≥ 2
Mechanical stress			
σ_i	Low	High	High
σ_T	High	Low	Low
Preparation conditions			
T_s (°C)	$\leq$ Ambient	$\sim$ Ambient	≥ 500

Table 2. Thin film layers used in lead-alloy logic circuits.

Layer	Material	Thickness (nm)	Function
1	Nb	300	Ground plane
2	Nb_2O_5	25 to 35	Ground plane insulation
3	SiO	145 to 275	Ground plane insulation
4	$AuIn_2$	30 to 43	Terminating, loading, and damping resistors
5	SiO	200	Logic interferometer isolation and resistor insulation
6	$Pb_{0.84}In_{0.12}Au_{0.04}$	300	Ground plane contacts and logic interferometer inductance
7	$Pb_{0.84}In_{0.12}Au_{0.04}$	200	Base electrodes, interconnections, and resistor contacts
8	SiO	275	Junction definition and insulation
9	PbO/In_2O_3	6.5	Tunneling barriers
10	$Pb_{0.71}Bi_{0.29}$	400	Counterelectrodes
11	SiO	100	Counterelectrode protection layer
12	SiO	500	Control line and interconnection insulation
13	$Pb_{0.84}In_{0.12}Au_{0.04}$	800	Control lines and interconnections
14	SiO	2000	Protective layer

high thermal conductivity and their suitability for use as the structural members in packaging large numbers of integrated circuits. Most of the superconducting layers are lead alloys which allow low-capacitance junctions to be formed. The alloy additions were selected to obtain films with improved chemical, mechanical, and superconducting properties compared to those of pure lead. The insulation layers are primarily SiO, selected because it can be prepared at the low substrate temperature and low stress levels desired for use with lead-alloy films. The tunnel barrier is an oxide grown on the Pb-In-Au base electrode alloy.

A scanning electron micrograph of a logic SQUID prepared by this process is shown in Fig. 3. The two $\sim$ 2.5-μm-wide control line loops (Fig. 3A) pass over the top of the device. The four junctions (Fig. 3B) of the same size are faintly visible beneath the inner control line. The counterelectrode (Fig. 3C) occupies the center region of the device, with the base electrode (Fig. 3D) extending beneath most of the region under the control lines. Logic devices of this type have been operated with an average logic delay of 1.3×10^{-11} second (10). Memory cells suitable for use in a memory with a 1×10^{-9} second access time have been fabricated using similar techniques and operated with a stored energy of only $\sim 6 \times 10^{-20}$ joules (watt-second) (11).

The experience gained to date in-

dicates that with further development this process has good potential for meeting the yield, tolerance, and reliability levels that will be required to make computers with large numbers of Josephson integrated circuits. The principal materials-related concern is the stability of the devices during thermal cycling between 300 K and 4.2 K. Devices similar to that shown in Fig. 3 will withstand $\sim$ 100 cycles before the first failures are observed in a population of 100 devices (12). Computers made with Josephson devices would probably not be subjected to very large numbers of thermal cycles (several hundred are estimated), but they would contain a very large number of devices. Thus, it is clear that further improvement in the thermal cycling stability of lead-alloy devices is needed. Investigation of the strain behavior of lead and lead-alloy thin films has recently resulted in the development of lead-alloy junctions with significantly improved thermal cycling stability, as discussed below.

For the case of niobium electrode materials, SQUID devices have been made with niobium base electrodes, niobium-oxide tunnel barriers, and counterelectrodes of either niobium (13) or lead alloys (14). At present, only those with lead-alloy counterelectrodes have been made with current-voltage characteristics with the low I_s values desired for integrated circuit applications. Much less experience has been gained with these devices than for the all lead-alloy

devices. However, arrays containing substantial numbers of individual devices have been made successfully. They exhibit excellent thermal cycling stability and the desired values of the Josephson current can be obtained with reasonable reproducibility and uniformity. These devices offer good potential for integrated circuit fabrication. Much of the experience gained during the development of circuits with lead-alloy junctions could be applied to the development of circuits using niobium electrodes. The principal concern with them is the $\sim$ 3.5 times higher values of their junction capacitance which would reduce computer performance by approximately a factor of 2 (15), according to estimates based on present circuit design concepts. It is not yet clear whether the effect of this capacitance can be reduced or circumvented.

For Nb$_3$Sn and other materials in the third group, small numbers of large area junctions have been made, usually with lead counterelectrodes. Some junctions with current-voltage characteristics of good quality have been obtained. The capacitance of Nb$_3$Sn:Pb junctions was found to be promisingly low (16), although this result may have been due to the presence of excess tin on the surface of the Nb$_3$Sn (17). However, too little experience has been gained to date to allow a realistic assessment of the potential of such junctions for integrated circuit applications. For example, the effect on the

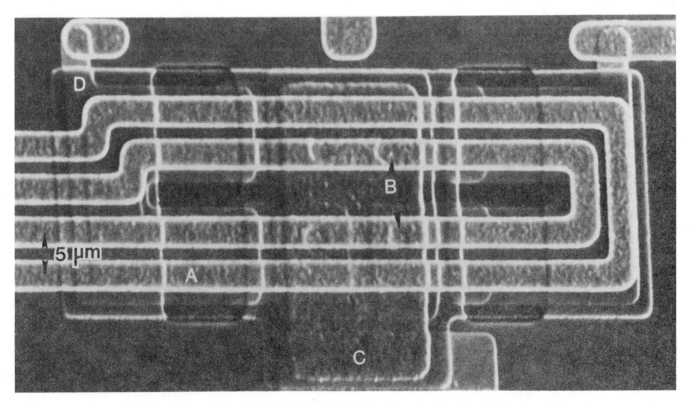

Fig. 3. Scanning electron micrograph of a Josephson logic interferometer containing 2.5-μm-diameter junctions.

junction properties of preparing the junctions by methods compatible with integrated circuit fabrication are not yet known. The development of a barrier formation process that would provide reproducible Josephson currents in the range needed for integrated circuits has not yet been attempted. A consistent set of materials and processes compatible with the ~ 700°C substrate temperatures needed for preparation of these materials has not yet been devised.

Lead-Alloy Junctions with Improved Thermal Cycling Stability

The nature of the cycling problem. The failure of lead-alloy junctions occurs as a short circuit that develops during thermal cycling because of the rupture of the tunnel barrier oxide. The rupture occurs abruptly after some number of cycles, giving rise to an additional nontunneling current in the current-voltage characteristic that grows with additional

cycling until it becomes dominant. This additional current has the properties of a superconducting microbridge of ~ 0.1-μm size shunting the tunnel barrier (*18*). The microbridge is believed to form because of stress relaxation in the junction electrodes during cycling.

Hillocks, micrometer-size protrusions that grow up from the film surface when a lead-alloy film on a substrate is repeatedly cycled to 4.2 K (*19*), are macroscopic evidence that stress relaxation is occurring. Such hillocks are suppressed when lead-alloy films are covered by additional layers, thus they are not observed in failed devices. Nevertheless it seems likely that device failures are due to the formation of "incipient" hillocks.

Hillocks can occur in a thin film when it is subjected to a compressive stress in the plane of the film. Such a compressive stress can develop in (initially stress-free) lead-alloy films during thermal cycling: the difference in thermal expansion coefficients of lead and the underlying silicon substrate are such that a lead-alloy film would be under tensile strain (elongation) during the cooling. If some relaxation of this strain occurs, the film would be compressively stressed upon rewarming to ambient temperature. According to this picture, hillock formation should be reduced if strain relaxation in the lead-alloy films could be reduced during cooling so that the films would return to ambient temperature in their original near-zero stress condition. We would thus like to determine the mechanisms by which strain relaxation can occur in lead-alloy films and to determine the amount of strain that such films will support elastically at 4.2 K, that is, without strain relaxation.

Strain relaxation. The amount of strain that a film actually supports elastically at 4.2 K, ε'_{33}, can be determined by using an x-ray diffraction technique to measure the spacing between atom planes in the film material after it has been cooled to 4.2 K (*20*). The total amount by which the film is strained during cooling, ε_{max}, can be calculated for lead by using known thermal expansion coefficient data for lead and silicon. The amount of strain relaxation is then $\Delta\varepsilon = \varepsilon_{max} - \varepsilon'_{33}$. The stresses corresponding to the strains can be calculated from the known elastic constants of lead.

The dominant strain relaxation mechanism in a material in a given region of stress and temperature can be identified by preparing a deformation mechanism map. The map constructed for 0.2-μm thick Pb-In-Au films (*21*) is shown in Fig.

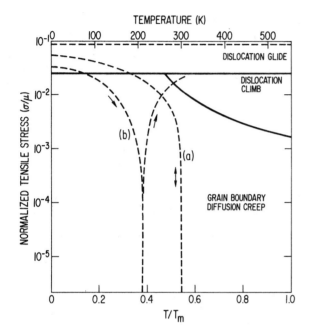

Fig. 4. Deformation mechanism map for Pb-In-Au thin films.

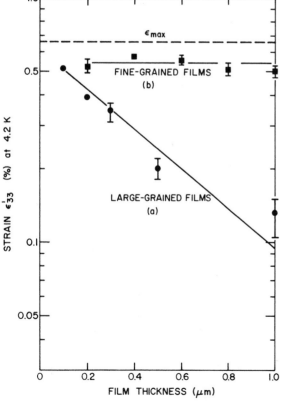

Fig. 5. Strain at 4.2 K for Pb-In-Au films with (a) large-grain and (b) small-grain sizes.

4. It was prepared by calculating for each stress σ (normalized by its shear modulus μ) and temperature, the strain relaxation rate for all relevant mechanisms to identify the dominant (fastest) mechanism. The σ-T map can then be divided into regions labeled by the dominant mechanism. The boundary between two regions is located at the set of (σ, T) values where the strain relaxation rates of the two corresponding mechanisms are equal. Three regions (mechanisms) are evident in Fig. 4. Also shown in Fig. 4, curve a, is the calculated (σ, T) trace that would be followed by a lead film (initial $\sigma = 0$) on a silicon substrate when cooled to 4.2 K, assuming no strain relaxation occurs. It is seen that the fields of two deformation mechanisms are traversed during cooling, that is, grain boundary diffusion creep at $T \geq 200$ K and dislocation glide at $\sigma/\mu > 2 \times 10^{-2}$. Because the cooling rates are very rapid near 300 K, the amount of strain relaxation by diffusion creep (strain relaxation by transport of atoms along the film grain boundaries) should be small. Thus, dislocation glide (strain relaxation inside grains of a film by shearing between planes of atoms) is expected to be the dominant deformation mechanism during cooling. If no strain relaxation occurs during cooling, the (σ, T) path during rewarming would reversibly follow the cooling curve, provided that the heating rate is sufficiently high (near 300 K) that diffusion creep is again negligible. If, however, dislocation glide does occur during cooling, then the path followed during rewarming would be along a curve such as curve b in Fig. 4: the tensile stress would reach 0 near 200 K, becoming compressive between 200 and 300 K and lying within the diffusion creep field, producing a driving force for hillock formation.

We would thus like to prevent dislocation movement in the electrode materials to improve the thermal cycling stability of lead-alloy devices. For bulk materials this is accomplished primarily by alloying or adding impurities to interrupt the almost perfect periodicity of the crystal structure within the grains of the material, thereby creating obstacles which inhibit or prevent dislocation motion. The stress that a material will support is $\sigma_c \propto l^{-1}$, where l is the spacing between the obstacles. Lahiri (19) added gold or indium or both to lead (to retard grain boundary movement), and this resulted in the formation of small intermetallic compound particles in the films. Both a lower incidence of hillock formation and an improvement in thermal cy-

cling stability were obtained (22). The elastic strains supported by these alloys as well as by pure lead are very similar and are below ε_{max}. Thus, the primary effect of such additions was to improve the uniformity of the strain relaxation, rather than reduce its average value.

For thin films, additional obstacles to dislocation movement are present, for example, the native oxide on the film surface, the substrate, and grain boundaries. Thus, the film thickness, h, and grain size, g, may be thought of as obstacle spacings where $l \simeq h$ or g.

The level of elastic strain ε'_{33} supported by $Pb_{0.84}In_{0.12}Au_{0.04}$ films of various thicknesses and grain sizes that were deposited on silicon substrates and cooled to 4.2 K is shown in Fig. 5. The dependence of ε'_{33} on film thickness for large-grained films (prepared by deposition on substrates held at 24°C) is given in curve a (18). For this case the film thickness controls the level of strain relaxation $\Delta\varepsilon = \varepsilon_{max} - \varepsilon'_{33}$. It is seen that ε'_{33} is less than ε_{max} and decreases with film thickness, that is, strain relaxation occurred during cooling at a level that decreased with decreasing film thickness. Decreasing the film thickness from ~ 0.45 μm to ~ 0.23 μm, which resulted in a decrease in $\Delta\varepsilon$ of ~ 30 percent, also resulted in a decrease in the level of hillock formation and a tenfold improvement in thermal cycling stability (23).

The effect of reducing grain size of $Pb_{0.84}In_{0.12}Au_{0.04}$ films is shown in Fig. 5, curve b (24), which gives the strain level supported by fine-grained films prepared at 77 K according to a process developed by Huang et al. (14) in order to obtain $g < h$. In this case the grain size controls the strain behavior: ε'_{33} is thickness-independent and nearly equal to ε_{max}, similar to the results previously obtained for pure lead films (25). At $h = 0.2$ μm (the minimum thickness currently used for lead alloy Josephson junction base elec-

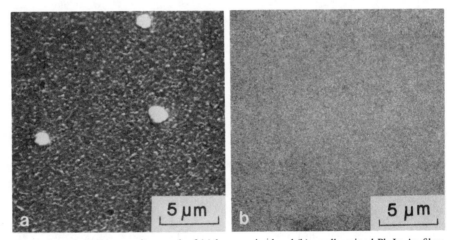

Fig. 6. Scanning electron micrograph of (a) large-grained and (b) small-grained Pb-In-Au films after repeated thermal cycling.

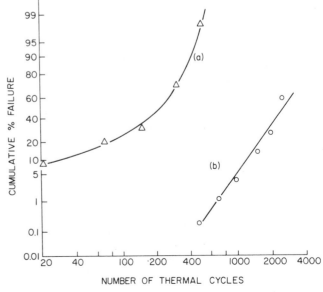

Fig. 7. Cumulative percentage failures for Josephson junctions made with (a) large-grained and (b) fine-grained base electrodes.

trodes, chosen to maintain $h > \lambda$), the strain relaxation upon cooling from 300 K to 4.2 K is much smaller in a fine-grained film than in the large-grained film. Thus, upon rewarming to room temperature, a compressive stress that can cause hillock formation is expected for the large-grained films, but not for the fine-grained films.

In Fig. 6, scanning electron micrographs are shown for two such 0.2-μm-thick films that have been cycled repeatedly to 4.2 K. In Fig. 6a a large-grained Pb-In-Au film (of the type used to prepare the device of Fig. 3) that was cycled 100 times can be seen to have several hillocks that developed during cycling. In Fig. 6b, a fine-grained film is shown that was cycled 725 times. No changes were visible. Nor were any cycling-induced changes evident in either the more sensitive x-ray analyses carried out on such films (24) or the more stringent tests to promote hillock formation (12). Thus, the fine-grained films exhibit significantly improved strain behavior.

Devices with improved thermal cycling stability. The stability during thermal cycling of devices fabricated with 0.2-μm-thick, fine-grained Pb-In-Au alloy base electrodes is substantially better than that of otherwise similar devices prepared with large-grained base electrode films (12). Figure 7 shows the best results obtained for each case from several controlled experiments in which ~ 1300 large area junctions were used. (The large junctions, chosen to enhance the probability of failure, were equivalent in area to ~ 375 devices of the type shown in Fig. 3.) The junctions with the fine-grained base electrodes appear to follow a log-normal statistical distribution with the first failures occurring only after 400 cycles between 300 K and 4.2 K. The companion junctions with large-

grained base electrodes showed ~ 100 percent failure at this point. The use of fine-grained base electrodes has thus improved the junction cycling stability by more than a factor of 100. Further significant improvements in the thermal cycling stability in lead-alloy junctions may be obtainable by using fine-grained films for the junction counterelectrodes. The potential for meeting the device stability levels required for computer applications by using such fine-grained lead-alloy film electrode materials appears to be very good.

Conclusion

We have considered the properties needed for superconducting metals to make them suitable for use as electrode materials in Josephson superconducting tunnel-junction devices and in integrated circuits operated at a temperature of 4 K. The properties of a $Pb_{0.84}In_{0.12}Au_{0.04}$ alloy, of Nb, and of Nb_3Sn have been compared as examples of three different groups of materials that are of interest. The Pb-alloy and Nb cases have been explored sufficiently to show that they have good potential for use in integrated circuits containing large numbers of devices. At present, the most important differences between the two appear to lie in their potential for achieving the high levels of reliability and performance desired for computers. Development of devices with Nb electrodes is less advanced, but it is clear that the properties of Nb favor good stability, and excellent device thermal cyclability has been obtained in initial experiments. However, it would be desirable to identify ways to reduce the capacitance of Nb junctions or to circumvent its anticipated reduction of computer performance. Lead-

alloy devices have already been used successfully in high-performance logic and memory integrated circuits. For these devices, the principal materials-related concern has been about their ability to withstand repeated cycling between 300 and 4.2 K without failure. Advances in understanding the strain behavior of Pb-alloy materials have led to the development of fine-grained base electrode alloys which provide an additional hundredfold improvement in device cyclability. Use of such alloys may allow the reliability needed for computer applications.

References and Notes

1. W. Anacker, *IEEE Spectrum* **16**, 26 (1979).
2. J. Matisoo, *IBM J. Res. Dev.* **24**, 133 (1980).
3. T. Gheewala, in preparation.
4. A. V. Brown, *IBM J. Res. Dev.* **24**, 167 (1980).
5. J. H. Greiner et. al., *ibid.*, p. 195.
6. A. T. Fromhold, Jr., *Theory of Metal Oxidation* (North-Holland, New York, 1976).
7. H. H. Zappe, *Jpn. J. Appl. Phys.* **16** (Suppl. 16-1), 247 (1977).
8. D. F. Moore, R. B. Zubeck, J. M. Rowell, M. R. Beasley, *Phys. Rev.* **20**, 2721 (1979).
9. M. Klein, *IEEE Trans. Magn.* **13**, 59 (1977).
10. T. Gheewala, *IEEE J. Solid-State Circuits* **14**, 787 (1979).
11. W. H. Henkels and J. H. Greiner, *ibid.*, p. 794.
12. H-C. W. Huang, S. Basavaiah, C. J. Kircher, E. P. Harris, M. Murakami, S. Klepner, J. H. Greiner, *IEEE Trans. Electron Devices*, in press.
13. R. F. Broom, R. B. Laibowitz, Th. O. Mohr, W. Walter, *IBM J. Res. Dev.* **24**, 212 (1980).
14. S. I. Raider and R. E. Drake, private communication.
15. H. H. Zappe, private communication.
16. R. E. Howard, D. A. Rudman, M. R. Beasley, *Appl. Phys. Lett.* **33**, 671 (1978).
17. D. A. Rudman, R. E. Howard, D. F. Moore, R. B. Zubeck, M. R. Beasley, *IEEE Trans. Magn.* **15**, 582 (1979).
18. S. Basavaiah, M. Murakami, C. J. Kircher, *J. Phys. (Paris)* **39** (Suppl. C6), 1247 (1978).
19. S. K. Lahiri, *J. Appl. Phys.* **41**, 3172 (1970); *J. Vac. Sci. Technol.* **13**, 148 (1976).
20. M. Murakami, *Acta Metall.* **26**, 175 (1978).
21. _____ and C. J. Kircher, *IEEE Trans. Magn.* **15**, 443 (1979).
22. S. Basavaiah and J. H. Greiner, *J. Appl. Phys.* **48**, 4630 (1977).
23. S. K. Lahiri and S. Basavaiah, *ibid.* **49**, 2880 (1978).
24. M. Murakami, private communication.
25. _____ *Thin Solid Films* **59**, 105 (1979).
26. We thank our colleagues S. Basavaiah, E. Harris, H-C. W. Huang, and S. Raider for allowing us to use their unpublished results and P. C. Long for preparation of the manuscript.

Index

high-temperature
    research programs for, 41
    structural, 37–43
nondestructive testing of, 42
production, scale-up problem in, 42
silicon carbide, 38
silicon nitride, 39
thermal shock-resistant, 37
Ceravital, 26
CF_4, action as etchant, 120
charge-transfer mechanism, 90
chemical
    breakdown, controlled, 23
    plasma etching, 120
    vapor deposition, 80, 107, 109
        compared with LPE method, 107
Chevrel
    phase compound, 78
    phases, 80
cholesteric-type liquid crystal, 128
chlorophosphazine, 6
chopped fiber composites, 33
Chromindur alloy, 86
chromium, in coatings, 49. See also Cr
circuits, integrated, 138, 139
    electrooptic, 113
    very large scale, 117
cluster theory of metallic glasses, 54
CoAl intermetallic compounds, 48
coating compositions, ductility of, 49
coatings, 48
    overlay, 49
    thermal barrier, 49
cobalt
    cemented carbide pistons, 74
    -chromium alloys, 23
    magnets, 82
    molybdate catalyst, 91
    molybdenum-sulfide catalyst, 91
    price rise, 84
    promoter effect of, on MoS_2, 91
    shortages of, 47, 50
cobalt-based alloys, 45, 46
    heat-resistant, 43
    See also alloys, superalloys
coercivity, 82, 85, 133
    mechanism for in Cr-Co-Fe alloys, 85
coextrusion process, 13
coherence length (ξ), 76
    superconducting, 139
cold-cathode gas discharge cells, 125
cold discharge tube reactions, 119
collagen
    bonded in implant surface, 26
    fibers attached to Bioglass surface, 27
color television tubes, 122
columnar-grained structure, 50, 51
combustor components, 47
combustors, reaction-sintered silicon carbide, 41
communication links, large-band width optical fiber, 104
compatibilizer for immiscible polymers, 12
Compax
    diamond die performance, 73
    machining performance, 70
composite
    biomaterials systems, 26
    materials, market potential, 29
    multipolymer systems, 9
    processing, 33
composites
    advanced, 44
    advanced graphite fiber, 35
    in automotive industry, 34
    boron/aluminum, 48
    carbon-epoxy, 44
    E-glass, 33
    in experimental automobile, 35
    fiber-reinforced plastic, 29–36

gas turbine, 44, 48
graphite fiber-reinforced, 28
graphite/Kevlar, 29
graphite/polyimide, 48
high current density, 79
hybrid fiber, 33
reinforced thermoplastic, 33
silicon carbide fiber-silicon, 39
superconducting, 81
composition, alloy, of metallic glasses, 55
compositions
    allowable, of groups III and V elements, 103
    eutectic, distribution of, 54
compounds
    A 15, superconducting, 80
    binary and pseudobinary, superconducting, 78
    intermetallic, 48, 79
    III–V group elements, 102
    transition metal, 75
compression
    explosive, 78
    molding, 34
compressive stress in thin films, 142
computer
    capability, 37
    graphics and modeling, 35
    terminals, 121
    ultrahigh-speed, 138
computer-aided design, 35
conductive
    coatings, transparent, 123
    polymers, 15–21
conductivity
    dark, 19
    photoinduced, 19
    polyacetylene, 19
conductor
    magnet, high-field superconducting, 77
    Nb-Ti alloy, production of, 79
    Nb-Ti composite, multifilamentary, 81
confined carrier quantum states, 115
control line, Josephson device, 138, 141
controlled
    chemical breakdown, 23
    surface-active materials, 26
controlling biomaterials interfaces, methods of, 23
cooling steel, rapid, 62
copolymer
    acrylic acid-methylmethacrylate, 23
    block, 10
    graft, 12
    tetrafluoroethylene, 5, 6
copper compounds. See Cu
copper-substituted magnets, 83, 84
    hafnium in, 84
    iron in, 84
    zirconium in, 84
corrosion resistance of metallic glasses, 55
Corvette, GlFRP usage for, 34
coulomb well, 21
counterelectrodes, 141, 144
couplers, double heterostructure, 113
Cr-Co-Fe alloys
    applications, 85
    coercivity in, 85
    demagnetization curves, 87
    electron micrographs, 85
    fabrication of, 85
    metallurgy and magnetic behavior, 84
    permanent magnets, 84
creep
    grain boundary diffusion, 143
    strength, 46, 50
crimp, fiber, 13
critical
    field, high upper, 78
    state magnetization, collapse of, 77

temperature (T_c), 75
    vs. electron-to-atom ratio, 80
    highest among elements, 77
crystal
    growth, bulk, 105
        liquid encapsulation method for, 106
        Stockbarger method for, 106
    lattice, imperfections in, 76
    structure
        A 15, 76, 140
        A 15, superconductivity in, 78
        hexagonal, 82
crystalline morphology, control of, 9
crystallinity, eliminating in copolymers, 6
crystallization, time-temperature diagram, 53
crystals
    extended-chain, 9
    ferrite, 58
cube-on-edge texture, 86
$CuInSe_2$ cells with CdS, 101
Cu_2S, 99, 123
Curie temperature, 135
curing agents, epoxy, 32
current
    density, critical, vs field, 77
    gains in CuCd solar cells, 100
    photogenerated, 97
CVD (chemical vapor deposition), 107
cyclohexane to methylcyclopentane, isomerization of, 90
cyclohexene, hydrogenation of, 90
Czochralski technique, 106, 134

DARPA (Defense Advanced Research Projects Agency), 41
defects, cation vacancy, 93
deformation
    aging of alloys, 85
    mechanism of Pb-In-Au thin films, 142
    twinning of sintered diamond, 70
demagnetization curves
    Cr-Co-Fe alloy, 87
    permanent magnets, 84
dendritic web process, silicon by, 99
dental
    implants, 22
    reconstruction, 25
deposition, chemical vapor, 107
Dexon suture, molecular breakdown of, 25
DH (double-heterostructure) laser, 104, 113
diamond, 67
    chemical reactivity of, 72
    cleavage, 67
    compacts, sintered
        formation of, 69
        thermal properties of, 71
    hardness, 67
    mechanical properties, 69
    polycrystalline, 67
    properties, 68
    single-crystal, mechanical properties of, 69
    sintered
        on cemented tungsten carbide, 71
        deformation twinning, 70
        hardness vs. microstructure, 70
        hot hardness, 70
        machining performance of, 70
        mechanical properties of, 71
        photomicrograph of, 70
        in piston apparatus, 72
        for ultrahigh-pressure apparatus, 72
        wire-drawing dies, 71
Diarylide yellow, 129
dichroism of the K polarizer, 18
dielectric anisotropy, 127
diesel engine, adiabatic turbocompounded, 42
diffusion
    brazing, 51